To my students and to the Biomeasurement team at Anglia Ruskin University

Biomeasurement

Biomeasurement

A Student's Guide to Biological Statistics

THIRD EDITION

Dawn Hawkins
Reader
Anglia Ruskin University

OXFORD
UNIVERSITY PRESS

OXFORD
UNIVERSITY PRESS

Great Clarendon Street, Oxford, OX2 6DP,
United Kingdom

Oxford University Press is a department of the University of Oxford.
It furthers the University's objective of excellence in research, scholarship,
and education by publishing worldwide. Oxford is a registered trade mark of
Oxford University Press in the UK and in certain other countries

First edition 2005
Second edition 2009

Impression: 2

Published in the United States of America by Oxford University Press
198 Madison Avenue, New York, NY 10016, United States of America

British Library Cataloguing in Publication Data
Data available

Library of Congress Control Number: 2013951465

ISBN 978-0-19-965044-6

Printed in Great Britain by
Ashford Colour Press Ltd, Gosport, Hampshire

Preface

This textbook is based on a course at Anglia Ruskin University of the same name, the introduction of which represented a paradigm shift in the teaching of statistics to biologists. We moved from a math-centred approach to a 'by biologists for biologists' approach. Building confidence and motivation, as well as knowledge and skill, became explicit aims. It worked—failure rates plummeted while understanding and interest blossomed.

Biomeasurement, the course, runs over a twelve-week semester during which we cover the material in Chapters 1 to 11 plus 14 in this third edition of *Biomeasurement*, the book. Students on the course are timetabled for three hours per week. Each week starts with a lecture and then the class is broken down into groups of around twenty-five. In some cases, those leading the tutorials have had almost as much to learn about statistics as the students, but this does not matter: the tutorial materials are provided and a sense of going through them and learning together works well. The hands-on 'active learning' aspect of the tutorial is very important. We assess the students through coursework including a test. Lecture, tutorial, and assignment material are available through the Online Resource Centre. The material in Chapters 12 and 13 is covered in second and third year courses.

online resource centre

While *Biomeasurement* explains how to carry out techniques and why they are needed, it does not explain how the underlying math work. Mathematicians often disapprove of this approach. To them I say that unless you start with this approach then you will recruit very few biologists to the study of statistics, and biology desperately needs such sons and daughters. With *Biomeasurement* I hope to leave young scientists chomping at the bit for more—both wanting and able to read more mathematically orientated texts. It might not seem like it to them at first, but actually I am on the mathematicians' side.

Aim and audience

This is a textbook for the statistically challenged or uninitiated bioscientist. Aimed primarily at first year undergraduates, this book is about building motivation and confidence as well as ability. Philosophical underpinnings are mixed with simple 'recipes' for descriptive and inferential statistics. This gives the reader a firm grounding in why statistics are needed and how to choose and carry out a range of the most useful and widely used techniques. While this book may well be initially read from cover to cover as a companion to a taught course or self-teaching aid, I hope it will stay on the reader's bookshelf as a well-used and familiar source of reference for many years. Certainly I have noticed that even quite experienced researchers find this type of material useful to dip into from time to time.

Scope and style

This book is a broadly bioscience-based introductory text with examples coming from biomedicine, genetics, cell biology, biochemistry, behaviour, ecology, and conservation. The techniques covered span all the topics that the majority of readers will encounter during their degree course. For some it may see them through their entire careers. It is my hope that all readers will complete this text with the confidence to be better able to tackle more advanced texts, or to ask a statistician for help and have the insight required to understand the answer.

The style is deliberately relaxed, leading the reader through a series of statistical facts, concepts, and procedures in an unintimidating style. The book contains mathematical formulae and shows how they are used but avoids explanations of how they are derived. It's like driving a car: knowing what a car looks like, how to drive it, and why that might be useful is all that's needed. Being able to explain how the internal combustion engine works is not essential. The driver has faith that the mechanic behind the system did his or her job well. Similarly biologists don't need to understand the mathematical principles behind statistical equations to be good scientists. They merely need faith in the mathematicians who created the statistical formulae. If after studying this book the reader wants to find out more about the math, that would be wonderful, but it's not essential to their becoming a successful bioscientist.

Associated projects

Biomeasurement, both the course and the book, has led to two associated Higher Education Academy (HEA) funded projects. Funding from the HEA Centre for Biosciences enabled the development of a departmental web-based resource centre to support the teaching and learning of all numerical techniques, including statistics. This project, known as NuMBerS (Numeric Methods for Bioscience Students) can be accessed at <http://web.anglia.ac.uk/numbers/>. In turn this spawned associated projects called SUMS (Students Upgrading Maths Skills) and NUMBAT (<http://numeracy-bank.net/>) funded by JISC, UKOER, and the HEA. These projects aim to identify teaching and learning challenges which are underpinned by weaknesses in student numeracy skills throughout Higher Education and to explore ways in which these challenges might be overcome. For more information on NUMBAT and SUMS, contact Toby Carter (<Toby.Carter@anglia.ac.uk>).

In June 2012, Jenny Keonig and I launched the Biomaths Education Network (<http://biomathed.wordpress.com/>). As a special interest group of the Higher Education Academy it is targeted at the teacher rather than the student. Its purpose is to bring together all those interested in maths and stats education within the biosciences, from school through undergraduate degree programmes to postgrads and scientists working in industry.

Additional acknowledgements for the third edition

Since the second edition, the following newcomers have contributed to the teaching of the Biomeasurement course at Anglia Ruskin University: Abby Stancliffe-Vaughan Alex Dittrich, Fabrizio Manco, Jeremy Thomson, Joab Omondi, Kat Karlsson, Leah Collett, Loveena Delori, Mya Fekry, Nicola Rodgers, and Sarah Sharrat. Nicola and Leah have also contributed to managing the course—a less fun but equally important task. Dimitrios Kiourtsoglou and Chris Holmes have provided great support on the technical side. Paul Elliott continues to provide much appreciated back-up, but Deb Clements, who was there for me on many occasions over many years, is now enjoying a happy retirement and is much missed.

At OUP Jonathan Crowe, who got me started writing this textbook, continues to be a source of ideas and encouragement on matters of teaching statistics to biologists. However, the task of overseeing this edition has gone to Dewi Jackson to whom I am most grateful for help and tolerance. Through Jonathan, Dewi, and other colleagues, OUP continues to be a great publisher with which to work.

The old material in this edition has benefited from useful comments and corrections from a number of readers, especially Abby Stancliffe-Vaughan and Julie Horrocks, and colleagues from the University of Guelph. The new material for this edition benefited from ideas and encouragement from 10 reviewers organised by OUP, Jonathan Amory, Nieky van Veggel, Will Hoppit, Julia Mackenzie, Nancy Harrison, Martin Gammell, Cat Gallacher, Guy Norton, and Micheal Weale. Julia Mackenzie also generously supplied me with the data for the GLZMs in Chapter 13, with Brampton Wood data provided through her by Shelley Hinsley, Centre for Ecology and Hydrology. Toby Carter has again taken on the role of producing the final versions of the updated screenshots and figures and co-authoring the online resource centre.

online resource centre

Meeting and working with Jenny Koenig on the BioMaths Education Network has been wonderfully inspiring and has added new dimensions and motivations to my work on communicating the uses and abuses of statistics to biologists. Working with Martin Gammell and Cat Gallacher (née Karlsson) on our statistical power 'project' has also been stimulating and rewarding, including our first research paper together (Hawkins *et al.*, 2013) contributing, we hope, to more use and less abuse.

All my thanks again to my family—Guy, Garrett, Lilian, Alan, and Janet—for slowing me down and speeding me up.

Additional acknowledgements from the second edition

Since the first edition, the following newcomers have contributed to teaching of the Biomeasurement course at Anglia Ruskin University: Adrian Scruton, Alvin Helden, Blaise Martay, Daisy Berthoud, Helen MacRobie, Mattieu Vizuete-Forster, Ngaio Richards, and Paul Elliott. Some of current teaching team deserve special long service awards for their contribution to the teaching year after year: Alison Thomas, Anne Spiess, Deb Clements, Charlie Nevison, Toby Carter, and Mark Kennedy. Paul Elliott took great care of the module while I went off on maternity leave to produce and enjoy the early months at home with my daughter Lilian Rosa and, as ever, the course has benefited from the presence of Deb Clements in moments of need and uncertainty. Jody Gunn's support with course management and teaching has been invaluable.

Jonathan Crowe and colleagues at OUP have continued to provide exceptional support and the seven reviews for the first edition they organized all made positive contributions to the evolution of this text. The new material for this edition benefited from encouragement and feedback from Alvin Helden, Barry Shoesmith, Deb Clements, Franc Hughes, Gemma Taylor, Katreen Ruckstuhl, Susan Lingle, two anonynous reviewers, and, in particular, Nancy Harrison. Franc Hughes generously provided the raw data for the GLM in Chapter 12. Toby Carter has repeated his key roles of coming to my aid to produce updated screenshots and figures and co-authoring the online resource centre. Working with Jacqui McCary, Julian Priddle, Toby Carter, and Phillipa Priddle on associated projects has been stimulating and rewarding.

My family—husband Guy Norton, descendents Garrett and Lilian and ancestors Alan and Janet—once more provided diversion and support.

Acknowledgements from the first edition

I am fortunate enough to have a team of colleagues at Anglia Ruskin University who share my vision of how to teach statistics to biologists. Their professionalism and energy have been integral to the success of the course on which this text is based and to my abilities and enthusiasm in writing the text itself. Over the years the following people have all contributed to the teaching of this course: Alison Thomas, Andrew Smith, Anna Spiess, Betty Tomova, Caroline Fowler, Charlie Nevison, Deb Clements, Deb Ottway, Emma Ormond, Franc Hughes, Fredrick Mofulu, Guy Norton, Helen Roy, Iain Brodie, Jo Kelly, Julian Doberski, Mark Kennedy, Nancy Harrison, Sandy Preston, Sheila Pankhurst, Toby Carter, and Viki Gutierrez-Diego. Deb Clements in particular has helped share the administrative burden of running the course. Our students, too numerous to name individually but no less important for this, have contributed enormously through their questions, struggles, successes, and compliments.

The role that Nancy Harrison has played deserves special mention. Without Nancy's exceptionally strong commitment to improving the educational experience of students, the course on which this text is based would never have made its way on to the Anglia Ruskin University books. Toby Carter's energetic personality and technical expertise have also played a special role, particularly in his co-authoring of the companion website. Although not involved directly in the teaching of Biomeasurement, Denis Wheller as our supportive and understanding Head of Department, has played a vital role in the Biomeasurement team.

I was able to take a three-month sabbatical from teaching at the start of writing this book through the Central Sabbatical Fund and the support of my colleagues at Anglia Ruskin University. Tim Clutton-Brock found me space in his research group in the Department of Zoology at the University of Cambridge. This provided an excellent environment for making the most of the opportunity this sabbatical gave me to get some momentum going with the writing process. An added benefit was that I met more generous researchers willing to share their raw data with me. Kathreen Ruckstuhl and Peter Neuhaus provided especially welcoming and stimulating company.

Jonathan Crowe, my commissioning editor at OUP, has provided encouragement, feedback, and ideas from which the process of writing and the book itself have benefited tremendously. The numerous reviewers he found for the proposal for this book provided motivation and guidance. The comments from the two reviewers he found for the manuscript provided more encouragement and led to further improvements in the text. The input from Nik Prowse, my copy-editor, was also helpful and constructive.

Nancy Harrison, Deb Clements, Alison Thomas, Guy Norton, Mike Weale, Amy Stelman, Sarah Ashworth, Kathreen Ruckstuhl. and Peter Neuhaus provided useful feedback on draft chapters. My mother, Janet Hawkins, nobly read through final drafts as they appeared and caught numerous errors of English and communication.

During those final weeks of getting the manuscript ready for submission in particular, I received much appreciated extra help. My husband, Guy Norton, and my parents, Janet and Alan Hawkins, kicked in with extra shifts looking after our young son, Garrett. Toby Carter provided great practical support by tidying up figures and getting them into their final form and working on statistical tables for Appendix 2. Alison Thomas gave me lots of useful advice as an experienced author and helped maintain motivation. Phillip Pugh gave me one especially memorable pep talk and made sure the manuscript was safely sent off to OUP while I went off on holiday.

Three people in particular have influenced my thoughts on statistics and how to teach them: Guy Norton, Mike Weale, and Nancy Harrison. Guy introduced me to SPSS, in the days when it was a command-driven package run on mainframe computers, and to nonparametric statistics, through the excellent Siegel (1956) in the days before the edition co-authored by Castellan (Siegel and Castellan, 1988). Our discussions have continued to be a great source of inspiration and especially instrumental in helping me formulate useful questions. Mike has been generous in allowing me to draw on his extensive statistical expertise for many years to help answer the more challenging of these questions. His insightful and humorous explanations of statistical concepts are always illuminating. Nancy has been a constant source of inspiration and support. Our many discussions of statistical issues and how to communicate them have been a lot of fun.

Data, Tables, and Figures

I gratefully acknowledge those who have given me access to their raw data, especially Fredrick Mofulu and colleagues from Tanzania National Parks (TANAPA), Peter Neuhaus, April Ereckson, Kathreen Ruckstuhl, Deb Clements, Justin Brashares, Helen Roy, Liz Smith, and Pinakin Gunvant.

Tables 2.2, 6.6, 6.9, 6.10, 7.5, 7.8, 8.6, 2.3, A2.4, A2.6, and A2.8 plus Figures 3.5 and 8.6 all appear with the kind permission of the publisher and/or authors.

Contents

Using this book

Order and organization

The text starts by placing the role of data analysis in the context of the wider scientific method (Chapter 1) and introducing students to key terms and concepts (Chapters 1 and 2). Then it goes on to introduce descriptive statistics (Chapter 3) and inferential statistics, both estimation and null hypothesis significance testing (Chapter 4). The next eight chapters focus on null hypothesis statistical testing (NHST). First points of procedure, terms, and concepts common to all techniques are covered (Chapter 5). Next, a range of different techniques are presented (Chapters 6 to 13). How to choose between the different null hypothesis significance testing techniques is explained in the final chapter (Chapter 14) and guidance is given on choosing and producing supporting graphs.

The null hypothesis significance testing techniques are divided into three groups: tests on frequencies (Chapter 6), tests of difference (Chapters 7, 8, and 9), and tests of relationship (Chapters 10 and 11). The chapter on tests on frequencies (Chapter 6) features one-way and two-way chi-square tests. Tests of differences are covered in three chapters: two unrelated samples (*t*-test and Mann–Whitney U test; Chapter 7), two related samples (paired *t*-test and Wilcoxon signed-rank test; Chapter 8), and three unrelated samples (one-way Anova and Kruskall–Wallis Anova; Chapter 9). Tests of relationship are covered in two chapters: linear regression (Chapter 10) and correlation (Pearson test and Spearman test; Chapter 11). For tests of difference and for correlation, parametric and nonparametric alternatives are described. The Generalized Linear Model (GLZM) is introduced through the General Linear Model (GLM) in Chapter 12 and expanded to the binary logistic and loglinear models in Chapter 13.

Step-by-step worked examples are presented for all techniques. In most cases this is done showing both calculations by hand and the procedure adopted when using the statistical package SPSS. For reasons of economy of space, worked examples using SPSS only are presented for Anovas, regression, correlations, GLM, and graphs (Chapters 9 to 13).

The presentation of statistical techniques in Chapters 6 to 13 has a common format for each test. This format shows:

- When to use and when not to use.
- Reference to four steps introduced in Chapter 5.
- Worked example(s) by hand and/or using SPSS.
- An example from the literature.

The "choosing" chapter appears at the end because it relies on information and ideas introduced throughout the book. By the time the reader gets to Chapter 14, he or she should be using key words and concepts with ease and be able to focus on the job of understanding how to choose between techniques. In practice, of course, this is where the process starts, not finishes. In fact, choosing a statistical test should ideally be done at the planning stage to help reinforce good experimental design.

Learning features

In addition to the style and organization of the book there are six important pedagogical features:

- **Margin notes** give definitions and explanations of key terms and concepts, which appear in bold in the main text. Marginal notes can be located through the paged numbers in bold in the index.

> **Descriptive statistics** summarize the main characteristics of a sample of data.
>
> **Sample size (n)** is the number of datum points, that is pieces of information, in a sample.

- **Formulae and key rules** appear in boxes. Key rules include when and when not to use a test as well as decision-making rules in the null hypothesis significance testing process.

> **Box 4.2** Formulae for calculating the the mean
>
> Confidence interval = mean $\pm\, t\, s_{\bar{y}}$
> Lower confidence limit = mean $-\, t\, s_{\bar{y}}$
> Upper confidence limit = mean $+\, t\, s_{\bar{y}}$

- **Worked examples** demonstrate the application of techniques both by hand and using SPSS. These worked examples enable the reader to complete the analysis of their data without assuming any prior experience of the technique. Just pick up this text and go.

> With expected according to a 1:1:1:1 Mendelian
>
> **STEP 1** State the null hypothesis (H_0)
>
> H_0: There is no difference between the observed frequency distribution based on a 1:1:1:1 ratio.
>
> **STEP 2** Choose a critical significance level (α)
> We will use 5% (0.05).

- **Data** used in worked examples are explained in some detail. Explanations emphasize the characteristics of the data that make them suited to analysis using the particular technique under consideration. Real data are also included to motivate: all the data sets are provided by working biologists and have been, or are being, used to help address exciting biological questions.

> 4.3 **Example data: ground squir**
>
> In Section 4.4 I will be showing you how to ge tervals, and errorplots. In this section, I am goi I will be using to illustrate these procedures.
> The data come from a long-term study b Columbian ground squirrels. These animals ar and the north-western United States. Peter's st Sanctuary in the foothills of the Rocky Mou studies of wild vertebrates are essential for le biology and factors affecting the growth of the
> In their native undistributed habitat, like th burrowing activities of ground squirrels are b selves are important food for bird and mamm are present on agricultural land they often ca

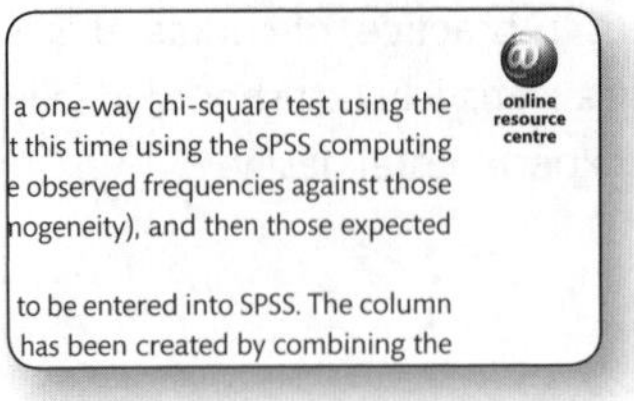
a one-way chi-square test using the
t this time using the SPSS computing
e observed frequencies against those
nogeneity), and then those expected

to be entered into SPSS. The column
has been created by combining the

- **Screencasts** going through all worked examples in both SPSS and R are available through the Online Resource Centre and are linked to the relevant section of text through the screencast callout icon.

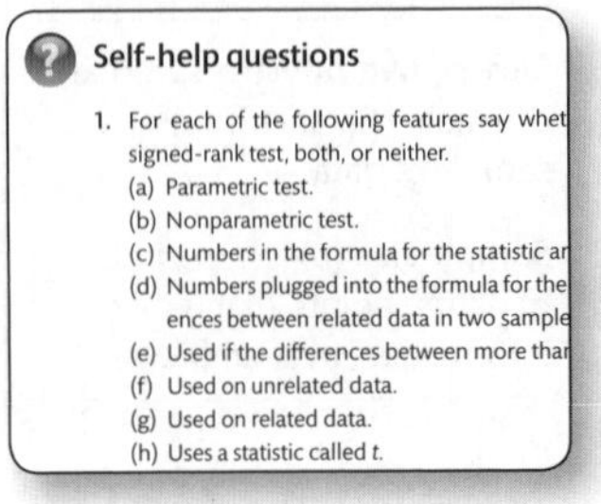
Self-help questions

1. For each of the following features say whet
signed-rank test, both, or neither.
(a) Parametric test.
(b) Nonparametric test.
(c) Numbers in the formula for the statistic ar
(d) Numbers plugged into the formula for the
ences between related data in two sample
(e) Used if the differences between more tha
(f) Used on unrelated data.
(g) Used on related data.
(h) Uses a statistic called *t*.

- End-of-chapter **self-help questions** are important in engaging the reader in the material and czhecking understanding. Along with the worked examples these provide ideal opportunities for 'active learning'.

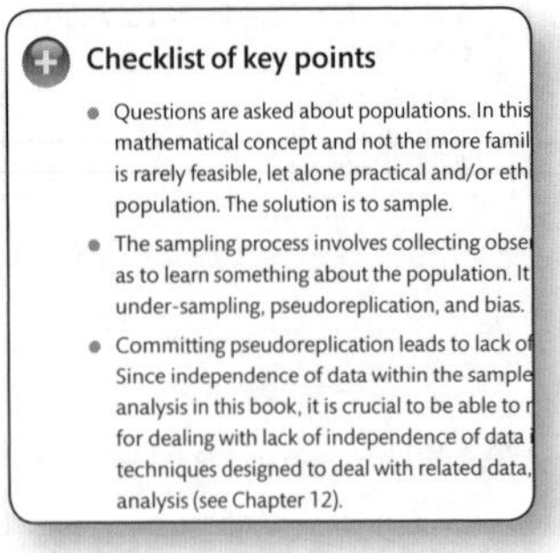
Checklist of key points

- Questions are asked about populations. In this
mathematical concept and not the more famil
is rarely feasible, let alone practical and/or eth
population. The solution is to sample.
- The sampling process involves collecting obse
as to learn something about the population. It
under-sampling, pseudoreplication, and bias.
- Committing pseudoreplication leads to lack o
Since independence of data within the sample
analysis in this book, it is crucial to be able to r
for dealing with lack of independence of data
techniques designed to deal with related data,
analysis (see Chapter 12).

- At the end of every chapter there is a **Checklist of key points** summarizing the main points.

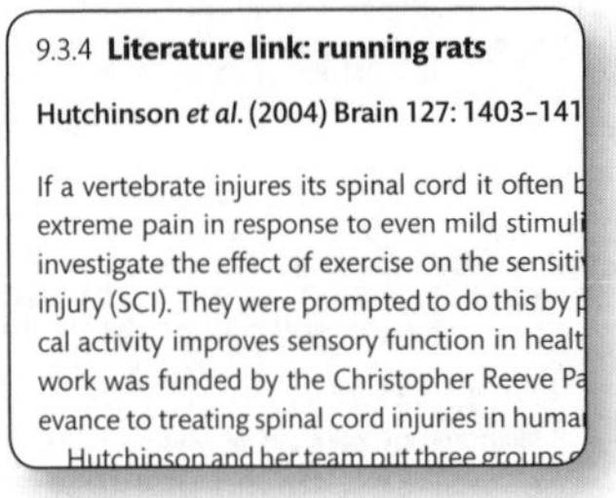
9.3.4 **Literature link: running rats**

Hutchinson *et al.* (2004) Brain 127: 1403–141

If a vertebrate injures its spinal cord it often b
extreme pain in response to even mild stimuli
investigate the effect of exercise on the sensiti
injury (SCI). They were prompted to do this by p
cal activity improves sensory function in healt
work was funded by the Christopher Reeve Pa
evance to treating spinal cord injuries in huma
Hutchinson and her team put three groups

- **'Literature links'** sections describe the use of techniques in published papers, and demonstrate just how important statistical tests are in research. These sections also show how to report findings in a professional manner. Full text versions of these papers are available on the companion website to download free of charge so students can see for themselves.

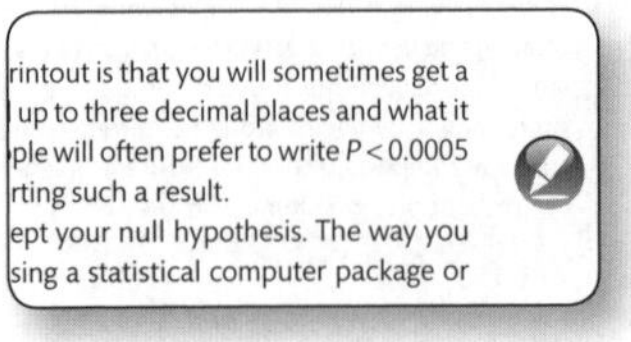
rintout is that you will sometimes get a
up to three decimal places and what it
ple will often prefer to write $P < 0.0005$
rting such a result.
ept your null hypothesis. The way you
sing a statistical computer package or

- The **reporting tip** icon indicates sections of text especially relevant to learning how to report results.

Statistical packages

This text primarily supports SPSS, now, more correctly, IBM SPSS (<http://www.spss.com>): The screen shots come from version 20.0. The Online Resource Centre has support for other versions of SPSS and R.

Biomeasurement Online Resource Centre

online resource centre

Biomeasurement doesn't end with this printed book. There is also an Online Resource Centre at <http://www.oxfordtextbooks.co.uk/orc/hawkins3e> which is co-authored by myself and Dr Toby Carter and is designed to help develop hands-on experience of using statistical techniques. This website includes:

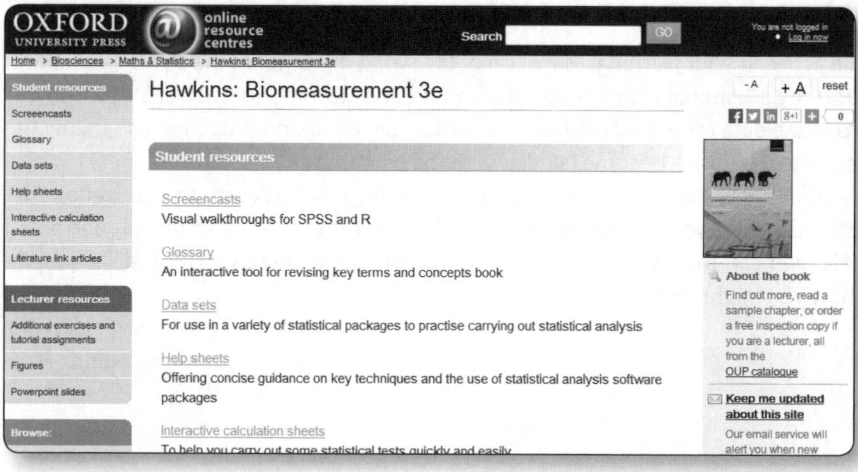

- *Overview, table of contents, and chapter summaries.* The book overview summarizes the progression of information through the book, and the table of contents shows the detailed structure of the different sections. The chapter summaries outline the main points covered in each chapter. Together these provide useful planning and revision materials.
- *Choosing charts and keys for tests and graphs.* Aids for choosing the right test and graph are given in two formats: flowchart and dichotomous key. They summarize and assist the decision-making process.
- *Example data in a number of formats including SPSS, Excel, Minitab, R, and CSV.* Actually going through the procedures, and not just reading about them, is essential to the learning process. Data used in the worked examples are therefore provided in electronic format so that readers can easily have a go themselves.
- *Screencasts and Helpsheets for SPSS and R.* The worked examples that appear in the book are available as screencasts and standalone helpsheets for SPSS and R.
- *Interactive calculation sheets.* Customized Excel spreadsheets, which conduct the calculations for the different tests are available for readers to use with their own

data, as an alternative to using a statistical package such as SPSS. They emphasize the four-step method described in this book.

- *Literature link PDFs*. Full text versions of the papers featured in the literature link sections are available free of charge so that students can see the test in use by practising bioscientists.
- *Key terms and glossary*. Key terms are provided for each chapter and a glossary of key words is provided for reference and to assist in the learning of important vocabulary.
- *Teaching materials (lectures, tutorials, and assignments)*. Materials used for the course run at Anglia Ruskin University on which this book is based are available. These will be especially useful to lecturers preparing similar courses.

A final comment to all students

Be active in your learning: Make use of the worked examples, the self-help sections, and the additional resources on the companion website. To learn SPSS, just download the data for the worked examples from the companion website and follow the steps in the text to become familiar with the procedures.

Above all, I want you to come away from using this book with the realization that statistics make your life easier rather than harder. They are a powerful tool—and a tool that you can master. It really is important that you acquire statistical skills, despite how you might feel!

Dawn Hawkins
Cambridge, August 2013

1 Why am I reading this book?

Book and Chapter Aims

The aim of this book is to help you to be able to use information, or data, to answer questions about biological systems. In other words it's about handling bioscience data correctly and effectively. Since data are usually in the form of counts and measures, this means doing something with numbers. You need to be able to process these numbers and other information—that is, analyse data—and present your findings clearly.

To this end this book seeks to give you grounding in the following skills:

- The ability to choose the appropriate analysis technique for the circumstance.
- The ability to perform and interpret analyses correctly.
- The ability to communicate the results of analyses honestly and clearly.

The role of this first chapter is to

- Emphasize the importance of these skills and why you need to acquire them.
- Put handling data in the context of the big picture, the process of doing science.
- Start to give you a feel as to what data analysis involves.

1.1 My lecturer is a sadist!

Let us consider the likely scenario that you are a student of the biosciences. Whether you are a biomedic, a physiologist, a behaviourist, an ecologist, or whatever, you like learning about living things—you enjoy learning about the human body, bugs, and plants. Now, lo and behold, you have been forced to take a course that will make you do things with numbers and, dread-o-dread, even do something with numbers using a computer. You have probably decided that the people who are making you do this are mindless sadists.

This is not the case, I assure you. Three crucial reasons why you must learn how to handle data are explained below. Do not lose sight of these reasons for learning how to do things with numbers, even if it involves using a computer! They will keep you motivated if you find that the going gets tough.

Reason 1: you'll be able to explore bioscience actively, rather than just being knowledgeable about it.

Think about the biological information you enjoy learning about in other lectures. Now ask yourself: where did it come from? A scientist somewhere posed and answered a question. In other words they didn't just 'imbibe' information, they generated it. So, the first reason for studying this book is so that you can do your chosen bioscience rather than just be knowledgeable about it.

This book will help you select and perform (by hand or using a computer) an appropriate technique to analyse data which, as we will see later in this chapter, will allow you to answer questions about biological issues. Students typically need to do this as part of a practical exercise or project. For those who become research scientists this activity will dominate their lives, but many career paths, biological or otherwise, require people to use data to answer questions.

Reason 2: you'll be able to effectively communicate the biological research that you have done to others.

Once you have carried out some science yourself it's going to be essential for you to be able to tell other scientists about your findings. This might be the lecturer marking your coursework, the reviewer of your first scientific paper, or your fellow students in the bar at lunchtime. The second reason then for studying this book is so that you can communicate the biological research that you have done effectively.

You can get through a lot of paper doing the working for any particular analysis, whether by computer or by hand. In spite of the amount of work it takes, only a few key bits of information need to be reported. This book tells you exactly which bits of information are key for the different techniques and gives you examples of how this information should be presented.

Communication of the results of an analysis often benefits from the use of a supporting illustration—a graph or a table. This book will help you decide which type of supporting material to use and will help you to get a computer to produce it for you.

Reason 3: you'll be able to understand and evaluate the work of others.

Scientists are fallible—probably more so than you currently assume—and if you are unable to detect poor practice it will mislead you. You need to be able to assess whether the analyses used by others are appropriate. In other words, can you trust the conclusions and explanations you are given? Therefore, the third reason for studying this book is so that you can understand and critically evaluate what other scientists are telling you.

If you know the right way to select, perform, and communicate data analyses you will be able to spot when other scientists are doing it the wrong way. Since this book will assist you to do this it will help prevent you from getting duped.

1.2 Doing science: the big picture

Hopefully Section 1.1 will have started to convince you that reading this book will not just build character; it will give you skills that will be vital to you throughout your career as a student and beyond. But, where exactly in the process of doing science does data analysis come in? To put the skills of analysis that you will hear about in subsequent chapters in context, I am going to start with a story about Einstein, a physical scientist rather than a life scientist, but a clever chap all the same!

> Einstein claimed that he was very late to speak and even after he did, he found whole sentences tricky, rehearsing them in an undertone before speaking them out loud. That delayed development, Einstein said, meant that he went on asking childlike questions about the nature of space, time and light long after others had simply accepted the adult version of the world. (*New Scientist*, 26 June 1999)

You should find this encouraging. You can ask questions—in fact you have been doing it all your life! You have what it takes—curiosity; it might have killed the cat but it makes the best scientists! Reading, listening, observing the world, even watching TV, all fuel the curious mind.

Questions can be divided roughly into two main types: descriptive questions and questions which require you to answer them by testing research hypotheses. As we shall see in the following sections, answering both these types of question will require you to use your data-handling skills.

1.2.1 Descriptive questions

The answers to many questions in the biosciences require a simple description of the natural world. I'll call these **descriptive questions**. The following are examples of descriptive questions.

A **descriptive question** requires an answer that's a simple description of the natural world.

- Does the size of elephant groups change with the season?
- How fast do kangaroos hop?
- Does regular exercise lower blood pressure?
- Do high levels of aluminium ions in soil inhibit root growth in wheat?
- What is the rate of decay of dead bodies in deserts?

Getting answers to descriptive questions is the first of two times in the process of doing science where the data-analysis techniques introduced in this book are used. For example, to answer the first question in the list above, you would be looking at a bunch of numbers representing the sizes of elephant groups in different seasons. If you study these data in their raw, unanalysed form, you are unlikely to be able to see the answer to your question. Data-analysis techniques are designed to help you to see the wood (the answer to your question) for the trees (the raw data).

1.2.2 Questions answered using a research hypothesis

The **hypothetico-deductive approach** is the three-stage process needed to answer **questions about a cause, mechanism, or function**.

Answers to descriptive questions will frequently lead you to ask further questions. For example, if you find that elephant group size varies with season then you are likely to become curious as to why this happens. To get answers to these sorts of questions—that is, **questions about causes, mechanisms, and functions**—scientists go through a particular series of stages designed to promote objectivity as detailed below. The fancy name for the series of stages most commonly used by modern biologists is the **hypothetico-deductive approach**.

Stage 1: developing research hypotheses

A **research hypothesis** is an educated guess at the answer to a question about cause, mechanism, or function.

Alternative hypotheses are different educated guesses to the same question.

The first stage in this approach is to generate a **research hypothesis**. That sounds like a very sophisticated and special thing to do, but it's not much more than having an educated guess at the answer to your question. Your educated guess will be derived from your general understanding of the world based on your learning and experience. This general understanding is the theoretical framework or model within which you are working. Often you might think of more than one possible answer and in doing so you are generating **alternative research hypotheses**.

Let us use a specific example to help you get your head around this process: a study of burrowing owls by Smith and Conway (2007). Unlike any other owls, burrowing owls nest underground in burrows. As if this wasn't strange enough, they also scatter mammalian excretment around the entrance to these burrows. You might ask yourself, as Smith and Conway did, why they do this. Why do burrowing owls scatter mammalian dung around the entrance to their nesting burrows?

Smith and Conway explored four hypotheses but for illustration we shall just consider three here. Firstly, Smith and Conway thought that perhaps the answer was related to mating. They argued that the phenomenon could function in mate attraction and assessment, based on the observations that the scattering behaviour was done just by males and that some individuals did not scatter dung, but other materials, like grass and dried moss. This sort of behaviour is certainly not unheard of in the animal kingdom: male bower birds, for example, build elaborate structures to attract females and the females use these to assess the quality of the males. Perhaps female burrowing owls have a thing about mammalian dung!

> Hypothesis 1: burrowing owls scatter mammalian dung around the entrance to their burrows to attract females (mate-attraction hypothesis).

Secondly, Smith and Conway thought that perhaps the answer was related to predation. They argued that badgers, coyotes, and striped skunks are all potential predators of burrowing owl nests and all use smell to help them find their prey. What might be important therefore is that the smell of mammalian dung hides the smell of burrowing owls, thus helping to prevent predators detecting them and their nests.

> Hypothesis 2: burrowing owls scatter mammalian dung around the entrance to their burrows to hide their smell from predators (olfactory-camouflage hypothesis).

Thirdly, Smith and Conway thought that perhaps the answer was related to food. Insects and other arthropods like dung and burrowing owls like arthropods. Instead of going out and foraging for arthropods, perhaps the burrowing owls lure them to their burrows. A bit like having your pizza delivered rather than going out to eat!

Hypothesis 3: burrowing owls scatter mammalian dung around the entrance to their burrows to attract arthropods which they then eat (prey-attraction hypothesis).

Stage 2: generating predictions

The second stage is to take each hypothesis and work out what you would expect to find happening if it were true. In other words you make **predictions**. Smith and Conway (2007) predicted that (Table 1.1):

A **prediction** is what you would expect to find if a particular hypothesis is true. A **general prediction** is a prediction expressed in broad general terms, while a **specific prediction** is a prediction expressed in term of things you can measure. An **exclusive prediction** applies to only one of a set of alternative hypotheses. Predictions need to be **testable**.

- If the mate-attraction hypothesis were true, scattering behaviour would only occur before mates paired up.
- If the olfactory-camouflage hypothesis were true, removing dung from burrow entrances would make them more likely to be predated upon.
- If the prey-attraction hypothesis were true, sites with dung would have more arthropods than sites without dung.

The trick comes in constructing predictions, or a suite of predictions, that distinguish between alternative hypotheses. To do this they need to be **mutually exclusive**. For example, the prediction that scattering behaviour only occurs before males and females pair up, and doesn't happen once pairs are formed, applies to Smith and Conway's (2007) mate-attraction hypothesis, but to neither the olfactory-camouflage nor the prey-attraction hypothesis. The prediction is therefore exclusive to the mate-attraction hypothesis but does not help us distinguish between the other two hypotheses.

Predictions also need to be **testable**. It helps in this regard to express predictions as specifically as possible. A **general prediction** of the prey-selection hypothesis is

Table 1.1 Summary of Smith and Conway's (2007) hypotheses, predictions, and data-analysis results.

Predictions (stage 2)	Hypotheses (stage 1)			Results (stage 3)
	Mate attraction	Olfactory camouflage	Prey attraction	
Scattering only occurs before mates pair.	✓	✗	✗	✗
Predation higher on nests from which dung removed.	✗	✓	✗	✗
Sites with dung have more arthropods than sites without dung.	✗	✗	✓	✓

that sites with dung would have more arthropods than sites without dung. A more **specific prediction** would be that sites with dung would have a greater biomass of arthropods that burrowing owls eat than sites without dung. Specific predictions give you a better idea about what data need collecting. If these data cannot be collected the prediction is not testable. For example if it were not possible to identify which arthropods burrowing owls eat and which they do not, or if it were not possible to measure the biomass of arthropods, then our specific prediction for the prey-attraction hypothesis could not be tested.

Stage 3: testing predictions

The third stage is to go forth into the world and find out if your predictions hold. If they do not then you can reject your hypothesis. If they do then you have found support for your hypothesis.

Look at the column on the far right of Table 1.1, which represents the outcome of this stage in Smith and Conway's study. You can see that their results match the pattern of predictions for the prey-attraction hypothesis, but not the other two. It is thus the prey-attraction hypothesis that Smith and Conway concluded that their work supported.

Let us consider what was involved in stage 3 for one of Smith and Conway's predictions. To assess the prediction that 'sites with dung would have a greater biomass of arthropods that burrowing owls eat than sites without dung', Smith and Conway added dung to some sites and removed dung from other sites. They then set pitfall traps at all of the sites for at least six days, after which time they collected up and identified their contents and found the biomass of everything burrowing owls are known to eat. This would have generated a whole bunch of numbers that they would only have been able to make sense of through analysis.

Here then we come to the second time in the process of doing science where the skills you can learn from this book play a crucial role. The techniques of data analysis introduced are essential tools for testing predictions, which allow us to assess research hypotheses. Just as when answering descriptive questions, if you look at your study data in their raw, unanalysed form, you are unlikely to be able to see what is going on. Again, data analysis is designed to help you see the wood (in this case the prediction) for the trees (the raw data).

1.3 The process in practice

There are several problems that you will encounter when you read scientific papers and try to make sense of the research process, in terms of the framework of asking and answering the kind of questions outlined in Section 1.2. For a start, authors do not generally make it explicit whether they are doing descriptive or hypothesis-driven analyses. Partly this is because the two types of question are really either end of a continuum of question types and the distinction is not always as clear as I have presented here. Moreover, the different types of question feed back to each other, the answers to one determining and shaping future questions. In general, the main phase of a biological research project is likely to be hypothesis-driven, but descriptive analyses have an important role in the development of a good research project. Indeed purely descriptive studies do have their merit, especially as foundations for future research. For example, a necessary precursor to the why-do-burrowing-owls-scatter-dung study we looked at in Section 1.2 was descriptive work to answering the question: what behaviours do burrowing owls exhibit? Only once the answer to this question revealed that their behaviour included scattering mammalian dung around their burrows, did the question of why do they do this arise and thereby spawn hypothesis-driven work.

Another problem you will encounter is that questions, hypotheses, and predictions are often phrased or treated interchangeably. Furthermore, even when hypotheses are clear their predictions may not be stated explicitly, or general and specific statements of predictions not distinguished. It is also quite a common weakness for there to be only one hypothesis under consideration and any alternatives ignored. For example, if Smith and Conway (2007) had only tested their prey-attraction hypothesis and ignored the alternatives, the work would not have been so convincing.

Given all this, it would not be terribly surprising if you got a headache trying to grasp the big picture of the scientific process. The important thing is not to get discouraged. In terms of what you are trying to get from this book the key point is to appreciate where, in the process of doing science, the skill of data analysis that you will learn come in. In short, this is firstly for answering questions of a descriptive nature, which may be a study in itself, but more likely part of the development of a bigger investigation. And secondly, to test predictions generated by hypotheses in hypothesis-driven studies or phases of studies.

1.4 Essential skills for doing science

Analysing data

It is essential that you are able to analyse data if you are going to be a bioscientist rather than just learn about biology. This is what this book is primarily about. Other essential skills required to answer questions about the natural world come under the general headings of Developing hypotheses and predictions; Experimental design; Taking measurements; Critical evaluation; and Health, safety, and ethical assessment. None of these are the remit of this book but all are related to its content. I will briefly review these areas and suggest where you can go to discover more.

Developing hypotheses and predictions

This is part of the three-stage hypothetico-deductive approach I outlined in Section 1.2.2. It is pretty easy to generate hypotheses as part of a good student project. However, the consistent ability to develop robust hypotheses to answer the most sophisticated and exciting questions in biology is perhaps the defining characteristic of a great scientist. There is a definite art to constructing the hypotheses themselves and to generating easily testable predictions. This comes with experience and practice. It is not something that is easy to teach but Ruxton and Colegrave (2010) and Barnard *et al.* (2007) provide useful guidance to the beginner.

Experimental design

Hypotheses and predictions define what information you need to collect to answer your question. Experimental design is about from where you should collect this information and how much you should collect. Experimental design is also about minimizing factors that might mess up your results or complicate their interpretation. In the next chapter I will explain the terms sample and control in more detail but, for reference, what we are talking about when we say experimental design is the *sampling process* and *controlling for confounding factors*.

Experimental design is closely related to data analysis. For example, if you design an experiment with a particular statistical analysis in mind, you will make your life a lot easier in the long run and do much better science. Ruxton and Colegrave (2010) provide an excellent introduction to the general principles of experimental design. (See Appendix IV.)

Taking measurements

The process of actually obtaining information requires you to take measurements. Ruxton and Colegrave (2010) give some general advice on taking measurements. It might be as simple as counting the number of burrowing owl burrows, but might involve special procedures and apparatus. The equipment and techniques you will need for actually obtaining accurate, precise, and reliable information will tend to be subject-specific. For example, Martin and Bateson (2007) is the place to go for more information if you are an ethologist, while Jones *et al.* (2012) covers a range of biological techniques.

Critical evaluation

A sound critical faculty is a very important attribute for a good scientist. In contrast to our legal system, which holds anyone innocent until proven guilty, you should regard any research as having to convince you of its innocence! It's also very important that you apply the same scepticism to yourself as to others. Ruxton and Colegrave (2010) urge you to be your own devil's advocate. This is excellent advice.

Every stage and aspect of the scientific process should be under your scrutiny. Is the logic behind the hypotheses and predictions sound? Have alternative hypotheses been considered? Are the predictions testable? Do the predictions distinguish between alternative hypotheses? Does the description of the experimental design provide us with enough information? Does it minimize potentially confusing factors? Are the analyses used appropriate? And so on...

Like developing hypotheses and predictions, an ability to evaluate critically is hard to teach but a healthy dose of scepticism and using your common sense will get you a long way. Experience will do the rest.

Health, safety, and ethical assessment

Not only is it a moral requirement for scientists to assess the health, safety, and ethics of what they are doing, it has over the last few years increasingly become a legal requirement. Of particular relevance to the bioscientist are the laws and guidelines laid down for taking measurements from human and non-human animals (see, for example, under ethics at http://www.asab.org). The safe use of chemicals is also of relevance in many areas of the biosciences. The legal aspects of this are embodied in the Control of Substances Hazardous to Health (COSHH) regulations (http://www.coshh-essentials.org.uk).

Not only can designing experiments carefully save time and resources, it often has positive ethical implications. For example, good experimental design can allow you to use fewer data in your analysis, involving fewer subjects, while still getting a convincing result.

1.5 Types of data analysis

In this chapter we are looking at what this book is about and why this subject area is so important. So far we have only dealt with its contents in fairly general terms. That is, this book is about conducting and communicating data analyses. In this section I am going to give you a little bit more of a feel for what we are talking about when we say that we are analysing data. I cannot go too far without the background in the subsequent chapters, but I can try to relate what we will be doing in this book to things you might already have heard about.

We use statistical techniques, supported by tables and graphs, to analyse data. Tables and **graphs** come in various designs. I use the term graph to cover a range of pictorial techniques used to assess and communicate patterns in data, including pie charts, bar graphs, and scatterplots. Graphs are often called **charts** by computer programs and **figures** in scientific papers and books.

A **graph, chart,** or **figure** is a visual method of analysing and communicating data. Types of graph include pie charts, bar graphs, and scatterplots.

There are two main categories of statistical technique that you will use: **descriptive statistics** and **inferential statistics**. My guess is that you will have heard of the term average. You probably calculate the average of a set of numbers by adding them all up and dividing them by the number of numbers. In statistical parlance you are actually calculating a type of average called a mean. A mean is an example of a descriptive statistic. Descriptive statistics are used to organize, summarize, and describe data.

Descriptive statistics are used to organize, summarize, and describe data. **Inferential statistics** are used to look for differences and association in your data. **Estimation** and **null hypothesis significance testing** (NHST) techniques come under inferential statistics.

Inferential statistics are used to assess patterns in data. You might, for example, want to know if the probability of dung-scattering behaviour occurring is related to the timing of mate-pair formation. Or, you might be comparing predation rate on nests with dung removed or dung added. Or, you might be be interested in the difference between the amount of arthropods at sites with and without dung.

There are a range of inferential statistical techniques. In this book we look at the two most common historically: **estimation** and **null hypothesis significance testing (NHST)**. An estimation technique you may have heard of is constructing confidence intervals. There is also a good chance that you have heard of a *t*-test or a chi-square test, both of which are null hypothesis significance tests.

Null hypothesis significance tests involve using statistical hypotheses. It is easy to confuse a statistical hypothesis with a research hypothesis, but they are not usually directly comparable. Research hypotheses are developed to answer questions using the hypothetico-deductive approach outlined earlier. Statistical hypotheses generally relate to specific predictions rather than the research hypothesis from which these predictions are generated.

Checklist of key points

- You need the skills covered in this book in order to:
  1. answer questions about the natural world yourself and understand the limitations of your answers.
  2. communicate your answers to other scientists.
  3. evaluate the answers other scientists give you.
- The data-analysis techniques covered in this book will allow you to answer questions:
  1. directly, if they are of a descriptive nature.
  2. indirectly, by testing predictions generated by hypotheses about cause, mechanism, or function.
- The three stages of the hypothetico-deductive approach used to answer questions about cause, mechanism, and function are:
  1. generate research hypotheses.
  2. develop predictions.
  3. test predictions.
- The attributes of a really good prediction are that it is:
  1. exclusive.
  2. specific.
  3. testable.
- Other important interrelated skills for answering questions in the biosciences, which are not dealt with in detail in this book, are:
  1. developing robust hypotheses and predictions.
  2. designing good experiments.
  3. taking accurate and reliable measurements.
  4. critically evaluating everything.
  5. assessing health, safety, and ethical implications.
- There are two types of statistical technique that we use to analyse data:
  1. descriptive statistics are used to organize, summarize, and describe data.
  2. inferential statistics are used to assess patterns in data. Inferential statistical techniques include estimation and null hypothesis significance testing.

Self-help questions

Link back: Section 1.1

1. You should study this book so that you can (choose one or more of the following):
   (a) find out the answers to questions yourself.
   (b) communicate your findings to others.
   (c) understand what other scientists are trying to tell you.
   (d) evaluate your own work.
   (e) evaluate the work of others.

Link back: Section 1.2

2. What are the two main types of question you can ask and how should you go about answering each type?

Link back: Section 1.2.2

3. List the three stages of the hypothetico-deductive (hypothesis-driven) approach.

Link back: Section 1.2.1 and Section 1.2.2, Stage 3

4. When in the process of doing science do the skills of data handling come into play?

Link back: Section 1.4

5. In addition to being able to handle and communicate data, what other skills does a scientist need?

Link back: Section 1.5

6. What are the two main types of statistical procedure used to analyse data and what different roles do they play?

2 Getting to grips with the basics

Chapter Aims

Having read Chapter 1 you should be very aware of the crucial importance of understanding, performing, and communicating data analyses in biosciences. You should be raring to go and get these skills under your belt, but first I must introduce you to some basic terms and concepts. If you can become familiar and comfortable with these at an early stage it will make your life a whole lot easier in the long run, because they recur explicitly and implicitly throughout this book. Introducing you to some key ideas and jargon is one aim of this chapter.

While you may be happy with the importance of data analyses, you may still be wondering why this is not a matter of just comparing a bit of information here with a bit of information there. The reason for this should start to sink in as you read this chapter. Addressing any lingering concerns you have about the need to pay attention to techniques of data analyses is a second aim of this chapter.

2.1 Populations and samples

In a nutshell, the message from Chapter 1 is that you need to be able to handle data competently in order to find, evaluate, and communicate answers to questions about the natural world. Questions, whether they are stated explicitly or not, are the starting point. They define what is being talked about or, in technical terms, they define the **population**. For example, Sackey *et al.* (2003) carried out an investigation entitled 'Predictors and nutritional consequences of intestinal parasitic infections in rural Ecuadorian children'. One of their questions was, what proportion of children living in rural Ecuador have intestinal parasites?

A **population** is the entire set of observations relevant to a research question. Since it is usually impractical or impossible to take records from every item in a population we normally work with a subset of observations, known as a **sample**.

This is a descriptive question about all the children in all the rural parts of the country at the time of their study. The presence or absence of intestinal parasites in each child living in rural Ecuador represented the population in which Sackey *et al.* (2003) were interested.

The word, population, is being used here in a mathematical not a biological sense. For a bioscientist, the word population typically conjures up an image of antelope migrating through the African landscape, or some similar collection of human or non-human bodies. You'll have to let that image go. In terms of data analysis you need to think about populations in a different, more abstract, way. In our Ecuadorian parasite example the population is the collection of observations on whether the children are infected or not. The population is not the children as such.

Of course it would have been quite impractical for Sackey and her colleagues to check the health of all these children given that there would have been between two and three million of them! Moreover, it would also have been quite unnecessary for them to do so because they could get a sufficiently good answer by looking at a sensibly chosen **sample** of children.

I am sure the idea of sampling is second nature to you. For example, I might ask you to go to the Emirates Stadium in north London for the Arsenal football team's next home game to answer the question, 'how tall are the spectators watching Arsenal's next home match at the Emirates Stadium?' My guess is that you would not attempt to measure all 60 000 people in a capacity crowd but that you would select perhaps 200 individuals. I think both you and I would be reasonably happy that the heights of these 200 individuals would give a pretty good indication of the kinds of numbers that would have been generated by measuring all of the individuals. The population in this football-watching example is made up of the heights of all 60 000 spectators. The sample is made up of the heights of the subset of 200 people.

The question, 'how tall are the spectators watching Arsenal's next home match at Emirates Stadium?', is quite specific to a particular match and therefore the population is restricted to the heights of the spectators at this game. Even though it would be impractical to do so, it would be possible to make observations on the entire population in this particular case. The population is limited in time (next match) and space (Emirates Stadium). This need not be the case. In fact moreinteresting or useful questions tend to be less specific. For example, we are more likely to be interested in the heights of spectators at the Emirates Stadium in general. In this case the population is not clearly defined in space or time and it would be impossible to take records on all the relevant heights; sampling is not only practical but essential.

2.1.1 The sampling process

You will typically be using samples taken from the population in order to answer your questions. How you actually select your sample is very, very important. In this section I am going to introduce you to three important issues associated with this process: sample size, pseudoreplication, and bias. Considering these issues in any detail is beyond our needs here, but I recommend that you study them further before embarking on research of your own. They come under the umbrella of experimental design and you should refer back to Section 1.4 for guidance and further reading (also see Appendix IV).

Sample size

One of the first decisions you will have to arrive at is how many observations you need to make. In other words, you will need to decide on your **sample sizes**. For example, Sackey *et al.* (2003) looked at intestinal parasites in 244 children.

The number of items in a sample is known as the **sample size. Power analyses** can be used to determine the smallest sample size on which it is worth doing data analyses.

Returning to our football spectators, I would expect that you would be happy that a sample of 200 could give you a fair idea about the population. But could you be certain this is enough? Conversely, could you have got away with less? You would be unlikely to think it sensible just to take one measurement, but would you really need to spend time taking all 200? In general, the bigger the sample size the safer in terms of your data analyses but there are often practical and ethical limitations to consider. If you collect a sample that is very small it may turn out to be worthless. You can just go with a gut feeling on how large your sample should be and see how it works out. However, there are techniques known as **power analyses** that will help you with this choice.

Replication and pseudoreplication

Repeating the same process of information collection until you have acquired the sample size you have decided on, is referred to as **replication** and the information produced termed **replicates**. For example, Sackey *et al.* (2003) had 244 replicates of their presence/absence of parasites information.

Replication is repeating the same process of data collection, producing **replicates**, until you have your desired sample size.

You need to be careful from whom, or what, you obtain your replicates. You would not go to the Emirates Stadium and measure the same person 200 times, rather than 200 different people once each. This is obviously a stupid way to get a sample to answer our question on spectator heights. If you did, your replicates would lack independence and you would be committing **pseudoreplication**.

Pseudoreplication is when replicates are not **independent** of each other.

Two pieces of data are **independent** if the sampling of one does not influence the sampling of the other. In general, taking replicates from the same item under the same conditions is a major 'no no' in this context! Unfortunately it is not always as obvious as the example I have just used and avoiding committing pseudoreplication is something you need to be on your guard about (see Selected further reading: Ruxton and Colgrave (2010) for a more detailed insight). The statistical techniques that you will meet in this book rely on the assumption that data within samples are independent of one another; violate this assumption and your analyses will be unreliable.

What can you do if you realize that you have committed pseudoreplication and the data in your sample are not independent of one another? It depends on the circumstances but all may not be lost. One option is to condense data by taking averages of groups of data points that lack independence from each other. Alternatively it may be appropriate to do separate analyses for different time periods. At other times, separating the data out into different samples and using techniques designed to cope with related data (introduced in (Section 2.3.1) may be appropriate, the techniques described in Chapter 8 for example. In general linear models (GLMs), which you will meet in Chapter 12, pseudoreplication can potentially be addressed by including an additional independent variable in the analysis.

Random sampling and bias

I would also be very surprised if you went to the Emirates Stadium and selected the 200 tallest people to get the replicates for your sample. I am sure you would see this as creating a **bias**. The way to avoid biasing is to do **random sampling**. If you went to the Emirates Stadium and selected 200 people at random you would be conducting the process of random sampling. Due mainly to the foibles of being human, being truly random isn't as easy as it sounds.

Random sampling is the process of selecting items for a sample without **bias**.

You will intuitively recognize pitfalls like small samples, pseudoreplication, and bias and you are likely to take the basic steps to avoid these, instinctively. However, you should not rely on instinct alone. As I said at the start of this section, you will need to do your homework on this before starting a project of your own. You must give what, when, and how you sample careful thought if you are not going to find yourself opening a large can of worms!

2.1.2 **Sample error**

We will find out more detail in the next chapter on ways of summarizing sets of numbers, but one way with which I am sure you are already familiar is to take an average. In our football-watching example, the average height of the 60 000 spectators would be a single number, providing a summary description of the population and it is an example of a **parameter**. The average of the subset of 200 heights in the sample would be a single number, providing a summary description of the sample and it is an example of a **statistic**.

A **parameter** is a value that describes a population, while a **statistic** is a value that describes a sample. **Sample error** is the difference between a statistic and its corresponding parameter.

Earlier, in Section 2.1, I suggested that we would be happy that 200 heights in the sample would give a pretty good indication of the kinds of numbers that would have been generated by measuring all the spectators in the stadium. Similarly, the average height of the sample is likely to give us a good indication of the average height of the population. However, I would not have expected the average height of a 200-strong sample to be exactly the same as the average height of all 60 000, had we managed to measure them all. I suspect neither would you. It would be possible of course, but we would not be surprised to find a difference. The average of the population (a parameter) and the average of a sample (a statistic) could be the same, but it is unlikely. The sample infers something about the population: for example, we get an idea of whether the stadium is full of tall or short people. However, we don't expect the statistic and the parameter to be exactly the same. Most probably there will be a difference. This difference is called **sample error**. The bigger the sample error the less good an indication our sample gives us about what our population is like. However, we cannot state the exact size of the sample error of any particular sample unless we know the value of the parameter.

I am sure that, even if you have not thought about it directly before, this sample error phenomenon is something you intuitively know happens and are quite comfortable with. You might wonder why I have bothered to emphasize the point. Well, sample error is the reason that this book is 14 chapters long rather than four; it's the reason why we need to use the statistical techniques known collectively as **inferential statistics**. I introduced this term in the previous chapter as the name for one of two main types of statistical technique; the other being descriptive statistics.

Inferential statistics allow us to infer something about populations from samples.

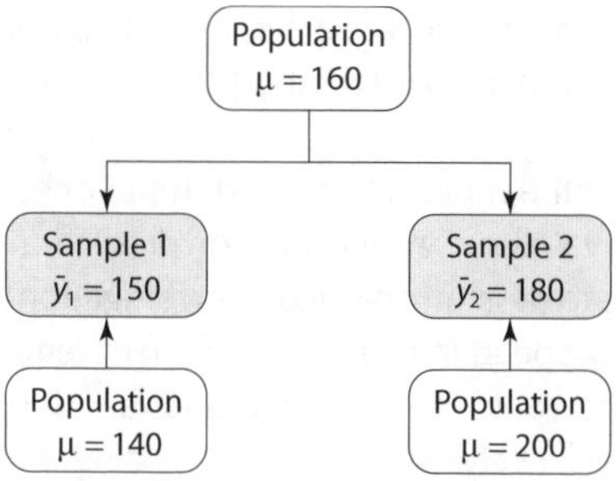

Figure 2.1 Dilemma caused by sample error. When two samples have a different mean, is this due to sample error alone (top half of the figure), or to sample error plus something of biological interest (bottom half of the figure)? The numbers in this figure could represent, for example, the height of football spectators in centimetres. $\bar{y}_1$, sample mean (sample 1); $\bar{y}_2$, sample mean (sample 2); μ, population mean.

Inferential statistics get their name by virtue of the fact that they use samples to infer information about populations. They provide procedures for assessing whether any patterns or trends we find are due to sample error alone or whether something interesting might be contributing. We will turn our attention to inferential statistics later, starting with an overview in Chapter 4. Here I am just going to outline where sample error comes into inferential statistics. Say our question was, 'are the heights of the home-team football spectators at the Emirates and Old Trafford Stadiums the same?'

As already discussed, you would want to take a sample at both the Emirates Stadium and at Old Trafford in Manchester and you would not expect the average of your two samples to be the same. But, is this due to sampling error or to sample error plus some biologically interesting phenomenon such as the genetics or diets of people in Manchester compared with north London? Do the heights of the home-team spectators belong to the same population of numbers, with the difference in the sample average due to sampling error alone? Or are the samples from different populations of numbers, with sampling error plus something else causing the difference? Take a look at Figure 2.1 which presents this dilemma visually. Inferential statistics do not allow us to distinguish absolutely between the two options, but they give us a decision-making process that has its limitations explicitly defined.

2.2 **Variation and variables**

Taking samples is something you undoubtedly already do automatically. The previous section should have just made you think a bit more about the need for and implications of sampling. The need for sampling stems from the impracticalities and cost of making observations on all the items to which a question refers. The implications of sampling arise from the phenomenon of sample error. There is, however, a more fundamental issue related to the need for samples, or at least for samples greater than 1, and that is variation.

Variation is a ubiquitous feature of biological systems and we will return to a fascinating feature of this variation in Section 3.3.3. For now, we are going to focus on what variation is and how it impacts on the process of answering biological questions. The heights of football spectators exhibit variation, but let us switch to something else. Frogs hop at different speeds: speed of hopping in frogs exhibits variation. The source of this variation is manyfold and includes psychological (e.g. motivation of the frog), physiological (e.g. muscle strength), anatomical (e.g. leg-bone length), ecological (e.g. ambient temperature in the environment), and methodical (e.g. inaccuracies in equipment) factors.

Most questions and predictions, and certainly the more refined and useful ones, will be seeking to identify specific sources of variation. A question or prediction defines what sources of variation are interesting to us in any particular investigation. For example, if we are investigating frog hopping speed in relation to ambient temperature then variation in relation to temperature would be interesting, and variation due to anything else would be a nuisance. Alternatively, we might be assessing the use of different methods of measuring the speed of frog hopping, such as a speed camera versus direct measurement using a stopwatch. In which case the variation we are interested in is that generated by the different data-collection techniques, and variation due to anything else, including temperature, would just get in our way. The factor causing variation that is not the focus of our investigation is technically known as a **confounding factor** or **confounding variable**.

A **confounding factor** or **confounding variable** is a cause of variation that is not the focus of a particular question. Confounding factors complicate interpretation of the results of data analyses.

Ideally we would remove the influence of all confounding factors in order to focus on the sources of variation pertinent to any particular investigation. However, the complexity of biological systems makes this almost impossible. Although with careful experimental design and data collection you will be able to control many confounding sources, you are never going to be able to eliminate them all. With our frog example, if you wanted to look at hopping speed at different temperatures you could arrange it so that the frogs are all well fed and measured at the same time of day using the same system. However, it will be far less easy for you to make sure that they all have had the same diet and exercise during their development, and errors in measurements will always be a potential source of variation.

The life of a biologist is therefore made difficult because biological systems are subject to multiple sources of variation and it is hard or impossible to standardize or indeed identify many of these sources. As part of their study on parasites in Ecuadorian children Sackey *et al.* (2003) asked, 'do haemoglobin levels differ between infected and non-infected children?'

Think about the potential sources of variation in haemoglobin levels between children in rural Ecuador—it's endless! The presence or absence of parasites is just one possibility.

Biological systems are highly complex and variation within them is caused by many factors. This is not a problem your colleagues in the physical sciences will encounter. Chemists and physicists study rather simple systems and in this respect have a much easier time of things! For example, the boiling point of water is always 100 °C at normal atmospheric pressure.

Variation in biological systems and its multiple sources has profound implications for data collection and analysis. Variation is the reason why we might even consider getting information from all items in a population and the reason why information from one item does not tell us about all items in a population. Furthermore, variation is the reason samples differ from one another and the reason we get sample error. Without variation we would not need the statistical techniques presented in this book, but biological systems do vary and so there is no getting out of it!

2.2.1 **Identifying variables**

Biological systems exhibit variation. A characteristic that varies is called a **variable**. Variables that we have already met in this chapter include the *presence of parasites* and *haemoglobin levels* in Ecuadorian children, *height* of football spectators, and *hopping speed* in frogs. Answering questions either directly or via predictions that test hypotheses requires you to identify variables and collect data on them. Normally you only collect samples of data rather than measuring the entire population.

A **variable** is a characteristic that varies between items in a sample. A **constant** is a characteristic that does not vary between items in a sample.

More formally defined, a **variable** is a characteristic that has different values for different entities; it is often denoted by the letters *x* or *y* in equations. This contrasts with a **constant**, which is the same for every entity and is generally represented by letters like *a*, *b*, or *c* in equations. While the *size of the African elephant group* varies, i.e. it is a variable, the species does not, i.e. it is constant. It is the characteristics of variables or the relationship between variables that you need to establish in order to answer questions in biology.

For a descriptive approach the characteristic or relationship is the answer. In the hypothesis-driven procedure, a **prediction** is a statement about the expected characteristics of a variable or the pattern connecting two or more variables. Data analyses are tools for assessing characteristics and patterns in observed data. Key variables will be obvious when a question or prediction is clear and specific.

A **prediction** is a statement about the expected characteristic of a single variable or pattern connecting two or more variables, if a particular research hypothesis is true (see also Section 1.2.2).

In Section 1.2.2 we looked at Smith and Conway's (2007) work. One of the predictions they tested was that the behaviour of scattering dung will occur before mates pair. The variables are *occurrence of dung scattering* and *time in relation to mate-pair formation*. To test this prediction they had to collect data on each variable and analyse the data to see if this prediction held.

2.2.2 **Dependent and independent variables**

When answering a question that involves looking at the relationship between two or more variables, you will generally be able to distinguish **dependent** variables (also known as **test, data**, or **response variables**) from **independent** variables (also known as **predictor, grouping**, or **explanatory variables,** or **factors** or **covariates**). In this book we will mostly stick to techniques involving up to two variables, one dependent and one independent, although we will look at situations with more than one independent variable in Chapter 12. Nevertheless, even in this simplest of situations you will find it very helpful if you can get into the habit of identifying which is which.

A **dependent** (**response, test**, or **data**) **variable**'s variation is suspected to be dependent on the variation of an **independent** (**explanatory, grouping**, or **predictor**) **variable** (or **factor**) or **covariate**.

With Sackey *et al.*'s (2003) question, 'do haemoglobin levels differ between infected and non-infected children?' the implication is that haemoglobin levels might be dependent on whether the child is infected or not. Therefore *haemoglobin level* is the dependent variable and the *infection status* of the child is the independent variable.

When Pfennig *et al.* (1999) predicted that the behaviour of avoiding eating relatives will be most pronounced in areas where more than one species of salamander coexists, they were considering the *behaviour of avoiding eating relatives* as dependent on the *number of species coexisting*.

It is important to refer back to your question or prediction when assigning the role of dependent or independent to a variable. What might be your dependent variable in one context could be your independent variable or confounding factor in another. Imagine a scenario in which you are testing the prediction, the number of species of salamander coexisting is related to altitude. In this context, *number of coexisting species* is the dependent variable and *altitude* is the independent variable. It all depends on the angle you are coming from and that is set by your question or prediction.

2.2.3 Manipulated versus natural variation in independent variables

In a **natural**, **correlational**, or **observational** study or experiment, the researcher exploits naturally occurring variation in the independent variable. In a **manipulative**, **planned**, **designed**, or **experimental** study or experiment, the researcher generates variation in the independent variable artifically.

Researchers can use either naturally occurring variation in their independent variable(s) or they can create it artificially: the former are often referred to as **natural, correlational**, or **observational** studies or experiments, and the latter as **manipulative, planned, designed**, or **experimental** studies or experiments.

In the examples from Sackey *et al.* (2003) and Pfennig *et al.* (1999) used in Section 2.2.2, the source of variation in the independent variables was natural. Sackey *et al.* (2003) measured haemoglobin levels in children that were infected and children that were not infected. Whether the children were infected was not determined by the researchers. Similarly Pfennig *et al.* (1999) exploited natural variation in the number of salamander species coexisting to assess their prediction about cannibalistic behaviour. There are often ethical, practical, and financial advantages to such natural experiments, but they suffer from the problem of being vulnerable to confounding factors and do not allow you to be absolutely sure in which direction the cause and effect is operating. For example, does parasite infection lead to lower haemoglobin levels or are those with lower haemoglobin levels more vulnerable to infection?

If you have ever investigated the effect of temperature on the rate of an enzyme reaction to see if there is an optimum, you will almost certainly have conducted a manipulative study. You will have set up a series of test tubes at different temperatures and measured the rate of reaction in each one. Confounding variables can be more easily controlled in manipulative studies and the direction of cause and effect is unambiguous. However, it may not always be possible or ethical to manipulate variation in your independent variable. You must also remember to make sure that the variation you generate is typical of biological systems and relevant to your question. There would be no point in doing the enzyme rate experiment with temperatures all over 50 °C!

Whether a study is natural or manipulative is often associated with the setting in which the data are collected. Typically natural studies are conducted in the field and manipulative studies are conducted in the laboratory. However, this is not fixed and manipulative studies in particular can be conducted in the field. For example, a study looking at whether fire attracts elephants to woodland could artificially set light to some areas of woodland and artifically exclude fire from other areas. This would be a manipulative study in a field setting.

2.2.4 Lack of independence between variables

The concept of dependent-independent variables is easily confused with that of lack of independence between variables. Two variables lack independence from one another, regardless of their status as dependent or independent, if one gives information about the other. If one gives a lot of information about the other they are said to show a high degree of dependency. For example *height* and *weight* in adult humans are likely to show a high degree of dependency. Other ways to say that two variables lack independence is to say that they are **correlated** (see Chapter 11) or that they are **non-orthogonal**. It is particularly useful to be able to identify a lack of independence between variables when you have multiple independent variables (see Chapter 12).

If one variable gives information about another they are said to be **dependent, correlated**, or **non-orthogonal**. If one variable gives no information about another they are said to be **independent, uncorrelated**, or **orthogonal**.

Conversely, two variables are independent of each other, regardless of their status as dependent or independent, if one gives no information about the other. For example the height of an adult human is unlikely to give us any information on how good they are at maths; the variables *height* and *mathematical ability* are likely to be independent. Other ways to say that two variables are independent is to say that they are **uncorrelated** (see Chapter 11) or that they are **orthogonal**.

2.3 Understanding data

For their study of kin recognition in tiger salamanders, Pfennig *et al.* (1999) had to make observations on the selection of variables needed to test their predictions and evaluate their hypotheses. The information they collected constituted **data**.

Data are information collected to answer a question. Notice that data is a plural. A single item of information would be a **datum**.

You might visualize data as numbers, but as we will see in Section 2.3.2 data can take text form. *Sex* is a variable among animals. Information on sex is typically recorded as male or female. The series of observations generated are not numerical but they are data.

It will make your life so much easier and your science so much better if you invest some time to get to grips with the nature of data. In particular, you must be able to (i) differentiate related data from unrelated data (Section 2.3.1); and (ii) recognize at what level the data are measured (Section 2.3.2).

Taking time to understand the data you have collected, or preferably plan to collect, is very important. Understanding relatedness of data and levels of measurement will dramatically enhance your ability to know which analytical techniques are appropriate for any given data set. Understanding data is vital if you are going to evaluate effectively other people's work.

2.3.1 Related and unrelated data

Experimental design is not the topic of this book, but it should be informed by and has consequences for, the techniques of data analyses, which are topics of this book. Being able to design good experiments is an essential skill which you must acquire if you are going to be an effective bioscientist (Section 1.4; Appendix IV). By designing your research carefully you will avoid sampling pitfalls such as bias, pseudoreplication, and inadequate size (Section 2.1.1). You will also identify, remove, or minimize the sources of variation on your dependent variable, in which you are not interested (confounding factors: Section 2.2), so that you can focus on the ones you are (independent variables: Section 2.2.2). When learning about experimental design you will come across designs described by terms such as blocked, paired, cross-over, split-plot, stratified, or latin squares. However, the key in terms of basic

Frog ID	Temperature (°C)	Speed (m/s)
Betty	21	2.1
Nora	17	1.9
Chloe	16	1.6
Mick	14	1.5
Ernest	15	1.6
Rick	10	0.5
Roger	20	2.3
Jo	12	0.6

Table 2.1 Related data: relationships. These data were generated by recording the hopping speed of eight different frogs and the temperature of the environment in which they were hopping.

data analysis is to understand whether you are dealing with a design that involves related or unrelated data.

Related data: repeated and matched

When data from different samples are linked in some way they are **related data**. This can be achieved through taking measurements on the same individuals for different samples, known as **repeated measures**. Alternatively, this can be achieved by linking data points based on the similarity of the characteristics of individuals contributing to different samples, known as **matched measures**. When just two samples are involved this is often called **pairing** rather than matching.

Let us start with **related data**. Relationships between variables, by their nature, involve related data. For example, in a study there is a piece of data for the dependent variable and a piece of data for the independent variable, for each item sampled. Imagine that we want to test the prediction that hopping speed for frogs will vary with ambient temperature.

This suggests we should look for a relationship between *hopping speed* and *ambient temperature*. Say that we exposed each frog in our sample to a different temperature and measured its hopping speed with a speed gun. For each frog we would have two numbers—one a measure of the variable speed, and the other of the variable temperature. Have a look at Table 2.1 to see what these data might look like.

Studies looking at differences between samples can involve either related or unrelated data. Related data are produced in studies of differences when the same item contributes to different samples by being measured under different conditions of the independent variable. This is known as **repeated measures**. Alternatively, data can be related by matching up different items in different samples based on their individual characteristics. This is known as **matched samples**. When just two samples are involved this is often called **pairing** rather than matching.

An investigation into the hopping speed of frogs might involve the following prediction: *hopping speed* for frogs will be different in hot compared to cold *ambient temperature*.

A repeated-measures design for our hopping frogs would go something like this. First we'd measure the hopping speed of eight frogs at room temperature and call the eight numbers generated our warm sample. Then we would measure the hopping speed of the same eight frogs in a controlled-temperature cold room and call this our cold sample. Actually we would ideally do half the frogs in the cold treatment first. In any event, each speed in the warm sample would be related to the number in the cold sample from the same animal; look at Table 2.2 to see how these data might look. Repeated measures has the advantage of helping to reduce sources of variation related to individual characteristics of the frogs such as *age, sex, size, weight*, and so on.

Frog ID	Speed (m/s) in cold environment	Speed (m/s) in warm environment
Ludwig	0.8	1.9
Blossom	0.6	2.1
John	0.5	1.8
Mona	0.5	1.7
Rollin	0.2	1.6
Ernie	0.0	0.8
Kat	0.7	1.9
Ron	0.9	2.0

Table 2.2 Related data: differences, repeated-measures design. These data were generated by measuring the hopping speed of eight frogs, first in a cold environment (5 °C) and then in a warm environment (20 °C).

Frog IDs	Speed (m/s) in cold environment	Speed (m/s) in warm environment
Kit and Bob	0.8	2.1
Mike and Robbie	0.4	1.8
Louise and Bess	0.9	1.7
Ojo and Laura	0.5	2.2
Will and Bill	0.2	1.6
Harry and Sam	0.1	0.7
Hilary and Jill	0.3	1.5
Louie and Don	0.8	1.3

Table 2.3 Related data: differences, matched design. These data were generated by measuring eight pairs of frogs matched for age and sex characteristics. One frog in each pair was in a cold environment (5 °C) and the other in a warm environment (20 °C).

For a matched design we'd have different frogs in the different samples, but we would partner them up according to age, sex and so on as much as possible; look at Table 2.3 to see how these data might look. This is not such a good way of controlling for variation generated by individual differences as doing repeated measures, but it could be the best option in some circumstances. For example you might decide for ethical reasons not to expose your study animal to more than one experimental condition.

Unrelated data

If data are not related then they are, not too surprisingly, **unrelated**. If we had measured the speed of eight frogs under cold conditions and eight different ones under warm conditions and had no ability to match them, we would have two unrelated samples of data. In contrast with related data, sample sizes do not have to be the same, so we could have had eight frogs in the cold and 11 frogs in the warm: Look at Table 2.4 to see how these data might look.

If data from different samples are not linked then they are **unrelated data**.

While related designs do have advantages in terms of controlling for confounding sources of variation, there are lots of reasons why you might have to go with an unrelated one. To answer their question about *haemoglobin* levels and *parasite infection*, status Sackey *et al.* (2003) collected two unrelated samples of haemoglobin measurements: one sample from uninfected children and the other from infected children.

Cold environment		Warm environment	
Frog ID	Speed (m/s)	Frog ID	Speed (m/s)
Alf	0.9	Ted	1.5
Nell	0.4	Dot	1.5
Lil	0.7	Jack	1.6
Ray	0.4	Jen	1.9
Bert	0.1	Bell	2.1
Fred	0.0	Nick	2.2
Flo	0.5	Doc	1.3
Oli	0.6	Rose	1.6
Sid	0.4		

Table 2.4 Unrelated data. These data were generated by measuring 17 different frogs: nine in a cold environment (5 °C) and eight in a warm environment (20 °C).

Trying to get their large samples to match up would have been impossible. To get a repeated measure they could have infected the children in the former sample and cured those in the latter. While their study did support a programme of treatment for the infected, obviously infecting the healthy was not an option!

Distinguishing related and unrelated data

In a **balanced design** sample sizes are equal. In an **unbalanced design** sample sizes are not equal. When data are related the design will be balanced. If the data are unrelated the design can be balanced or unbalanced.

I have already pointed out that if data are related, then samples will be the same size or, in statistical jargon, **balanced**. If they are unrelated, then sample sizes can be different or **unbalanced**, although a balanced design is always preferable in terms of data analyses. The tip to pick up from this is that if your sample sizes are different, you must be looking at unrelated samples, but if they are the same the data could be either related or unrelated.

Another way to distinguish related from unrelated data is to imagine moving the first data item in one sample to another position and asking if the order of the data in the second sample should be readjusted accordingly.

2.3.2 Levels of measurement

The levels at which data can be measured, from weakest to strongest, are **nominal** (categories with no intrinsic order), **ordinal** (ranks), and **scale** (counts and measures). For the scale level, counts produce **discrete variables** and measures produce **continuous variables**.

At the beginning of this section I pointed out that data can be recorded in text form, such as male or female, as well as in number form. This observation touches on the concept of levels of measurement. You must really make an effort to get your head around this concept as soon as possible. Doing so will stand you in very good stead. The first thing to remember is that there are three levels of measurement: **nominal** (categories), **ordinal** (ranks), and **scale** (counts and measures). Next, you need to learn the characteristics and some examples of each. Here they are, from weakest to strongest.

Nominal (categories)

At this weakest level of measurement, data are measured as simple categories, for example male or female. Unless you are sexist you will appreciate that these categories have no intrinsic order to them. Another example is *flower colour* in roses. This is usually measured as red, white, pink, or yellow. Sackey *et al.* (2003) used a nominal level to measure *parasite species* according the established taxonomic categories such as *Giardia intestinalis* and *Trichuris trichiura*.

Ordinal (ranks)

Data measured as categories that can be put into a logical order constitute a stronger level of measurement. The Beaufort *wind-force scale* ranges through 12 categories, from calm (level 1) to strong breeze (level 6) to hurricane (level 12). In their preliminary analyses Sackey *et al.* (2003) mention that they measured *parasite burden* as low, medium, or high. This is another example of using an ordinal level of measurement.

Scale (counts and measures)

Best, or strongest, of all is the scale level. The data are measured in such a way that you can tell the distance between values. *Temperature* in centigrade, *height* in metres, *age* in years, and *mass* in kilograms are all examples of scale level of measurement. Sackey *et al.*'s (2003) measurement of haemoglobin levels was on the scale level.

Within the scale level, mathematicians recognize interval and ratio data but this is not a crucial distinction for our purposes. More relevant to us is to appreciate that scale data generated by counting gives rise to **discrete variables,** and that scale data generated by measuring gives rise to **continuous variables**. This is exemplified by counting the *number of parasite worms* in a faecal sample compared with measuring the *length of these worms*. The length of a worm can fall anywhere along a continuum. In contrast, the number of eggs falls into discrete categories, for example four or five, but not anywhere in between.

To recap, there are three levels of measurement: nominal (categories without intrinsic order), ordinal (categories which can be ranked), and scale (counts and measures). In general people have little difficulty distinguishing nominal from ordinal or scale. What they find trickier to identify is the difference between ordinal and scale. The story of the first African woman to win the Olympic marathon should help.

In 1996 I sat enthralled in front of my television watching the women's marathon at the Atlanta Olympics. Before the twelfth mile Fatuma Roba moved to take up the front position then widened her lead with every mile. The commentators were not encouraging. She was ranked number 29 and had no hope. She wasn't drinking anything. She had over-extended herself and she would fade before the end.

But she did not fade. By the time she reached the Olympic stadium she was way, way ahead. As she entered, the crowd rose as one to their feet. The lone figure of Fatuma Roba ran the final lap of the 26 mile and 385 yard race to a standing ovation of thousands. It was very moving.

How does this help us distinguish ordinal from scale? Consider the result of this race, presented in Table 2.5. If you look at the places the athletes achieved (an ordinal-level measure) you get no indication of the impressiveness of this victory. Only by looking at the times of the runners (a scale-level measure) can you do that.

Place	Athlete	Country	Time (h:min:s)
1 (Gold)	Fatuma Roba	Ethiopia	2:26:05
2 (Silver)	Valentina Yegorova	Russia	2:28:05
3 (Bronze)	Yuko Arimori	Japan	2:28:39
4	Katrin Doerre-Heinig	Germany	2:28:45
5	Rocio Rios	Spain	2:30:50
6	Lidia Simon	Romania	2:31:04
7	Maria Machado	Portugal	2:31:11
8	Sonja Krolik	Germany	2:31:16

Table 2.5 Women's marathon result at the 1996 Atlanta Olympics.

Source of data: http://www.sporting-heroes.net

2.4 Demystifying formulae

There are two things about statistical formulae that are likely to make you panic. First, they look scary because they are full of squiggles, lines, and letters arranged in weird and wonderful ways. Dealing with this is a matter of staying calm and reading the meaning of the symbols and carrying out the calculations in the right order. Sections 2.4.1 and 2.4.2 are designed to help you with this.

Secondly, once you have got a particular formula sorted you will quite often find that someone else or some other book presents you with an entirely different-looking formula for the same thing. It is important that you do not automatically assume that you have done something wrong. There are different ways of writing the same formula. Some arrangements are easier to use when doing calculations by hand. Other arrangements provide insights into the underlying mathematics. Sometimes the arrangement used reflects the personal preference of the author or researcher. Applied correctly by translating the symbols accurately and doing the calculations in the right order, all arrangements of formulae for the same statistics will produce the same answer. We will revisit this using a specific example in Chapter 3.

2.4.1 Squiggles, lines, and letters

There are symbols such as ÷ or ×, meaning divide or multiply respectively, with which I am sure you are familiar. You may be less familiar with the use of a slash, /, or round brackets, (), as alternatives to these symbols. Table 2.6 reviews the symbols used for various mathematical operations in statistical formulae.

Another potential source of confusion is that the same symbol is not used for a statistic as its corresponding parameter. For example, big N is generally used to denote the number of items in a population and small n the number of items in a sample. This example is shown, with others, in Table 2.7. Do not worry, you have not missed anything. We have not yet even mentioned standard deviation and variance: I have included them for future reference.

A superscript to the right of a number tells you to multiply the number by itself that many times. When this is two times, the process has a special name, squaring. For example, y^2 means multiply whatever number y is by y; in other words, square y. The

Symbol	Meaning	Example
+	Add	6 + 2 is 8
−	Subtract	6 − 2 is 4
×	Multiply	6 × 2 is 12
()	Multiply	(6)2 is 12
÷	Divide	6 ÷ 2 is 3
/	Divide	6/2 is 3
>	Greater than	6 > 2
<	Less than	2 < 6
≥	Greater than or equal to	6 ≥ 2 or 6 = 6
≤	Less than or equal to	2 ≤ 6 or 6 = 6
=	Equals	6 + 2 = 8
≠	Not equal to	6 + 2 ≠ 18

Table 2.6 Symbols for some common mathematical operations.

Feature	Symbol for population	Symbol for statistic
Size	N	N
Mean	μ	$\bar{y}$
Standard deviation	σ	s
Variance	σ^2	s^2

Table 2.7 Symbols for some parameters and their corresponding statistics.

square of 4 (4^2) is 16. The reverse operation is indicated by a sort of extra-large tick (√). For example, to find $\sqrt{y}$, in other words to find the square root of y, you need to find the number that multiplied by itself would give y. The square root of 4 ($\sqrt{4}$) is 2 or −2. You cannot tell if it is 2 times 2 or −2 times −2, because whenever you multiply two signs together that are the same, it makes a positive. This does not matter as you can just say the square root of 4 is plus or minus 2, written as ±2. You can write this as $\sqrt{4} = \pm 2$. Most square roots are not whole numbers and are not so easy to work out. Fortunately your calculator is likely to have a square root button that you can use.

A particularly often-used squiggle is Σ. This squiggle is a Greek capital letter called sigma and it means *sum of*. You need to add everything up which lies directly to the right of the sigma. As I mentioned in Section 2.2.1, letters at the end of the alphabet are commonly used to represent variables, while those at the beginning are used for constants. Therefore Σy means add up all the values for your variable in your sample.

2.4.2 **Doing things in order**

The order you should do things in is important. First, if you see any brackets, you must do the calculations inside them before anything else. This means that if you meet $\Sigma(y+1)$, you need to add one to every value in your sample and then sum the numbers produced. This will give you a quite different answer from $\Sigma y + 1$, where you should sum up all your values and then add one to the total.

Second, perform any squaring or square rooting you can see, from left to right. Third, perform any multiplication and division, from left to right. Finally, you will need to do any addition or subtraction, from left to right. For example, for $\Sigma(1+4y^2)$ square *y*, multiply this by 4, add 1 for each value of *y*, and then add the results up. *Do not* add 4 to 1, to give you 5, then multiply 5 by *y* and square the result.

In short, the order you should do things is:

1. Brackets.
2. Squares and square roots.
3. Multiplying and dividing.
4. Adding and subtracting.

Checklist of key points

- Questions are asked about populations. In this context a population is an abstract mathematical concept and not the more familiar and easily visualized biological sort. It is rarely feasible, let alone practical and/or ethical, to answer a question using the entire population. The solution is to sample.
- The sampling process involves collecting observations from a subset of a population so as to learn something about the population. It must be done with care in order to avoid under-sampling, pseudoreplication, and bias.
- Committing pseudoreplication leads to lack of independence of data within a sample. Since independence of data within the sample is an assumption underlying all of the analysis in this book, it is crucial to be able to recognize lack of independence. Options for dealing with lack of independence of data include condensing data, using statistical techniques designed to deal with related data, or adding independent variables to the analysis (see Chapter 12).
- A parameter describes a population and a statistic describes a sample. The difference between a parameter and a statistic is known as sample error. Sample error underlies the need for inferential statistics.
- Variation is a feature of all biological systems and is the reason we need to use statistical techniques to analyse bioscience data. This variation has many sources. The question or prediction under consideration will define the sources of variation of interest in any given study. Confounding factors are sources of variation that are not of direct interest, but which need to be identified and controlled if they are not going to mess up your research.
- Characteristics that vary are called variables. Answering questions, either directly or via predictions which test hypotheses, requires variables to be identified and data to be collected on them. Characteristics whose variation we are seeking to explain are dependent variables and the variables we are investigating in relation to this are independent variables.
- Variables are independent from each other (uncorrelated or non-orthgonal) if information about one gives us no information about the other.

- In order to answer a question you can use natural variation in your independent variable, or you can create variation in your independent variable artificially. These two options have different advantages and disadvantages.
- Data are either:
  1. Related (matched or repeated).
  2. Unrelated.
- There are three levels of measurement:
  1. Nominal (categories).
  2. Ordinal (ranks).
  3. Scale (counts and measures).

  Within the scale level, counts produce discrete variables and measures produce continuous variables.
- Calmly translating symbols and carrying out calculations in a set order is essential to using statistical formulae successfully.

Self-help questions

1. List three reasons that might explain variation between the body length of 100 earthworms. *Link back: Section 2.2*
2. Fill in the missing words in the statements (a) to (i):
   (a) Although research questions typically concern a ____, a research study typically examines a ____. *Link back: Section 2.1*
   (b) The relation between a statistic and a parameter parallels the relation between a ____ and a ____. The difference between a statistic and its corresponding parameter is known as ____. *Link back: Section 2.1.2*
   (c) A selection process that ensures that each individual has an equal chance of being selected is called ____ sampling and avoids ____. *Link back: Section 2.1.1*
   (d) Classifying a species occurrence in the wild as rare, moderate, or abundant is an example of measurement on an ____ level of measurement. *Link back: Section 2.3.2*
   (e) Determining a person's reaction time in seconds would involve measurement on a ____ level of measurement. *Link back: Section 2.3.2*
   (f) Classifying people according to their country of origin would involve measurement on a ____ level of measurement. *Link back: Section 2.3.2*
   (g) In order to determine the size of the difference between measurements, a researcher must use a ____ level of measurement. *Link back: Section 2.3.2*
   (h) To test the prediction that tadpoles swim slower in polluted water, the speed of swimming of five tadpoles was first measured in clean water and then in polluted water. This is a ____ design and the data are therefore ____. *Link back: Section 2.3.1*
   (i) A researcher observed that students studying in a red room looked up more often than students studying in a blue room. For this study, the dependent variable was ____ and the independent variable was ____. *Link back: Section 2.2.2*
3. What sort of study is it when researchers make use of naturally occurring variation in their independent variable(s)? What's the alternative? *Link back: Section 2.2.3*
4. A researcher studying bird vocalization cannot tell if she has already recorded a bird or not. Why might this cause a problem with her data analysis? *Link back: Section 2.1.1*
5. What are the alternative names given to variables that lack independence from one another? *Link back: Section 2.2.4*

Link back: Section 2.4

6. For the set of scores 2, 0, 2, 4, 2 (*y* values) calculate (a) to (e) below:

(a) $\sum y^2$

(b) $\sum y + 1$

(c) $\sum (y + 1)$

(d) $\sum (y + 1)^2$

(e) $(\sum y)^2$

Link back: Section 2.4

7. How would the following mathematical operations be expressed using the sigma symbol and round brackets?

(a) Subtract 1 from each score and find the sum of the resulting values.

(b) Add 2 to each score, square the resulting value, then find the sum of the squared numbers.

3 Describing a single sample

Chapter Aims

Chapters 1 and 2 focused on the what and why of data analyses, but now it is time to get on with the how. The aim of this chapter is to show you how to conduct analyses to describe a single sample of data. You can do this by producing summary numbers, names, tables, or pictures (otherwise known as graphs, charts, or figures).

This chapter starts with descriptive statistics (Section 1.5), which are names or numbers used to summarize a single sample. It then moves on to a consideration of frequency distributions, which are a more visual way of looking at data in a sample, in table or graph form. After this, alternative ways of summarizing the information in a sample visually are considered.

The chapter ends by showing you how to get the SPSS software program to produce descriptive statistics, frequency distributions, and graphs for a single sample of data.

3.1 The single sample

Take a look at the numbers in the final column of Table 2.1, which represent the speed of frogs hopping at different temperatures and are an example of a single sample. Written out as a list the sample would look like this: 2.1, 1.9, 1.6, 1.5, 1.6, 0.5, 2.3, 0.6.

This is a fairly small sample, just eight numbers, so if someone asked you to describe your data you could list each number in turn. However, this is rather clumsy and the person told is unlikely to remember all eight numbers. You are unlikely to have communicated very much about what your data are like. Even if you wrote them down and gave them to your inquisitor, it is not really the most efficient way of doing things. This chapter covers techniques for describing data in a single sample in terms of numbers, text, and graphs. A graph is a visual representation of data (Section 1.5).

A single sample can be measured at any level of measurement. Our frog-hopping data are continuous scale data, but of course samples can consist of observations measured at any level. For example, a single sample of nominally measured gender data on humans might look like this: male, male, female, male, female, male, male. As you will see in this chapter, the appropriate technique to use for describing a sample often depends on the level of measurement.

3.2 Descriptive statistics

Descriptive statistics summarize the main characteristics of a sample of data.

Sample size (*n*) is the number of datum points, that is pieces of information, in a sample.

Descriptive statistics are numbers or names used to summarize the information in a sample. The simplest is the **sample size (*n*)**, which is the number of datum points, that is pieces of information, in a sample (Section 2.1.1) but others include measures of central tendency and measures of variability. We will cover the mean, median, and mode as ways of describing central tendency and the range, interquartile range, variance, and standard deviation as ways of describing variability. Below I will go through each of these techniques in turn, covering what they are, how to calculate them, what their characteristics are, what levels of measurement they can be used with, and examples of their use from the literature.

3.2.1 Central tendency

Measures of central tendency (or **location**) summarize what the data in a sample are typically like. They include the **mean, median,** and **mode**.

Measures of central tendency, or **location**, give us an indication of what the data generally look like. For example, if we have a room full of people over 1.85 m in height, a measure of central tendency would indicate that we were dealing with tall people rather than short people. The three principal ways of measuring central tendency are **mean, median,** and **mode**.

Mean ($\bar{y}$)

Some species of edible snail are more expensive than other species; being able to tell the difference is therefore of commercial as well as biological importance. Unfortunately telling them apart, other than by their taste, can be a little tricky. It can come down to the size of various bits of their genitalia. As part of a study on the genital system of snails, Van Osselaer and Tursch (2000) measured the penis length of 79 specimens of *Helix pomatia*. They added all the lengths up and divided by 79. The answer was 19.68 mm. They were calculating the type of measure of central tendency called a mean. The mean is what people typically understand by the average, although technically that is a broader term which can apply to any measure of central tendency.

I have no doubt that you are familiar with calculating means—you just add everything up and divide by the number of items. In other words the mean is calculated as follows:

- Sum all the values.
- Divide by the number of values.

The formula is given in Box 3.1.

Box 3.1 Formula for calculating a mean

$$\bar{y} = \frac{\sum y}{n}$$

Where n = sample size.

While the mean is very useful and is used widely it suffers two important constraints. Firstly, it cannot be used with nominal data and should only be used with great care for ordinal data; many say you should never calculate a mean for ordinal data. Secondly, it is affected by extreme values. An extreme value is one that is far outside the majority of values. Extreme values should be treated with caution as they may indicate that some problem has occurred. For example, if you have a sample of body lengths of great tits and you have one value that's 30 cm, it is likely there was an error in measurement, recording, or species identification!

Let us consider a very small and simple sample: 14, 16, 12. You can imagine these numbers representing anything you like. They could be the level of haemoglobin in people's blood in g/100 ml, or the length of time dogs spend wagging their tails each day in minutes; this does not matter for the purposes of the illustration. The sum of 14, 16, and 12 is 42; divide this by 3 and this gives you a mean of 14. Change the 12 to a 30 (an extreme value) and the mean changes from 14 to 20!

Median

The median is less sensitive to extremes and can be used freely with ordinal- or scale-level data. The median is the midpoint of the values, the value with as many values above it as below it. The median is calculated as follows:

- Put the numbers in numerical order.
- For an odd number of values, the median is the number in the middle.
- For even number of values, the median is half way between the two middle numbers.

Again, let us consider a very small sample, 2, 4, 6, 7, 20, representing anything you fancy. The median is 6. Alternatively, if there had been an even number of numbers, say 2, 4, 7, 20, the median would have been midway between 4 and 7, which is 5.5.

The median is less affected by extreme values than the mean. If we changed the 20 to a 100 in our sample data above, the medians would remain unchanged. It can be used for both scale and ordinal data.

Mode

For data measured at a nominal level you need to use the mode. The mode is the most frequently occurring value. Let's take the sample of flower colour recorded from 10 rose bushes: red, white, red, red, pink, pink, red, red, white, red. The mode is red.

The above example with flower colour shows the use of mode with nominal data. The mode has the advantage that it is the only measure of central tendency that can be used for data measured at the scale, ordinal, and nominal levels. If you had a sample of ordinal or scale measures—4, 7, 7, 8, 11—then the mode would be 7. The mode is not affected by extreme values.

3.2.2 **Variability**

Measures of variability (or **dispersion**) summarize how similar the data in a sample are. They include the **range, interquartile range, variance**, and **standard deviation**.

Measures of variability, or **dispersion**, give us an indication of the similarity or dissimilarity of the data. For example, if we have a room full of people, all around 1.9 m in height, a measure of variability would indicate that the data were not very variable.

As already mentioned, Van Osselaer and Tursch (2000) found the mean penis length of 79 specimens of *Helix pomatia* to be 19.68 mm. While this gives us some idea of what the 79 raw data records looked like, it leaves us with a somewhat incomplete picture. At one extreme all 79 records could have been around 19.68 mm. At the other extreme some could have been much smaller than 19.68 mm, while others could have been larger.

In short, we do not get an idea about the variability of the data from just the measure of central tendency. Two sets of numbers with the same mean can look very different depending on how variable the numbers are. For example, 8, 9, 10, 11, 12 look very different from the more variable 0, 5, 10, 15, 20, but both have a mean of 10.

Whenever you report a measure of central tendency you should report the relevant sample size. For ordinal and scale data, you should also get into the habit of reporting a measure of variability. The following paragraphs consider four measures of variability from which to choose: **range, interquartile range, variance,** and **standard deviation**. All can be used for scale data, but only the range and the interquartile range are appropriate for using with ordinal data. The variability of nominal data cannot be measured using any of these techniques—you have to rely on frequency distributions (see the next section) to get an idea of the variability of data measured at the nominal level.

Range

The range is the simplest way of expressing the spread of values in a sample. It is the difference between the smallest and the largest values. In other words: the range is the difference between the maximum and minimum values.

As part of their study of the genital system of edible snails, Van Osselaer and Tursch (2000) looked at the variation in their data due to variation in recording. To do this they repeated the same measurements of the same features on the same specimen 10 times. The smallest value, or minimum, they recorded for the penis length of specimen JD-0368 was 16.5 mm and the largest, or maximum, was 18.0 mm. The range was therefore 1.5 mm.

As an aside, you might like to note that this is a rare example of when taking replicates of the same item under the same conditions is not a case of pseudoreplication (Section 2.1.1), because of the context of the question being addressed.

You will be able to see that the range is sensitive to extreme values. If either the minimum or maximum is far away from the rest of the data the range could be misleading. For example, if one of the measurements of JD-0368's penis length had been recorded as 20.0 mm by Van Osselaer and Tursch, then this would have been reflected in the range even if most of the values were between 16.5 and 18.0 mm. The range can only be used with scale or ordinal data, not with nominal data.

Interquartile range

The term interquartile range sounds quite fancy but the concept and the calculation of this measure of variability are really quite straightforward. You already know about medians from Section 3.2.1. While a median splits a sample into two, a quartile splits it into four. The lower quartile is the midpoint of the values below the median and the upper quartile is the midpoint of the values above the median. For example, for the sample 2, 2, **7**, 7, **8**, 9, 10, 11, **12**, 14, 16, the median is 9, the lower quartile is 7, and the upper quartile is 12.

The interquartile range is the difference between the lower and upper quartile. For our example data, 2, 2, **7**, 7, 8, 9, 10, 11, **12**, 14, 16, the interquartile range is 12 minus 7, which is 5. If the maximum had been 50 rather than 16, then the interquartile range would have been the same, which illustrates that it is less affected by extreme values than the range. Of the measures of variability presented in this book it is in fact the least sensitive to extremes.

The interquartile range can be used with scale or ordinal data, but not nominal data. It is the most appropriate measure of variability to report in association with a median.

Variance (s^2)

Although the range and interquartile range do give an idea of the spread of values, most of the data items do not directly contribute to their calculation. It's easy to get the feeling that you are missing something! The variance is nice because all the numbers are involved. However, it can only be used with data measured at a scale level. Two alternative formulae for variance are given in Box 3.2. In short, the variance is the mean squared deviation from the mean.

Box 3.2 Formulae for calculating variance and standard deviation

	Formula A	Formula B
Variance	$s^2 = \dfrac{\sum (y-\bar{y})^2}{n-1}$	$s^2 = \dfrac{\sum y^2 - \dfrac{(\sum y)^2}{n}}{n-1}$
Standard deviation	$s = \sqrt{\dfrac{\sum (y-\bar{y})^2}{n-1}}$	$s = \sqrt{\dfrac{\sum y^2 - \dfrac{(\sum y)^2}{n}}{n-1}}$

Where n = sample size.

Following are the steps that you need to take to calculate the variance of a set of numbers using formula A in Box 3.2. It may sound a bit involved but it's really not difficult. Take a look at Box 3.3. This shows the calculation of the variance using formula A for a very simple sample of three numbers: 3, 4, 5. It will help if you go through Box 3.3 as you go through these steps:

- Find the mean of your numbers (Section 3.2.1, mean).
- Find $(y-\bar{y})$, the 'distance' between each of your data points and the mean by subtracting the mean from each item in turn. In other words you calculate the deviation from the mean, for each value of y.
- Find $(y-\bar{y})^2$ by squaring each deviation; that is multiply it by itself.
- Find $\Sigma(y-\bar{y})^2$ by adding all the squared deviations up so that you have the sum of the squared deviations (from the mean) which is otherwise known as the **sum of squares**.
- Divide the sum of squares by $n-1$, which is known as the **degrees of freedom**, and which in this case is one less than the sample size. This effectively finds the average sum of squares.

Sum of squares is the short way of saying that the sum of the squared deviations from the mean is a number related to sample size, $\Sigma(y-\bar{y})^2$.

Degrees of freedom is a number related to sample size. When calculating the variance of a sample, the degrees of freedom are 1 less than the sample size.

Because of the vagaries of sampling, which we touched on in Chapter 2, degrees of freedom rather than sample size are used to calculate statistics. For variance the degrees of freedom are one less than the sample size. Samples will tend to underestimate the variance of the population from which they are taken. By making the value that we divide by smaller by 1 than the sample size, we are counteracting this a bit.

Box 3.3 Calculating variance and standard deviation

Construct a calculation table with the column headings as shown below:

Column number	**1**	**2**	**3**	**4**
Column heading	y	$\bar{y}$	$(y-\bar{y})$	$(y-\bar{y})^2$
	3	4	−1	1
	4	4	0	0
	5	4	1	1
Σ				2

Put your data (y) in column 1: in this example $y = 3, 4, 5$.

Find the mean of y by dividing the sum of y by the sample size (n): in this example, $12/3 = 4$. Put this in column 2 for every value of y.

Find the deviation of each value from the mean: see column 3.

Calculate the squares of the deviations: column 4.

Add the squares of the deviations together to find the sum of squares: bottom of column 4.

Divide the sum of squares by the degrees of freedom (sample size minus 1; or $n-1$): $= 2/2 = 1$

This is the variance. To find the standard deviation take the square root of this number: $= \sqrt{1} = \pm 1$

There were a lot of new phrases in the last two paragraphs so your head may be spinning a bit. Here is a recap:

- Deviation from the mean (the difference between a datum and the mean of the data it comes from): $(y - \bar{y})$.
- Squared deviation from the mean (the deviation of the mean multiplied by itself): $(y - \bar{y})^2$.
- Sum of squares (the total of all the squared deviations of the mean for all the data): $\Sigma(y - \bar{y})^2$.
- Degrees of freedom (which in this case is one less than the sample size): $n - 1$.

Look back at Box 3.2, the choice of formulae illustrates a potentially confusing situation that I mentioned in Section 2.4. It is very important that if you see a different-looking formula elsewhere, you do not assume that you have not understood things after all. Formulae can be arranged to look quite different, but the key point is that they give the same answer if used properly. Different arrangements may reflect the context in which the formula is being presented. Some arrangements are better for giving an understanding of what is going on, while others may be easier for doing calculations with large sample sizes and reduce inaccuracies generated when rounding numbers up or down. Sometimes the formulae presented just reflect the personal preference or experience of the author. In summary, the take-home message is not to be thrown by seeing different formulae for the same statistic.

I favour formula A because it gives a good feel for why variance gives us a measure of variability of the data. Imagine that instead of 3, 4, 5, the data were 1, 4, 7. The mean would still be 4 but the data are more variable. The deviations from the mean for 1 and 7 are going to be bigger than for 3 and 5, leading to a bigger value for the variance. The more the data vary, the bigger will be the sum of the squared deviation and the bigger the variance.

Since, as we shall see later, it is very easy to get a computer package like SPSS to calculate variance, you need not worry about becoming adept at calculating variance by hand. However, it is worth considering the mechanism of calculation at least once so that you can get a feel of how variance provides a measure of variability.

If you've grasped the general principle of how variance works, then you might be wondering why it is necessary to square each of the deviations. Just for fun, you could calculate the sum of the deviations from the mean rather than the sum of the squares—you'll find it always comes to zero because the negative values cancel out the positive ones. Squaring the deviations gets rid of this problem because squared values are always positive. This is because a minus times a minus is a plus, as is a plus times a plus.

The need to square our values does leave us with a slightly tricky situation in interpreting the variance. The units of variance are those of the raw data squared. The bigger the variance the bigger the variability of the raw data, but direct comparison of the variance with the raw data is not possible. We shall see in the next section how using the standard deviation, which is derived from the variance, is much better for this. However, variance is worth knowing about in its own right, particularly because, although it is itself a descriptive statistic, it is widely used in inferential statistical procedures.

Standard deviation (s)

Variance takes into account all the data, which is good, but the units of variance are those of the raw data squared, which is not easily interpreted. The standard deviation is calculated in the same way as the variance, with a further step added of finding the square root. In other words, the standard deviation is the square root of the variance.

The standard deviation can only be calculated for data measured at a scale level and is reported in association with a mean. To calculate a standard deviation you follow the instructions on how to calculate a variance and find the square root of the result. Two alternative formulae are shown in Box 3.2. For a worked example using formula A, see Box 3.3.

Van Osselaer and Tursch (2000) reported the standard deviation for the mean of the penis length of their 79 snail specimens (Section 3.2.1, mean). The mean was 19.68 mm and the standard deviation was 3.12 mm. Strictly speaking the standard deviation could be plus or minus 3.12, because 3.12 times 3.12 equals 9.73, but so does −3.12 times −3.12. Typically this would be reported in the general form: mean ± standard deviation. For our example this is: 19.68 ± 3.12 mm.

This may seem strange at first, but if you think about it, it actually makes sense that a measure of variability should go both above and below the mean.

3.3 Frequency distributions

This chapter is all about how to describe a single sample. We have looked at ways of describing the general size or 'flavour' of the data (measures of central tendency), and ways of describing how much the data vary (measures of variability). This section shows how a sample of data can be described in terms of the frequency of the different values in the sample. Frequency literally refers to how often different values occur. **Frequency distributions** can be presented in table or graph form. Frequency distributions give us an idea of the central tendency and variability of our sample. However, they also summarize how many occurrences there are of each value in a sample. In other words, frequency distributions also allow us to describe the 'shape' of our data.

Frequency distributions show how many times different values occur in a sample. This information can be presented in tabular form as a **frequency-distribution table**, or graphically as a **frequency-distribution graph**. A frequency-distribution graph of continuous scale data has a special name, a **histogram**.

We are going to consider two basic procedures for constructing **frequency-distribution tables** and **graphs,** according to the level of measurement. In Section 3.4.1 we are going to deal with the procedure that covers data measured at nominal, ordinal, and discrete scale levels, as exampled by the nominal gorilla gender data in Table 3.1a. In Section 3.4.2, we are going to consider the other procedure, which you should use for continuous scale level, as exampled by the sperm-length data in Table 3.2a.

3.3.1 For nominal, ordinal, and discrete scale data

For nominal, ordinal, and discrete scale data, constructing a frequency distribution is fairly straightforward. Take a good look at Table 3.1b which shows how you make a distribution table by tallying up how many times each value occurs. The totals of these

(a) Male, Male, Female, Male, Female, Female, Female, Male, Male, Male, Male, Female, Female

(b)	Category	Tally	Frequency
	Male	IIIIIII	7
	Female	IIIII	5

Table 3.1 Gender in gorillas. (a) Raw observations (male/female). (b) Frequency-distribution table.

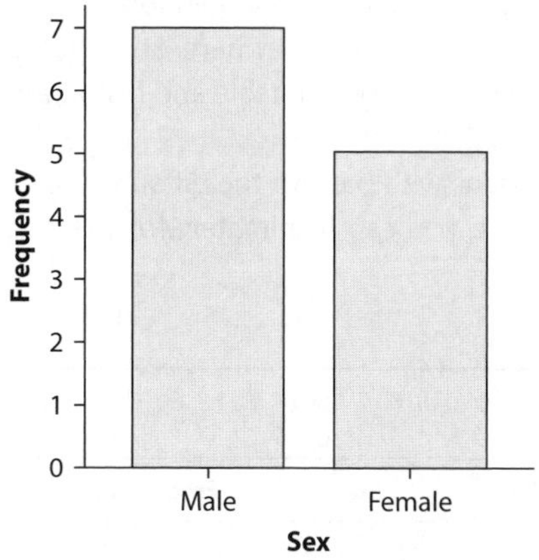

Figure 3.1 Column graph. Example of a column graph used to present the frequency distribution of data measured at nominal, ordinal, or discrete scale level. In this example the data are nominal. The graph shows that there are seven males and five females in the sample.

tallies, in the example seven males and five females, are the frequencies. You then plot these frequencies in a frequency-distribution graph, as illustrated in Figure 3.1. The convention for these three types of data is that the columns in their frequency graphs do not touch. For ordinal and discrete scale data the categories go in ascending order of their values' ranks. For nominal data, which do not have an intrinsic rank, the categories on the *x*-axis can go in any order.

3.3.2 **For continuous scale data**

To construct a frequency distribution for continuous scale data you have to go through the same procedure as for other types of measurement. However, the categories that you should use for tallying up values are not immediately obvious. What you have to do is define categories covering a range of values yourself. There are no hard-and-fast rules about this, but in general you should:

- Make your categories equal in size.
- Avoid using lots of categories with only a few values in each (no more than 10 categories as a rough guide).
- Avoid using so few categories that you cannot see any detail (no fewer than five categories as a rough guide).

By convention, the bars in the frequency chart for continuous data are drawn touching and the chart produced has a special name: a **histogram**. Unfortunately, this term is used very loosely, often being applied to graphs looking at the relationship between variables that involve more than one sample (Chapter 12). The main problem with this bad practice is that reading around may not reinforce this concept as it should. Therefore, you need to be confident in your understanding of what a frequency-distribution graph for continuous data is and what it should look like, in order not to get confused by what you read or you are told elsewhere. Study Table 3.2b and look at Figure 3.2: They show a frequency-distribution table and histogram for the sample of continuous scale data in Table 3.2a.

Having said that you should produce histograms with bars for the different *x* categories touching, there are circumstances where you can legitimately forsake this

Table 3.2 Sperm length in gorillas. (a) Raw observations. (b) Frequency-distribution table. Measurements shown are in micrometres (μm).

(a) 60.80, 60.92, 60.51, 60.62, 60.75, 61.19, 61.21, 60.71, 61.25, 60.74, 60.81, 60.83, 60.91, 61.15, 61.22, 60.83, 61.08, 60.93, 61.12, 61.34, 61.01, 61.05, 61.14, 61.44, 61.07, 61.37

(b)	Classes (range of values)	Tally	Frequency
	60.50–60.69	II	2
	60.70–60.89	IIIIIII	7
	60.90–61.09	IIIIIII	7
	61.10–61.29	IIIIIII	7
	61.30–61.49	III	3

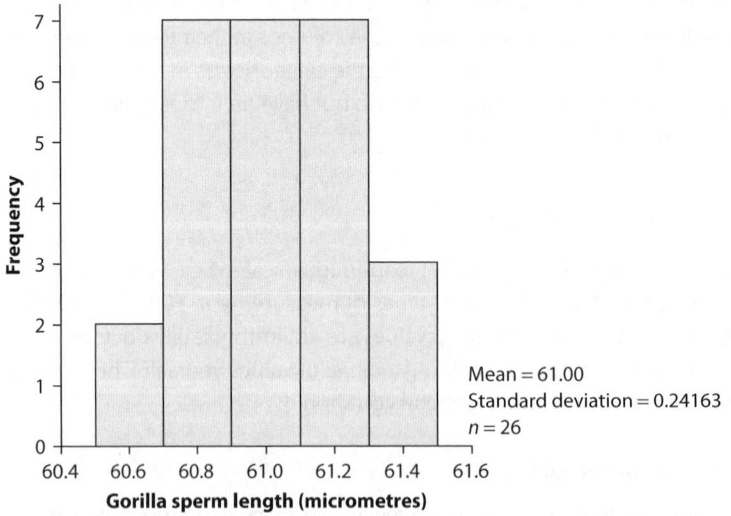

Figure 3.2 Histogram. Example of a histogram used to present the frequency distribution of data measured at continuous scale level. In this example, the frequency of length in a sample of gorilla sperm is shown ($n = 26$). The value marking each column on the *x*-axis shows the centre point of the class interval; for example, 60.6 marks the column for the interval 60.50 to 60.69.

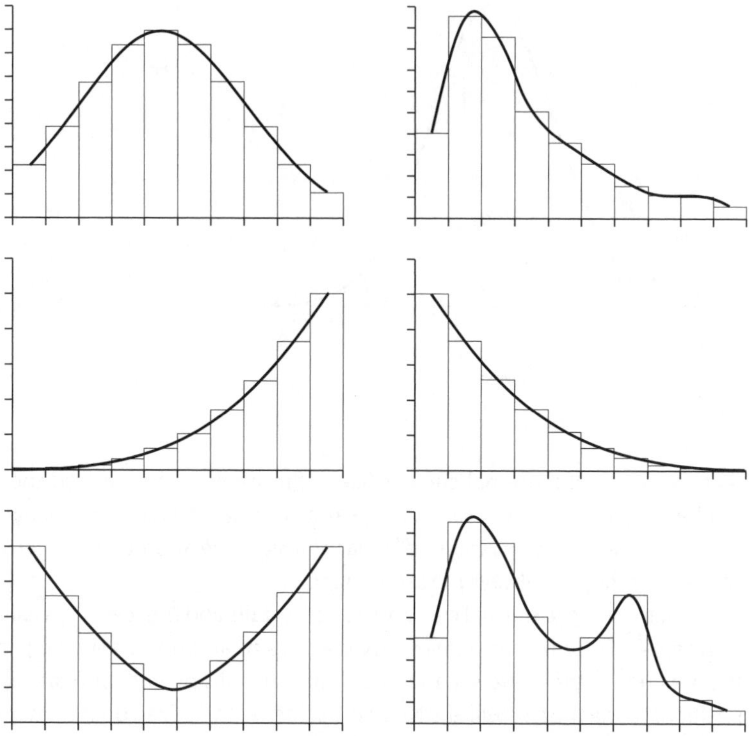

Figure 3.3 Variety of histogram shapes. This figure shows six different histograms. The lines drawn through the tops of the columns summarize the shape of each distribution.

convention. For example, Beasley *et al.* (2000) produced histograms of shell size in males and in females of the mussel *Paxyodon syrmatophorus*. In order that the histograms of the different sexes could be compared easily they drew the size categories apart, so that they could superimpose the female histogram on the male one.

Look at the range of example histogram shapes in Figure 3.3. A line has been drawn connecting the centres of each bar on each histogram. This is a common way of summarizing the shape of a histogram. The first histogram in Figure 3.3 depicts a symmetrical or bell-shaped curve, a particular example of which is the normal curve. The normal curve is special and I am going to devote the next section to telling you a little more about its importance and features.

3.3.3 **The normal distribution**

The **normal distribution** is a special sort of symmetrical bell-shaped frequency distribution. The extremes of a normal distribution are called its **tails** and the middle section is called its **body**. Approximately 95% of the area of a normal curve falls within two standard deviations either side of the mean.

Frequency distributions of biological data, measured at a continuous scale level, very often produce a normal curve. It is one of those situations where nature is weirder than science fiction. Whether it be shell length in snails, vigilance rates in birds, or rates of oxygen uptake in plants, a histogram of the data is likely to be normal.

The **normal distribution** is not any old symmetrical bell-shaped curve. The shape of a normal curve is very precise and described by a scary-looking formula. You can look this formula up in another text (see Selected further reading) if you are feeling curious, but it's not important that we know the details here. The important point is

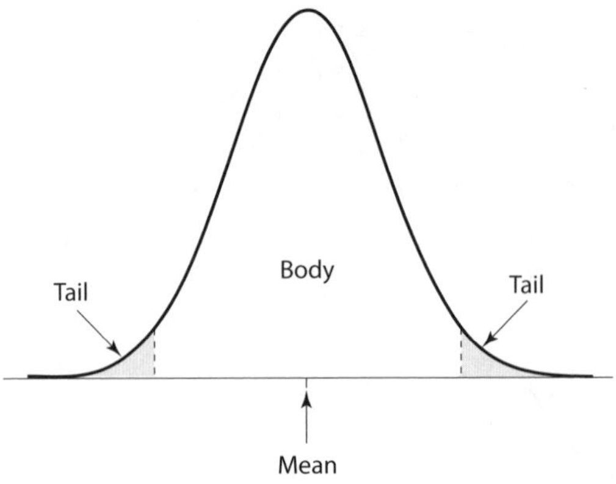

Figure 3.4 A normal curve. The areas in the tails beyond two standard deviations above and below the mean are shaded. This shaded area represents approximately 5% of the area under the curve.

that the shape of a normal curve is fixed. If more data are involved then the curve will be bigger, but still the same shape as a curve generated from a smaller amount of data. It is like those Russian dolls that fit inside one another. They get smaller and smaller but they are always the same shape.

The extremes of a normal curve are called its **tails** and the rest of it is called its **body** (Figure 3.4). Since normal curves vary only in size and not shape, the proportion of the area under the curve separated out into the tails by lines at relatively the same positions along the axis will always be the same. With Russian dolls the same percentage of the doll is above her neck, whatever size she is.

We can use the standard deviations of the data to identify the same relative points along the *x*-axis. For example, two standard deviations above and below the mean will always cut off 2.28% of the area of the curve in each tail (Figure 3.4). Another way of saying this is that 2.28% of the data falls above and 2.28% falls below two standard deviations either side of the mean. Yet another way of expressing this is to say that just under 5% of the values are more extreme than two standard deviations away from the mean.

The size of the sample is important. A histogram of a small sample, perhaps of eight values, is unlikely to look normal even if the population from which it is derived is normally distributed. The larger the sample size the more likely it is that the frequency distribution will come out as normal, assuming that the sample is taken from a normally distributed population.

Parametric statistics is a group of inferential statistical techniques that rely heavily on the assumption that the frequency distribution of the data is a particular shape, commonly **normal**, but other possibilities include **binomial** and **Poisson** distributions.

The normal distribution is an important phenomenon because of its application in inferential statistics. In particular, a very widely used group of techniques called the **parametric statistics** rely heavily on the special properties of the normal distribution. I am not going to seek to explain in detail how the normal distribution is used in inferential statistics, because this is not essential to conducting the basic data analyses appropriately or effectively. If you get keen then you can look this up in one of the texts listed in Selected further reading. What you do need to understand is that the normal distribution is a symmetrical bell-shaped frequency distribution of continuous scale data and that many, although not all, continuous scale data sets in biology form a normal distribution.

3.3.4 **Discrete data and the normal distribution**

In the last section I made the point that you can only get a normal distribution with continuous scale data. Certainly you cannot get a normal distribution with nominal or ordinal data, but we hit a bit of a grey area with discrete scale data on this. In theory, discrete data cannot produce a normal distribution. In practice, however, if discrete data produce a symmetrical bell-shaped frequency distribution, it can be regarded as approximating the normal distribution and you can proceed as though the data were continuous. This means that any scale-level data can potentially be treated as though it were normally distributed and used in parametric statistics.

3.3.5 **Other distributions: binomial and Poisson**

A couple of other distributions are worth mentioning: **binomial** and **Poisson**. Although these are not as commonly occurring as the normal distribution they are useful for working with certain types of data. The binomial distribution crops up where you have the sort of data created by tossing a coin and recording heads or tails; two nominal categories. More biologically relevant examples occur when data are collected as male or female, dead or alive, healthy or diseased, and so on.

The Poisson distribution is often a good description of count data (Section 2.3.2). Counts have the feature of being non-negative integers, in other words whole numbers zero or greater. The most famous example is the number of Prussian cavalry men killed each year over a twenty-year period through being kicked by a horse. However, you are more likely to collect, for example, counts of the number of eggs in a nest or the number of cells in a petri dish.

3.4 Pies, boxes, and errors

Frequency-distribution graphs are one way of inspecting and describing your sample visually. Three other useful methods to know about are pie charts, boxplots, and error bars.

3.4.1 **Pie charts as alternatives to frequency-distribution charts**

The relative number of values in different categories, or classes of values, can be displayed as pieces of pie on a pie chart. The angle at the centre of the circle, defining each piece of pie, is calculated by dividing the frequency of the category or class by the sample size and multiplying this by 360°. The different pieces of the pie should be clearly labelled or a key included using colour or shading to individualize them.

Pie charts should not contain more than about six sections, otherwise they become rather cluttered. For this reason they are not usually a good idea for continuous scale data and often are not the best idea for ordinal or discrete scale data. However, they are often very good for describing a single sample of data measured at the nominal level.

Morton and Britton (2002) conducted an investigation on a beach in Western Australia that included an assessment of scavenger feeding on dying, washed-up sea

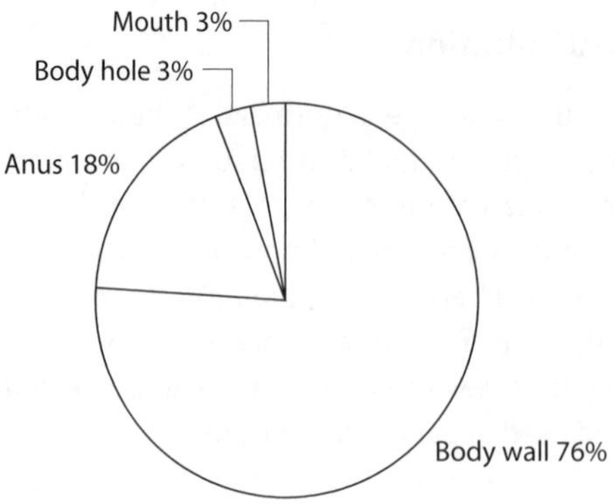

Figure 3.5 Pie chart of a single sample. This is based on data reported in Morton and Britton (2002), who recorded the feeding behaviour of 142 scavengers on sea cucumbers washed up on a beach in Western Australia. The feeding behaviour was measured as being on the body wall or through a hole in the body wall, mouth, or anus.

cucumbers. On 28th July 2000 they observed 142 scavengers and used a pie chart to display the proportion of this sample feeding on the body wall itself, or through a hole in the body wall, the anus or mouth. Take a look at Figure 3.5 to see what their data looked like in a pie chart. In this example, feeding method is being measured at a nominal level.

3.4.2 **Understanding boxplots**

A boxplot is a good way of describing a sample of scale or ordinal data visually, especially when data are not normally distributed. Another name for a boxplot is a box and whisker plot. Look at Figure 3.6 to see a generalized boxplot, which would appear relative to a *y*-axis of values in the sample. The thick horizontal line indicates the median.

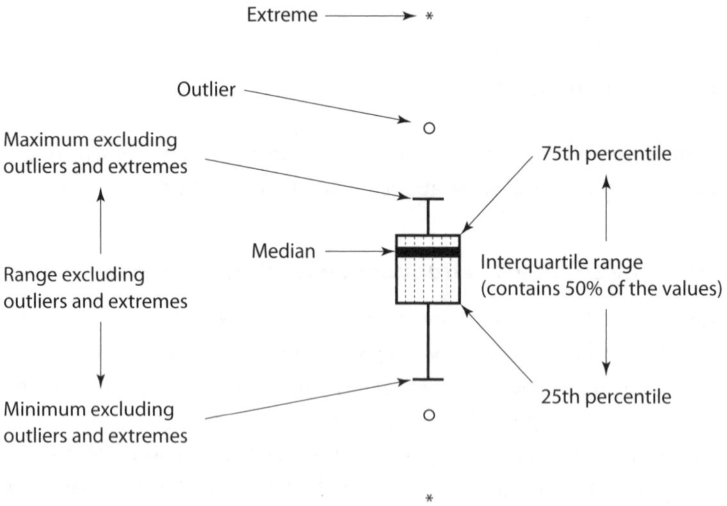

Figure 3.6 Boxplot of a single sample. The important features of a boxplot are labelled on this generalized diagram. The thick black line indicates the median, the surrounding box indicates the interquartile range, and the T-shaped bars indicate the range. In SPSS the T bars exclude outliers and extremes where outlying values are those between 1.5 and 3 interquartile ranges (box lengths) from the 25th or 75th percentiles (lower or upper edge of the box), and extreme values are those more than 3 interquartile ranges (box lengths) from the 25th or 75th percentiles (lower or upper edge of the box).

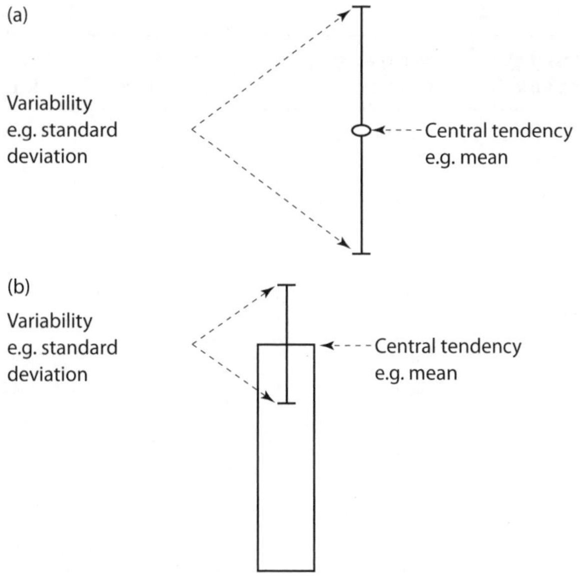

Figure 3.7 Error bars for single samples. Error bars can be used to depict the variability of a measure of central tendency, which can be marked as either points (a) or bars (b).

The box around the median indicates the interquartile range and therefore contains 50% of the values. The 'whiskers' indicate the range. In SPSS they indicate the range excluding outliers and extremes. Outliers are values between 1.5 and 3 box lengths from the upper or lower edge of the box. Extremes are values more than 3 box lengths from the upper or lower edge of the box.

3.4.3 **Introducing error bars**

Error bars are a way of depicting the variability of a sample on a graph. They are usually T-shaped and can extended above and/or below a point or a bar marking the central tendency of the sample. Typically, the measure of central tendency that error bars are used in conjunction with is the mean. The mean can be represented by a point, producing an errorplot as shown in Figure 3.7a, or a bar, as shown in Figure 3.7b. We will consider errorplots again in the next chapter.

Error bars are T-shaped marks that indicate the variability of a sample. They are drawn in association with points or bars indicating the central tendency of the data.

3.5 Example data: ranger patrol tusk records

In Section 3.6 I will be showing you how to generate descriptive statistics, frequency distributions, pie charts, and boxplots using SPSS. In this section, I am going to introduce the example data that I will be using to illustrate these procedures.

The data are records of elephant tusks found by ranger patrols in Mikumi National Park in 1983 are presented in Table 3.3. The data come from TANAPA via Fredrick Mofulu and his colleagues in the Ecology Department in Mikumi National Park. TANAPA is short for Tanzania National Parks, which is the authority that runs all the National Parks in Tanzania.

Table 3.3 Example data: ranger patrol tusk records.

Record	Number of tusks	Mass (kg): first tusk	Mass category: first tusk	Mass (kg): second tusk	Cause of death	Location found
2.004	2	9.03	Medium-heavy	9.03	Poaching	Daraja Mbili
2.005	2	12.03	Heavy	13	Control	Lodge
2.006	2	4.5	Medium-light	3.5	Poaching	Champgore
1.002	1	2.5	Light	-	Unknown	Champgore
2.007	2	1.5	Light	1	Unknown	Mwanambande
1.003	1	0.5	Light	-	Unknown	Mwanambande
1.004	1	2.5	Light	-	Poaching	Mgoda
2.008	2	1.5	Light	1.5	Poaching	Champgore
2.009	2	2.5	Light	2.75	Disease	Mahondo
1.005	1	3	Medium-light	-	Train	Mkemgumba
2.01	2	9	Medium-heavy	8.5	Control	Park HQ/Village
2.011	2	2.5	Light	3	Poaching	Mwanambande
1.006	1	3	Medium-light	-	Poaching	Mkemgumba
1.007	1	2	Light	-	Poaching	Champgore
1.008	1	1.5	Light	-	Poaching	Mgoda
1.009	1	20	Heavy	-	Unknown	Mgoda
1.01	1	3.5	Medium-light	-	Poaching	Kiraza-Kikwanza
2.013	2	3.5	Medium-light	3.03	Unknown	Ikoya
2.014	2	3.5	Medium-light	3.5	Unknown	Mgoda
2.015	2	2	Light	2.03	Disease	Mgoda
1.011	1	2	Light	-	Disease	Rungwa River
2.016	2	1	Light	1	Disease	Upper Mgoda
2.017	2	13	Heavy	14	Poaching	Mikwajuni
1.012	1	3	Medium-light	-	Unknown	Mahondo

Source of data: Mikumi National Park records (1983) courtesy of Tanzania National Parks (TANAPA).

Every time rangers on patrol come across a dead elephant they record the information for the variables underlined in the list below. The level of measurement of the data is given in brackets. Each variable has a sample of 24 data points, except the mass of the second tusk, as this is missing in about half the records. That is, the sample size is 24, except for mass of the second tusk for which the sample size is 13.

- *Number of tusks*. This tells us whether one or two tusks were found (discrete scale).
- *Mass (kg): first tusk*. This is the mass of the first tusk measured (continuous scale).
- *Mass category: first tusk*. This is the broad mass category of the first tusk measured (ordinal).
- *Mass (kg): second tusk*. This is the mass of the second tusk measured if a pair of tusks were found (continuous scale).
- *Cause of death*. This is the cause of death of the elephant (nominal). The categories are poaching (killed by poacher), control (killed by park management

in life-threatening situations for rangers, villagers, farmers, or tourists), train (hit by a train), disease (died of disease), or unknown.

- *Location found*. This is the area of the park in which the carcass was found (nominal). Example locations are Chamgore and Mahondo.

Sometimes the rangers only find one tusk. This may be because the elephant only had one tusk, or for some other reason, such as the carcass has been spread around and they could not find the second tusk. Although there are no examples in Table 3.3, rangers do occasionally find a carcass but no tusks. This can happen, for example, if the elephant was tuskless or the tusks have been removed by poachers. When the rangers find two tusks the first one to be measured is a random choice.

3.6 Worked example: using SPSS

We are going to use SPSS to produce descriptive statistics, frequency-distribution tables, and frequency-distribution graphs (Section 3.6.1), pie charts (Section 3.6.2), and boxplots (Section 3.6.3) of single samples from the tusk data introduced in Section 3.5. To do this the data must be entered into SPSS as shown in Figure 3.8. In newer versions of SPSS the **Data Editor** window has a **Variable View** tab and a **Data View** tab. Figure 3.8a shows the former and Figure 3.8b the latter. Appendix I has notes on how to enter data into SPSS.

In SPSS you can generate an errorplot, which is a graph with means plotted as points flanked by error bars depicting, for example, standard deviation stretching above and below the point. However, errorplots are generally used to compare the characteristics of two or more samples, rather than to explore the characteristics of a single sample. We are therefore going to leave a consideration of how to generate errorplots using SPSS until the next chapter.

3.6.1 Descriptive statistics and frequency distributions

In this section we are going to go through how to use the frequencies facility in SPSS to produce both descriptive statistics and frequency-distribution tables and graphs. As explained in Section 3.3, there are different approaches needed when constructing frequency tables and graphs for nominal, ordinal, and discrete, as compared to continuous scale data. Accordingly we are going to consider how to use SPSS's frequency facility separately for these two groups of data types.

For nominal, ordinal, and discrete scale data

To get SPSS to produce descriptive statistics and frequency tables or graphs for a single sample you must first open the data file. Then you must make the following selections:

Analyze
→ Descriptive Statistics
→ Frequencies...

(a)

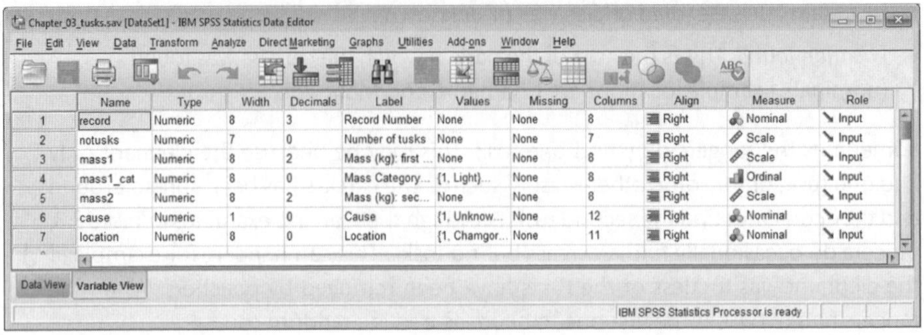

Chapter_03_tusks.sav [DataSet1] - IBM SPSS Statistics Data Editor

	Name	Type	Width	Decimals	Label	Values	Missing	Columns	Align	Measure	Role
1	record	Numeric	8	3	Record Number	None	None	8	Right	Nominal	Input
2	notusks	Numeric	7	0	Number of tusks	None	None	7	Right	Scale	Input
3	mass1	Numeric	8	2	Mass (kg): first ...	None	None	8	Right	Scale	Input
4	mass1_cat	Numeric	8	0	Mass Category...	{1, Light}...	None	8	Right	Ordinal	Input
5	mass2	Numeric	8	2	Mass (kg): sec...	None	None	8	Right	Scale	Input
6	cause	Numeric	1	0	Cause	{1, Unknow...	None	12	Right	Nominal	Input
7	location	Numeric	8	0	Location	{1, Chamgor...	None	11	Right	Nominal	Input

(b)

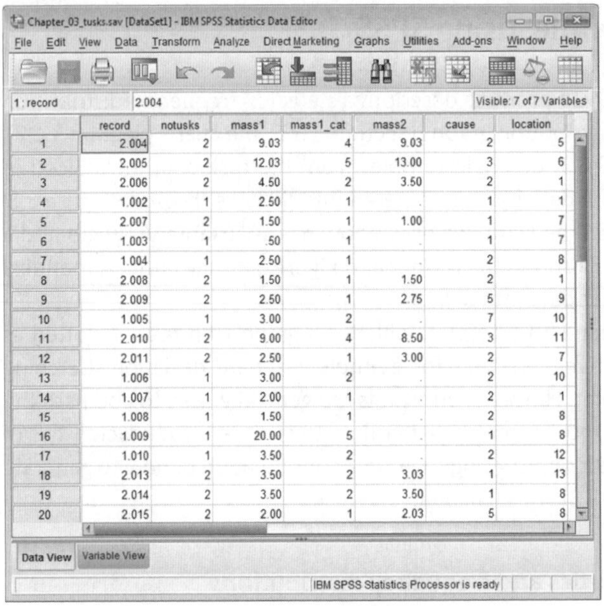

Chapter_03_tusks.sav [DataSet1] - IBM SPSS Statistics Data Editor

1 : record 2.004 Visible: 7 of 7 Variables

	record	notusks	mass1	mass1_cat	mass2	cause	location
1	2.004	2	9.03	4	9.03	2	5
2	2.005	2	12.03	5	13.00	3	6
3	2.006	2	4.50	2	3.50	2	1
4	1.002	1	2.50	1	.	1	1
5	2.007	2	1.50	1	1.00	1	7
6	1.003	1	.50	1	.	1	7
7	1.004	1	2.50	1	.	2	8
8	2.008	2	1.50	1	1.50	2	1
9	2.009	2	2.50	1	2.75	5	9
10	1.005	1	3.00	2	.	7	10
11	2.010	2	9.00	4	8.50	3	11
12	2.011	2	2.50	1	3.00	2	7
13	1.006	1	3.00	2	.	2	10
14	1.007	1	2.00	1	.	2	1
15	1.008	1	1.50	1	.	2	8
16	1.009	1	20.00	5	.	1	8
17	1.010	1	3.50	2	.	2	12
18	2.013	2	3.50	2	3.03	1	13
19	2.014	2	3.50	2	3.50	1	8
20	2.015	2	2.00	1	2.03	5	8

Figure 3.8 Example data in SPSS, ranger patrol tusk records. (a) Variable View. (b) Data View.

A window like that shown in Figure 3.9a will appear. Select the following three samples in turn and use the arrow to move it to the **Variable(s)** box: *Cause, Mass category: first tusk*, and *Number of tusks*. These samples contain data measured at nominal, ordinal, and discrete scale respectively.

Next click the **Statistics** button to bring up the **Statistics** dialogue window (Figure 3.9b). Select the statistics you want by clicking on the appropriate little boxes to either select or deselect the options. In this case, check that all the features we have mentioned in this chapter are selected: **Quartiles, Standard Deviation, Variance, Range, Minimum, Maximum, Mean, Median, Mode**, and **Sum**. Clicking the **Continue** button will take you back to the main dialogue window (Figure 3.9a). Now click the **Charts** button to bring up the Charts dialogue window (Figure 3.9c). You want **Bar charts** for **Chart Type** and **Frequencies** for **Chart Values**. Once again, click **Continue**

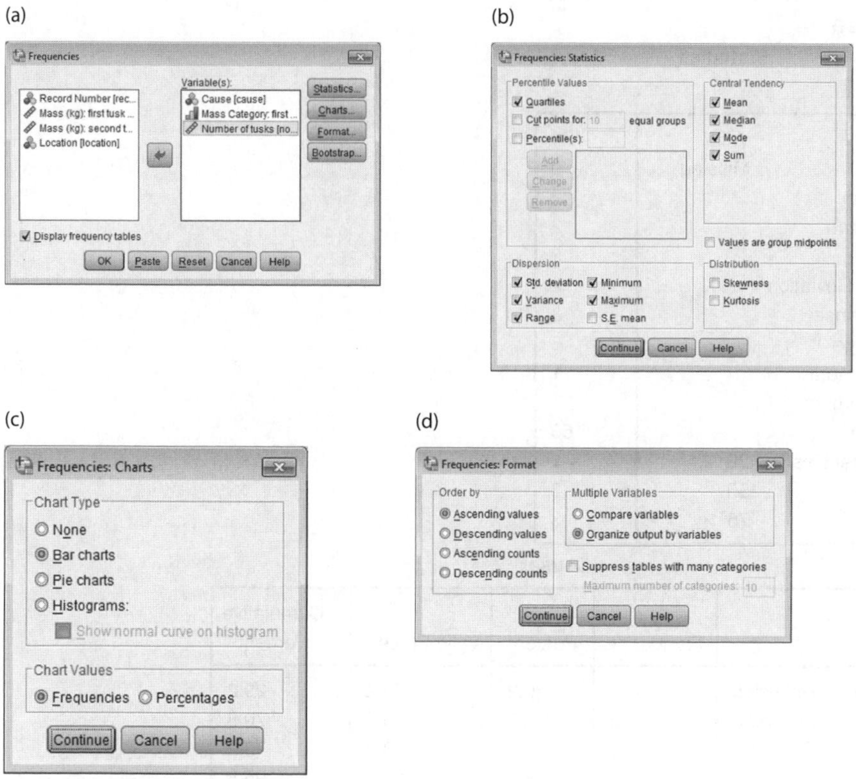

Figure 3.9 Producing descriptive statistics and column frequency-distribution graphs using SPSS. (a) Frequencies dialogue window. (b) Statistics dialogue window. (c) Charts dialogue window. (d) Format dialogue window.

to take you back to the main dialogue window. Now you need to click the **Format** button to bring up the **Format** dialogue window. Check that **Ascending values** is selected from the **Order by** choices and select **Organize output by variables** from the **Multiple Variables** options. Click **Continue** to go back to the main dialogue window (Figure 3.9a). Once you have done this you can click **OK** and, in the flash of an eye, you will get output like that in Figure 3.10.

For continuous scale data

To get SPSS to produce descriptive statistics and frequency tables and graphs for a single sample you must first open the data file. Then you must make the following selections:

Analyze
→ Descriptive Statistics
→ Frequencies...

A window like that shown in Figure 3.11a will appear. Select *Mass (kg): first tusk* and use the arrow to move it to the Variable(s) box. This sample contains data measured at continuous scale level.

Cause

Statistics

Cause

N	Valid	24
	Missing	0
Mean		2
Median		2
Mode		2
Std. Deviation		2
Variance		3
Range		6
Minimum		1
Maximum		7
Sum		60
Percentiles	25	1
	50	2
	75	3

NOTE:
Only N and mode useful information with nominal data

Cause

		Frequency	Percent	Valid Percent	Cumulative Percent
Valid	Unknown	7	29.2	29.2	29.2
	Poaching	10	41.7	41.7	70.8
	Control	2	8.3	8.3	79.2
	Disease	4	16.7	16.7	95.8
	Train	1	4.2	4.2	100.0
	Total	24	100.0	100.0	

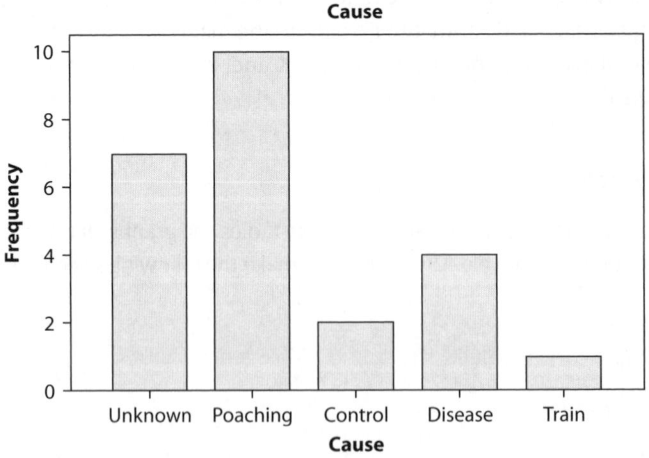

Figure 3.10 SPSS output for descriptive statistics and column frequency-distribution graphs.

Mass Category: first tusk

Statistics

Mass Category: first tusk

N	Valid	24
	Missing	0
Mean		2
Median		2
Mode		1
Std. Deviation		1
Variance		2
Range		4
Minimum		1
Maximum		5
Sum		49
Percentiles	25	1
	50	2
	75	2

Mass Category: first tusk

		Frequency	Percent	Valid Percent	Cumulative Percent
Valid	Light	12	50.0	50.0	50.0
	Medium-Light	7	29.2	29.2	79.2
	Medium-Heavy	2	8.3	8.3	87.5
	Heavy	3	12.5	12.5	100.0
	Total	24	100.0	100.0	

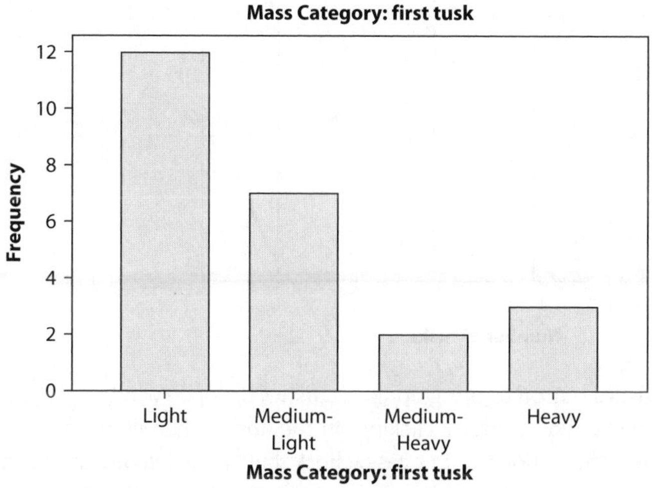

Figure 3.10 *(continued)*

Number of tusks

Statistics

Number of tusks

N	Valid	24
	Missing	0
Mean		2
Median		2
Mode		2
Std. Deviation		1
Variance		0
Range		1
Minimum		1
Maximum		2
Sum		37
Percentiles	25	1
	50	2
	75	2

Number of tusks

		Frequency	Percent	Valid Percent	Cumulative Percent
Valid	1	11	45.8	45.8	45.8
	2	13	54.2	54.2	100.0
	Total	24	100.0	100.0	

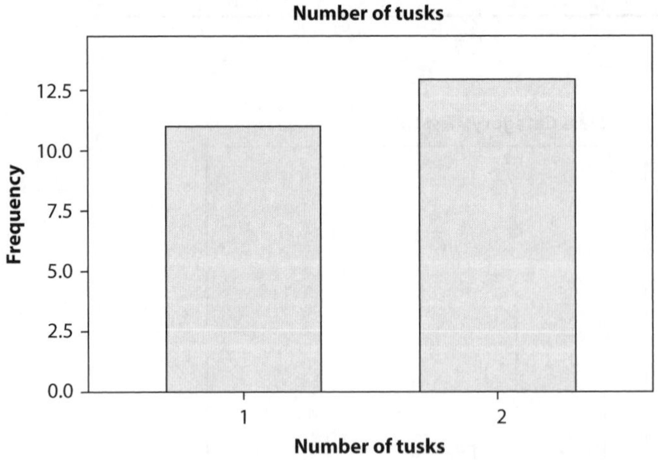

Figure 3.10 *(continued)*

Next click the **Statistics** button to bring up the **Statistics** dialogue window (Figure 3.11b). Select the statistics you want by clicking on the appropriate little boxes to either select or deselect the options. In this case, check that all the features we have mentioned in this chapter are selected: **Quartiles, Standard Deviation, Variance, Range, Minimum, Maximum, Mean, Median, Mode**, and **Sum**. Click the **Continue** button, which will take you back to the main dialogue window (Figure 3.11a). Now click the **Charts** button to bring up the **Charts** dialogue window (Figure 3.11c). This

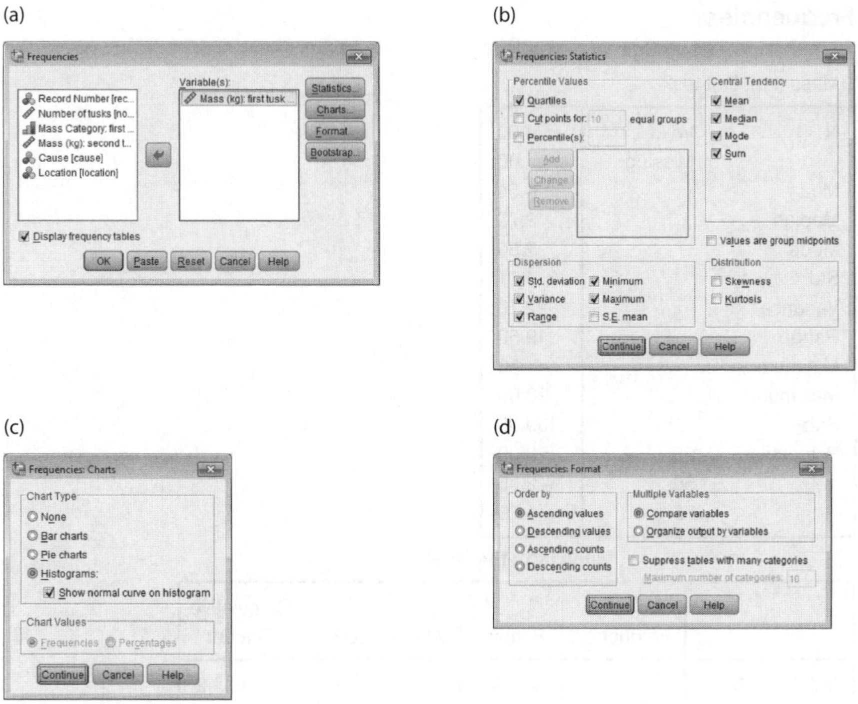

Figure 3.11 Producing descriptive statistics and histograms using SPSS. (a) Frequencies dialogue window. (b) Statistics dialogue window. (c) Charts dialogue window. (d) Format dialogue window.

time you want **Histograms** for **Chart Type**. It's a good idea to select the **With normal curve option** too. Once again, click **Continue** to take you back to the main dialogue window. Now you need to click the **Format** button to bring up the **Format** dialogue window (Figure 3.11d); check that **Ascending values** is selected from the **Order by** choices. Since we are only dealing with one variable this time, the **Multiple Variables** options are irrelevant. Click **Continue** to go back to the main dialogue window (Figure 3.11a). Once you have done this you can click **OK** and, in a flash, you will get output like that shown in Figure 3.12. Notice that the frequency-distribution table is done for individual values and not for classes covering ranges of values. Class intervals of 2.5 are used for the histogram. The first column, for example, covers 0–2.49, the second 2.50–4.99, and so on.

3.6.2 **Pie charts**

You can use the frequency facility on SPSS to produce pie charts rather than frequency graphs. This is only likely to be a useful alternative if you are dealing with nominal, ordinal, or discrete scale data where there are relatively few categories. After opening the data file, make the following selections:

Analyze
→ Descriptive Statistics
→ Frequencies...

Frequencies

Statistics

Mass (kg): first tusk

N	Valid	24.00
	Missing	.00
Mean		4.54
Median		2.75
Mode		2.50
Std. Deviation		4.71
Variance		22.15
Range		19.50
Minimum		.50
Maximum		20.00
Sum		109.05
Percentiles	25	2.00
	50	2.75
	75	4.25

Mass (kg): first tusk

		Frequency	Percent	Valid Percent	Cumulative Percent
Valid	0.5	1	4.2	4.2	4.2
	1	1	4.2	4.2	8.3
	1.5	3	12.5	12.5	20.8
	2	3	12.5	12.5	33.3
	2.5	4	16.7	16.7	50.0
	3	3	12.5	12.5	62.5
	3.5	3	12.5	12.5	75.0
	4.5	1	4.2	4.2	79.2
	9	1	4.2	4.2	83.3
	9.025	1	4.2	4.2	87.5
	12.025	1	4.2	4.2	91.7
	13	1	4.2	4.2	95.8
	20	1	4.2	4.2	100.0
	Total	24	100.0	100.0	

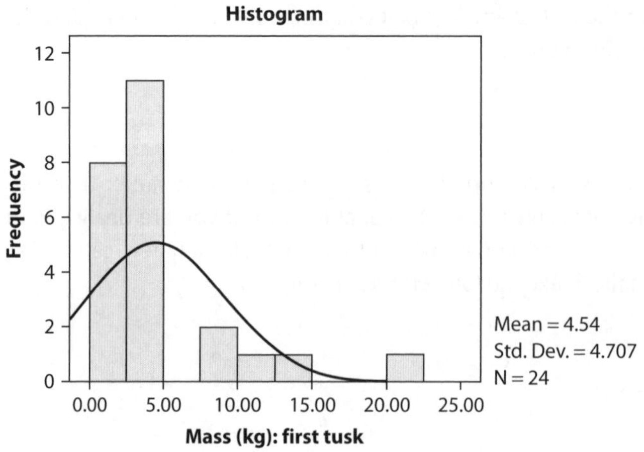

Figure 3.12 SPSS output for descriptive statistics and histograms.

Proceed as for nominal, ordinal, and discrete scale data in Section 3.6.1, but in the **Charts** dialogue window, select **Pie charts** rather than **Bar charts** (Figure 3.13). You can select either **Frequencies or Percentages** for the **Chart Values**. I have gone with the default, **Frequencies**, for the figures here. The output produced will include pie charts like those shown in Figure 3.14.

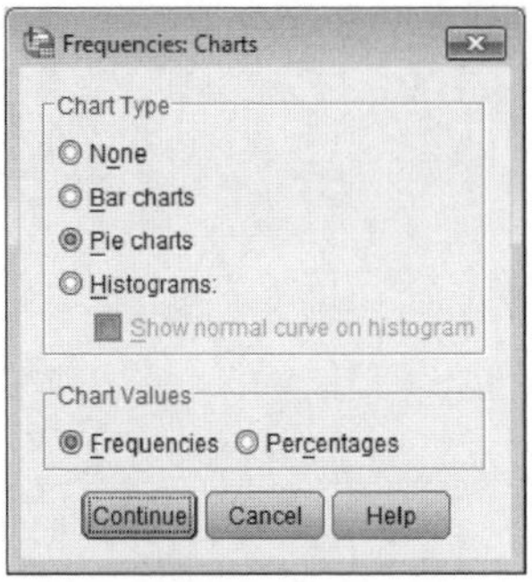

Figure 3.13 Producing a pie chart for a single sample using SPSS. Charts dialogue window showing Pie chart option selected.

3.6.3 **Boxplots**

SPSS will produce a boxplot of a single sample if, after opening the data file, you make the following selections:

Graphs
→ Chart Builder...

This will bring up the **Chart Builder** dialogue window (Figure 3.15a). (If a window called **Element Properties** is in the way just close it.) On the gallery tab in the bottom half of the window, select **Boxplot** from the **Choose from** list, then select the single boxplot to the far right and drag it to the **Chart preview...** box at the top right of the window. Now select the variable called *Mass (kg): first tusk* from the **Variables** box on the left and drag it to the area marked **x-axis?** in the **Chart preview...** box (Figure 3.15b). You might have to wiggle it around a bit until it 'connects', which is indicated by the blue rectangle going black. Click **OK** and the output shown in Figure 3.16 will appear.

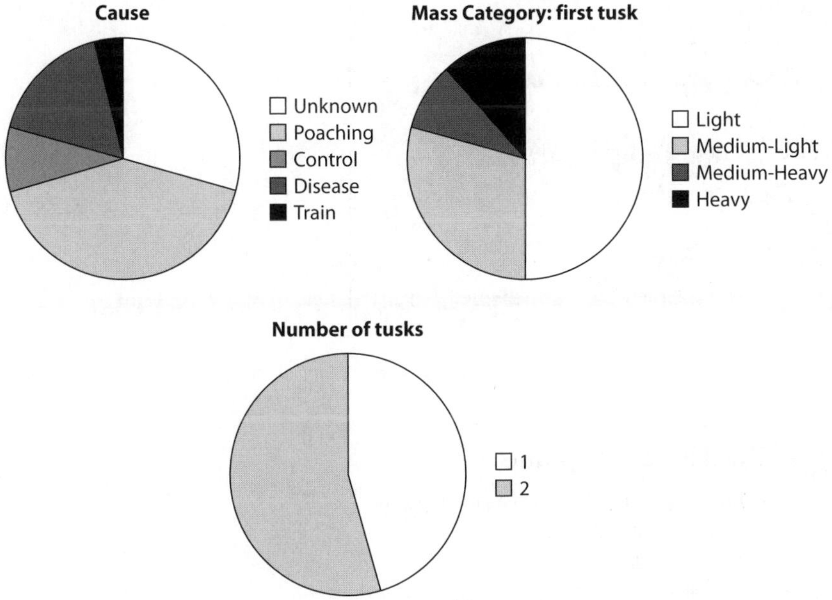

Figure 3.14 Pie charts from SPSS.

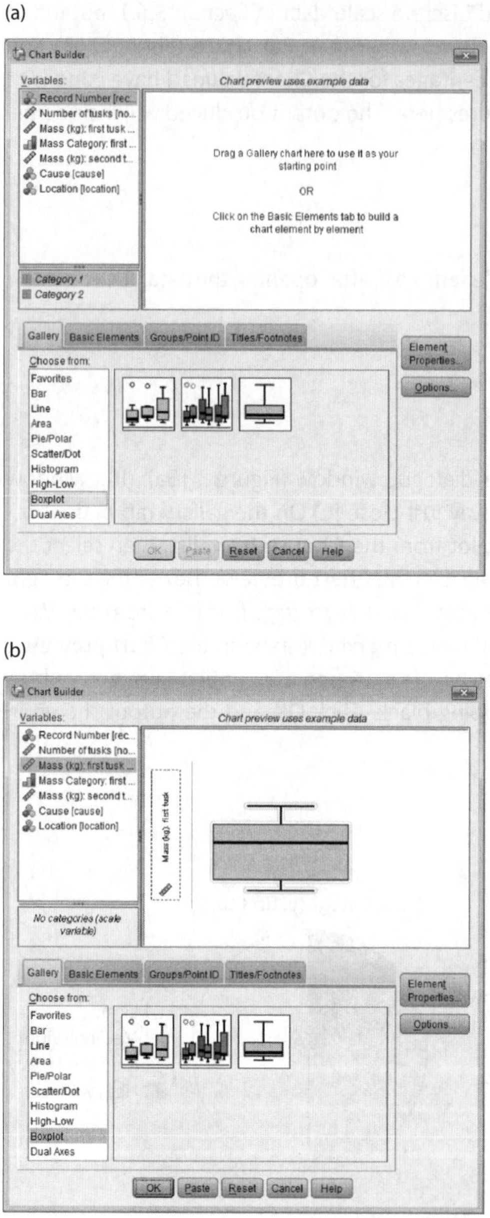

Figure 3.15 Producing a boxplot for a single sample using SPSS. (a) Chart Builder dialogue window. (b) Chart Builder dialogue window with single boxplot with mass(kg): first tusk selected.

Checklist of key points

- Three important features of a sample are its:
  1. Central tendency;
  2. Variability;
  3. Shape.

Explore

Case Processing Summary

	Cases					
	Valid		Missing		Total	
	N	Percent	N	Percent	N	Percent
Mass (kg): first tusk	24	100.0%	0	.0%	24	100.0%

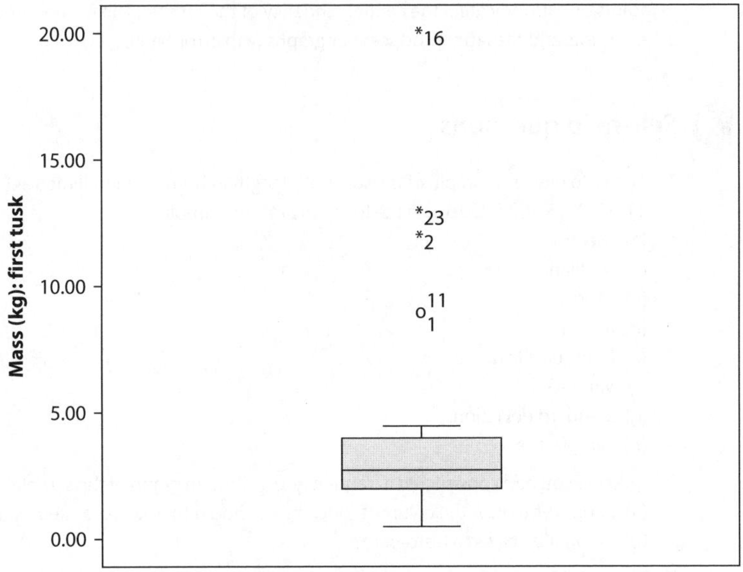

Figure 3.16 SPSS output for boxplots.

- You can summarize a sample's central tendency and variability using descriptive statistics and you can assess its shape by looking at its frequency distribution.
- Measures of central tendency include the:
  1. Mean.
  2. Median.
  3. Mode.
- Measures of variability include the:
  1. Range.
  2. Interquartile range.
  3. Variance.
  4. Standard deviation.
- Frequency distributions indicate central tendency, variability, and shape of sample.
- The bars on frequency-distribution graphs should not touch for nominal, ordinal, and discrete scale data, but should touch for continuous scale data. A frequency-distribution graph for continuous data has a special name: a histogram.
- The normal distribution is a special bell-shaped histogram. Most, but not all, samples of continuous biological data are normally distributed. Irrespective of sample size involved,

approximately 95% of the area under a normal curve lies within two standard deviations either side of the mean. Other distributions that sometimes describe biological data are the binomial distribution (for two nominal categories) and the Poisson distribution for count (discrete scale) data.

- Theoretically, the normal distribution cannot apply to data measured at the discrete scale level. However, this is a grey area and it does not usually create problems to treat discrete data as normally distributed if they are symmetrical. The normal distribution cannot apply to nominal or ordinal data.
- Pie charts and boxplots are alternative ways of presenting visually the data in a single sample.
- Error bars are T-shaped marks indicating sample variability. They are drawn around points or bars indicating the central tendency of the sample. The former produces errorplots and the latter produces bar graphs with error bars.

Self-help questions

1. The following is a sample of incisor tooth length in humans in millimetres (mm): 8, 9, 11, 11, 13, 12, 6, 9, 11. Calculate the following for this sample.
   (a) Mean. *Link back: Section 3.2.1*
   (b) Median. *Link back: Section 3.2.1*
   (c) Mode. *Link back: Section 3.2.1*
   (d) Range. *Link back: Section 3.2.2*
   (e) Interquartile range. *Link back: Section 3.2.2*
   (f) Variance. *Link back: Section 3.2.2*
   (g) Standard deviation. *Link back: Section 3.2.2*
   (h) Sample size. *Link back: Section 3.2*
2. Imagine you had constructed a frequency-distribution graph of the sample in Question 1. *Link back: Section 3.3.2*
   (a) Should you draw the columns touching or should they be separated by a small gap?
   (b) Could it be called a histogram?
3. The DAFOR scale is an ordinal-level measuring system, which ranges from 5 (dominant) to 1 (rare). It is used to record the occurrence of plant species in quadrats. The following is a sample of grass cover measured using this system: 4, 3, 4, 5, 2, 2, 1, 3. *Link back: Sections 3.2 and 3.3*
   (a) Select the most appropriate measure of central tendency and the most appropriate measure of variability for these data. Justify your choice.
   (b) Calculate the measures of central tendency and variability that you selected in (a).
4. Imagine you had constructed a frequency-distribution graph of the sample in Question 3. *Link back: Section 3.3.1*
   (a) Should you draw the columns touching or should they be separated by a small gap?
   (b) Could it be called a histogram?
5. List ways of graphically presenting data in a single sample. *Link back: Section 3.4*

4 Inferring and estimating

Chapter Aims

We are now ready to make the move from descriptive to inferential statistics. This chapter deals with general issues relating to inferential statistics, introducing both estimation and null hypothesis significance testing (NHST).

The ideas and procedures for estimation will then be explored in more detail. In particular you will find out how to calculate the standard error and confidence intervals of the mean and present them graphically as errorplots. Using estimation to compare samples is discussed. In the last chapter you were told that the mean should only be used for scale data. I will mention briefly about estimation techniques relevant to other data measured at other levels.

The chapter ends by showing you how to get SPSS to produce standard errors, confidence intervals, and errorplots to compare samples.

4.1 Overview of inferential statistics

I think that before we go on, it would be good to remind you of why you need to know about inferential statistics and to touch briefly on the concept of probability and how it fits into inferential statistics.

4.1.1 Why we need inferential statistics–a reminder

The numbers of freshwater mussels have declined rapidly over the last 50 years. One conservation activity to help counter this trend is to relocate vulnerable colonies. Unfortunately many mussels die during the relocation process, which is obviously not desirable. One possible reason for this is that exposure to air has a detrimental effect on the health of the mussels. Mussel health can be assessed biochemically: increased lipid concentrations are, for example, a sign of deteriorating health. Greseth *et al.* (2003) monitored the biochemical composition of mussels after exposure to air.

They found that the mean concentration of lipids in the mantle tissue of pimpleback mussels exposed to air for between 15 and 45 minutes was 25.0 mg/g of dry weight, compared with that in the control sample (mussels that had not been exposed to air) of 22.9 mg/g of dry weight. They used samples of 30 specimens in each case.

You may be thinking that since 25.0 is greater than 22.9, lipid levels are higher in mussels exposed to air, which suggests that the health of these individuals has been compromised. If so, you have forgotten about sample error (Section 2.1.2 and Figure 2.1)! The difference in these means might be entirely due to the phenomenon of sample error and the two samples may in fact be from the same population. In other words, they might be different due to chance alone. There may be something of biological interest going on but we cannot tell this just by looking at the means. This is why we need inferential statistics.

Inferential statistics are so named because they allow us to infer something about populations from samples. Knowing something about populations allows us to assess whether patterns apparent in samples are just a product of sample error, or not. We use inferential statistics in two ways: **estimation** and **null hypothesis significance testing**.

Descriptive statistics are estimates of their corresponding parameters. **Estimation** involves assessing how good these estimates are.

In estimation we use descriptive statistics (which describe samples) to estimate parameters (which describe populations). For example, in Section 4.2 we will see how information on population means can be inferred from sample means using confidence intervals.

A **null hypothesis significance test (NHST),** e.g. the *t*-test, is a procedure which allows you to assess if the results you have are just a result of sample error, that is, chance.

We use null hypothesis significance testing procedures, such as a *t*-test, to decide whether our results have been obtained by chance alone. For example, we could use a null hypothesis significance test to decide whether lipid levels in mussels exposed to air are different from those in mussels remaining immersed in water. Chapter 5 deals with general issues relating to null hypothesis significance testing.

4.1.2 **Uncertainty and probability**

We use inferential statistics to infer things about populations from samples. Notice that the sample only infers something about a population; it does not tell us with absolute certainty what it is like. You need to get used to the idea that science involves uncertainty. You only have to look at well-publicized debates about the transmission of bovine spongiform encephalopathy (or mad cow disease) to humans and the supposed link between the MMR vaccine and autism to realize that we live with a lack of certainty about scientific findings. People like yes/no answers but this is not how science works. What we can do is define and be explicit about the level of uncertainty that we are dealing with. This brings us to probabilities.

There are whole textbooks, even whole degree courses, just about probability. Nevertheless, I believe that you have enough understanding of the subject from your everyday life to appreciate, in general terms at least, the role of probability in statistical analysis. Commonly used words such as risk, likelihood, and chance all embody the concept of probability. What is the risk of you catching malaria during a visit to east Africa? Is there a chance of that attractive person asking you out? What are the chances of Arsenal winning their next game? What is the likelihood that the bus will arrive on time? All these questions reflect both an acceptance of uncertainty and an idea about what it means. I have already relied on your intuitive understanding of probability to explain the idea of sample error in Section 2.1.2. I used phrases such as 'likely to give a good indication', 'not be surprised to find a difference', 'not have expected', and 'most likely' to convey the role of chance in generating sample error.

Probabilities can be expressed as percentages or proportions. For example, a 5% probability can be written as 0.05. The probability that a coin will land heads up when flipped is 50% or 0.5. You should make sure that you are comfortable with expressing probabilities as percentages and proportions interchangeably.

In short, probability can be used to link samples and populations and is therefore at the foundation of inferential statistics. Your understanding of probability from everyday life should be enough to get you through the basics of data analysis. It is just a matter of realizing this and becoming comfortable with the idea that, even as a scientist, it's OK to talk in terms of probabilities rather than certainties.

4.2 Inferring through estimation

A parameter is a value that describes a population and a statistic is a value that describes a sample (Section 2.1.2). Descriptive statistics can be thought of as estimates of their corresponding parameters. For example, the mean of a sample provides an estimate of the mean of the population from which the sample was taken. For any particular statistic it would be nice to know how good an estimate it was. In other words, how precise is the estimate? **Standard errors** and **confidence intervals** are used to answer this question.

Standard errors and **confidence intervals** are used in estimation and can be calculated for any statistic. The **standard error of the mean** and the **confidence interval of the mean** are examples. Confidence intervals have an associated probability: **95% confidence intervals** are commonly used. The 95% confidence interval of the mean of any one sample has a very good chance of including the population mean between its **upper** and **lower confidence limits**.

Every statistic has an associated standard error. Standard errors give us an indication of how reliably a statistic estimates its parameter: the bigger the standard error the less reliable. We are only going to deal with the most commonly used of these: the **standard error of the mean**. This type of standard error is so common that if you see standard error written without reference to any particular statistic, then you can assume that it is referring to the standard error of the mean.

Standard errors can be used to calculate confidence intervals. Although we can set any level of confidence, it is very common to work with a probability of 0.95, producing what is called a **95% confidence interval**. Again, although confidence intervals can be produced for any statistic, the **confidence interval of the mean** is the most commonly used and what is meant when the term is used without specifying the statistic.

As for standard errors, confidence intervals give us an indication of how reliably a statistic estimates its parameter: the bigger the confidence interval the lower the reliability. This is all you really need to know, but in case you are interested, a 95% confidence interval is saying that if we took 100 samples we would expect 95 of those samples to have confidence intervals including the population mean. The confidence interval ranges above the statistic to the **upper confidence limit** and below the statistic to the **lower confidence limit**. The following section explains how to calculate and present standard errors and confidence intervals of means.

4.2.1 Standard error (of the mean, $s_{\bar{y}}$)

Usually we work with a single sample and do not take samples repeatedly of the same variable under the same conditions. For example, Greseth *et al.* (2003) had just one sample of lipid concentrations in the mantle of pimpleback mussels exposed to air for between 15 and 45 minutes. Nevertheless, conceptually we could keep repeating our sampling procedure and generate an endless number of samples. If we then constructed a frequency distribution of the means of these samples rather than of the raw data, we would produce a **sampling distribution** of the means. The standard error of the mean is the standard deviation of this distribution.

If you take lots of samples from the same population, calculate a statistic (for example the mean) for each sample, and then create a frequency distribution of these statistics, you will have created a **sampling distribution**.

The standard error of the mean of a sample can be estimated from the standard deviation (Chapter 3, Box 3.2) of that sample, using the formula in Box 4.1. Divide a standard deviation by the square root of the sample size and you get the corresponding standard error. For example, if you have a sample of 144 numbers with a standard deviation of 120 you would calculate the standard error as follows:

- 120 divided by the square root of 144, which is either +12 or −12;
- 120 divided by +12 is +10 and 120 divided by −12 is −10;
- The standard error of this sample is therefore ±10.

Box 4.1 Formula for calculating the standard error of the mean ($S_{\bar{y}}$)

$$S_{\bar{y}} = \frac{s}{\sqrt{n}}$$

Where s = standard deviation and n = sample size.

The need for the sampling distribution of the mean to be normal is not as restrictive as you might first think because of the following three characteristics of the sampling distributions.

1. If the raw data are normally distributed, the sampling distribution of the means will be normally distributed.
2. The means of discrete scale data will be continuous, so the sampling distribution of the mean of any scale-level data can be normally distributed.
3. Even if the raw data are not normally distributed, the sampling distribution of the mean will tend to be so. This tendency increases as the size of the samples used to generate the means increases. This particular fun fact about the sampling distribution of the mean is known as the **central limit theorem**.

Central limit theorem: the bigger the sample size the more likely the sampling distribution of the means of these samples will be normally distributed, even if the population from which they are drawn is non-normal.

Calculating standard error using the formula in Box 4.1 is therefore valid with just about any scale-level data, even if the raw data themselves are discrete or are not normally distributed. You will need to consult one of the advanced texts in the Selected further reading section, for example Quinn and Keough (2002), regarding techniques for calculating standard error of the mean for scale data on the rare exceptions that this does not apply (or cannot be assumed to apply), or for data measured at the nominal or ordinal level. You will need to learn about fun-sounding procedures like **jackknifing** and **bootstrapping**.

Jackknifing and **bootstrapping** are methods used in estimation when the sampling distribution of the statistics being estimated is not known. They involve repeatedly sampling a sample.

Since standard error incorporates standard deviation it is essentially another way of measuring variability. As for standard deviation it only makes sense to report it in relation to a mean. For example, Greseth *et al.* (2003) found that the mean concentration of lipids in the mantle tissue of pocketbook mussels was 39.3 mg/g of dry weight with a standard error of 1.9, compared with 22.9 mg/g of dry weight with a standard error of 0.7 for pimpleback mussels. They actually report these statistics in a table with the standard error in brackets after the mean. An alternative to the brackets would have been to use plus and minus (±) signs, for example, 39.3 ± 1.9 mg/g of dry weight. Sample size was 30 in all cases.

4.2.2 **Confidence intervals (of the mean)**

When the sampling distribution of a mean is normally distributed, which as we reviewed will apply to just about all scale data, confidence intervals can be calculated from standard errors using something called *t*. A confidence interval is bound by an upper and lower limit. The **lower confidence limit** is the mean minus *t* multiplied by the standard error. The **upper confidence limit** is the mean plus *t* multiplied by the standard error. This is summarized in Box 4.2. As for standard error, you will need to consult Quinn and Keough (2002), or another of the advanced texts listed in the Selected further reading section regarding techniques for calculating confidence intervals when the sampling distribution of the mean is not normally distributed or if you are dealing with a statistic other than the mean.

Confidence intervals have an associated probability: 95% confidence intervals are commonly used. The 95% confidence interval of the mean of any one sample has a very good chance of including the population mean between its upper and lower confidence limits.

Box 4.2 Formulae for calculating the confidence interval of the mean

Confidence interval = mean $\pm\, t\, s_{\bar{y}}$
Lower confidence limit = mean $-\, t\, s_{\bar{y}}$
Upper confidence limit = mean $+\, t\, s_{\bar{y}}$

It is a complicated procedure to calculate *t*, and it takes on different values depending on the sample size you are dealing with and the level of confidence that you want. Fortunately there are tables of *t* values for different sample sizes and probabilities are widely available. To look up a value of *t* in a *t* table (Appendix II, Table AII.2) you need to decide what level of confidence, or critical significance level, you want to work with, and how many degrees of freedom you have. Degrees of freedom are related to sample size, in this case one less than the sample size. As mentioned already, we typically use 95% confidence intervals for which we take the value listed in the column labelled $\alpha = 0.05$ in the *t* table.

Here is how to calculate a 95% confidence interval for a sample of size 15 with a mean of 100 and a standard error of 10. The value of *t* with 14 degrees of freedom and a significance level of 0.95 is 2.145. The confidence interval and limits for this example are therefore:

- 95% confidence interval = mean ± *t* multiplied by $S_{\bar{y}} = 100 \pm 2.145 \times 10 = 100 \pm 21.45$;
- Lower 95% confidence limit = mean − *t* multiplied by $S_{\bar{y}} = 100 - 21.45 = 78.55$;
- Upper 95% confidence limit = mean + *t* multiplied by $S_{\bar{y}} = 100 + 21.45 = 121.45$.

For an example from the literature, Lohar and Bird (2003) counted the number of galls on wild-type (genetically 'normal') Japanese lotus plants, two weeks after inoculation with a parasitic worm. They calculated the mean to be 30.9 with a 95% confidence interval of 25.9–35.9. The sample size was between 8 and 12; we are not told exactly.

Since confidence intervals are mathematically related to standard deviation and standard error, confidence intervals are yet another measure of variability. However, they tell us something more. A 95% confidence interval tells us that if we take 100 samples, the chances are that the confidence interval of the mean of 95 of these will include the population mean. This is subtly different from saying that a particular confidence interval has a 95% chance of including the population mean. Perhaps the best statement for you to remember and to work with is that there is a very good chance that 95% confidence intervals of your statistic include the value of the parameter.

4.2.3 Error bars revisited

Error bars are T-shaped marks around measures of central tendency, such as means, on graphs. They can be used to represent the range, interquartile range, standard deviation, standard error, or confidence interval and they indicate how variable the data are. In the case of standard error and confidence intervals, they give a visual indication of how reliably the statistic estimates its parameter.

I introduced you to **error bars** in the previous chapter (Section 3.4.3). They are T-shaped marks around a central tendency that show the variability of the sample. For example, they can be used to show range, standard deviation, standard error, or confidence intervals around a mean. But which of these options should you go for? The most appropriate to use depends on your context. You need to ask yourself whether you are trying to show variation within a sample, or to compare the variability of different samples in order to make inferences about the population(s), as summarized below.

- If what's important is the variability within a sample, then the range or the standard deviation are most appropriate. Generally the standard deviation will be best because it is less affected by extremes and takes into account all the data.
- If what's important is the variability of different samples in relation to each other, then the standard error or confidence intervals are more appropriate. This is the case if you are using inferential statistics to try to evaluate whether samples are different or correlated due to chance alone, or chance plus something else.

It is easier to get a feel for what the error bars are telling you if you use confidence intervals rather than standard error. However, standard error has traditionally been reported far more frequently than confidence intervals. In almost all cases this preference reflects the fact that standard errors are smaller than confidence intervals and are therefore easier to plot and look better! A standard error will always be smaller than the corresponding standard deviation. This is because the calculation involves dividing the standard deviation by a number, which will always be larger than 1. A 95% confidence interval will always be bigger than the corresponding standard error, because t is always bigger than 1 at this level of confidence.

Either columns or points can be used to indicate the mean around which error bars can be drawn. In the example I used in the previous section, Lohar and Bird (2003) used a column to show the mean number of galls. If they had used a point we would call it an errorplot. I will show you how to construct an errorplot with confidence intervals error bars using SPSS in Section 4.4.2.

4.2.4 **Comparing samples**

The extent to which the confidence intervals of two samples overlap gives an indication of how likely it is that the two samples are from the same population. Look at Figure 4.1a. This shows the one extreme: there is almost complete overlap and therefore the two samples are very likely to come from the same population, any difference being due to sample error alone. Now look at the other extreme in Figure 4.1b. There is no overlap and therefore it is unlikely that the two samples come from the same population. Finally, turn your attention to Figure 4.1c, which shows an intermediate situation where the samples might or might not be from the same population.

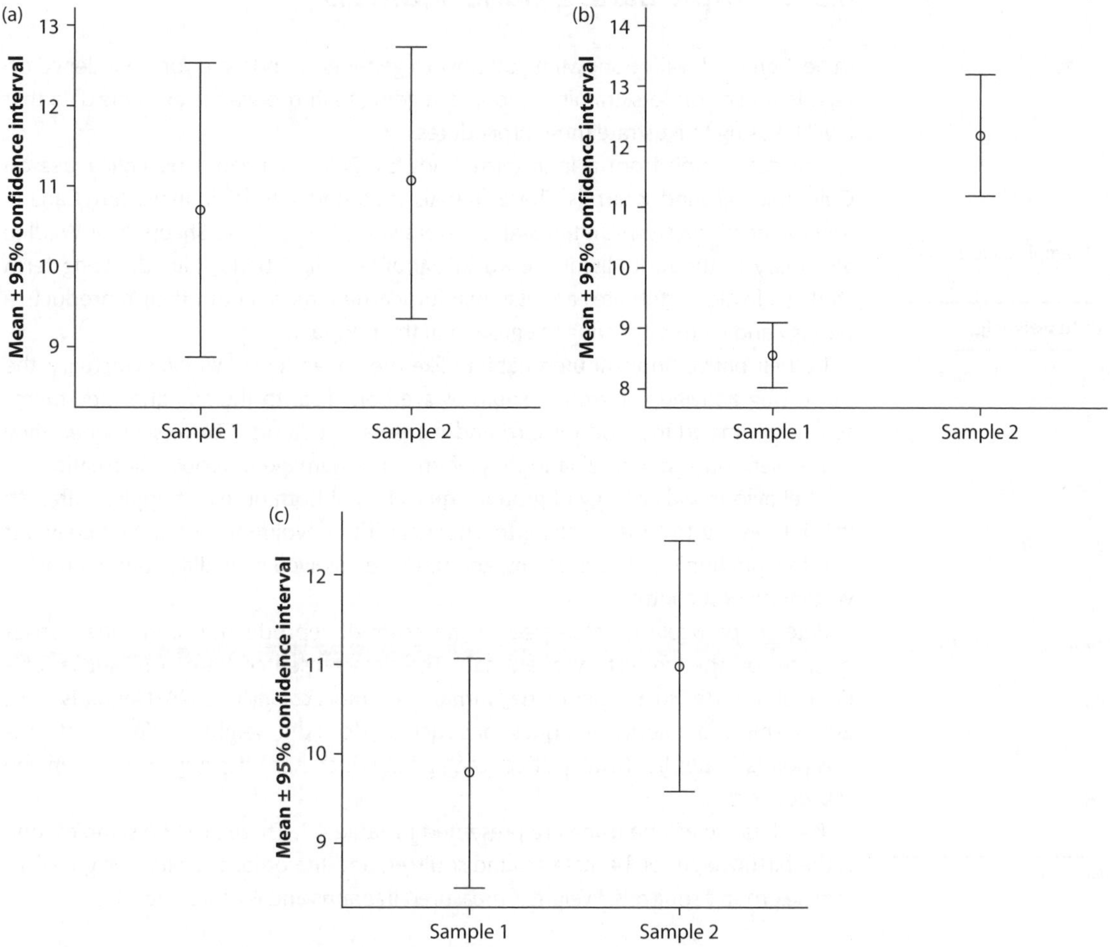

Figure 4.1 Errorplots comparing two samples. (a) Almost complete overlap. (b) No overlap. (c) Partial overlap.

Picking up on the lotus plant example once more, Lohar and Bird (2003) actually presented the column with error bars for the wild type next to two other columns: one for a sample of a mutant strain called *har1-1* and the other for a sample of a mutant strain called *etr1-1* (see their Figure 3). The confidence intervals for the wild-type plants and the *etr1-1* mutants overlapped almost completely, but the *har1-1* mutants overlapped with neither of these. This suggests that the number of galls on the *har1-1* strain differs by more than could be accounted for by chance alone.

We are starting to use confidence intervals and visual inspection of errorplots to make judgements about whether our samples are different due to chance alone or chance plus something of biological interest. The null hypothesis significance testing procedures that are the subject of the next seven chapters allow us to make such judgements in a more objective, clearly defined fashion. There are different tests suited to looking at data structured in different ways and measured at different levels.

4.3 Example data: ground squirrels

In Section 4.4 I will be showing you how to generate standard error, confidence intervals, and errorplots. In this section, I am going to introduce the example data that I will be using to illustrate these procedures.

The data come from a long-term study by Peter Neuhaus and colleagues on Columbian ground squirrels. These animals are found in the wild in western Canada and the north-western United States. Peter's study site is in the Sheep River Wildlife Sanctuary in the foothills of the Rocky Mountains in Alberta, Canada. Long-term studies of wild vertebrates are essential for learning more about their reproductive biology and factors affecting the growth of their populations.

Table 4.1 Example data: ground squirrels.

Birth weight (g)	
Males	**Females**
12.32	10.88
11.51	8.91
10.15	8.92
9.39	6.09
7.20	8.56
9.09	8.55
8.88	9.22
9.19	9.10
8.82	8.75
13.69	11.36
13.51	14.00
13.16	
14.66	
13.42	

Source of data: Courtesy of Peter Neuhaus.

In their native undistributed habitat, like the Sheep River Wildlife Sanctuary, the burrowing activities of ground squirrels are beneficial to the soil and they themselves are important food for bird and mammal predators. Conversely, when they are present on agricultural land they often cause damage to crops. Information on the behaviour and ecology of ground squirrels is of both pure and applied interest: it is interesting to those seeking to understand the evolution of animal behaviour and the functioning of ecosystems, and to those involved in wildlife conservation or vertebrate pest control.

Peter is particularly interested in the costs of reproduction and their consequences on the life history of animals. The data presented here are a subset of a data set reported in a paper in the *Journal of Animal Ecology* in 2004 (Neuhaus *et al.*, 2004). This paper addressed questions such as, does the weight of a female at various points in her life history affect the age at which her offspring wean when she reproduces?

The data we will be using are presented in Table 4.1. There are two samples: one is the birth weight of 14 male ground squirrels and the other the birth weight of 11 female ground squirrels. Weight is measured in grams and is at scale level.

4.4 **Worked example: using SPSS**

We are going to run through how to get SPSS to calculate standard error, confidence intervals, and errorplots using Peter Neuhaus' data, introduced in Section 4.3. To do this the data must be entered into SPSS as shown in Figure 4.2. In newer versions of SPSS (version 10 onwards), the **Data Editor** window has a **Variable View** tab and a **Data View** tab. Figure 4.2a shows the former and Figure 4.2b the latter. Appendix I has notes on how to enter data into SPSS.

Notice that the data are set out differently from how they are in Table 4.1. In Table 4.1 there is a column of weights for males and a column of weights for females. In SPSS all the weight data goes into one column, with whether a data point applies to a male or a female recorded in a second column. In other words there is a row for each ground squirrel, with a column indicating its sex (code 1 for males and code 2 for females), and a column containing weight data. This is how you arrange data in SPSS when they are unrelated (Section 2.3.1). As we shall see in later chapters, a different arrangement is used if the data are related.

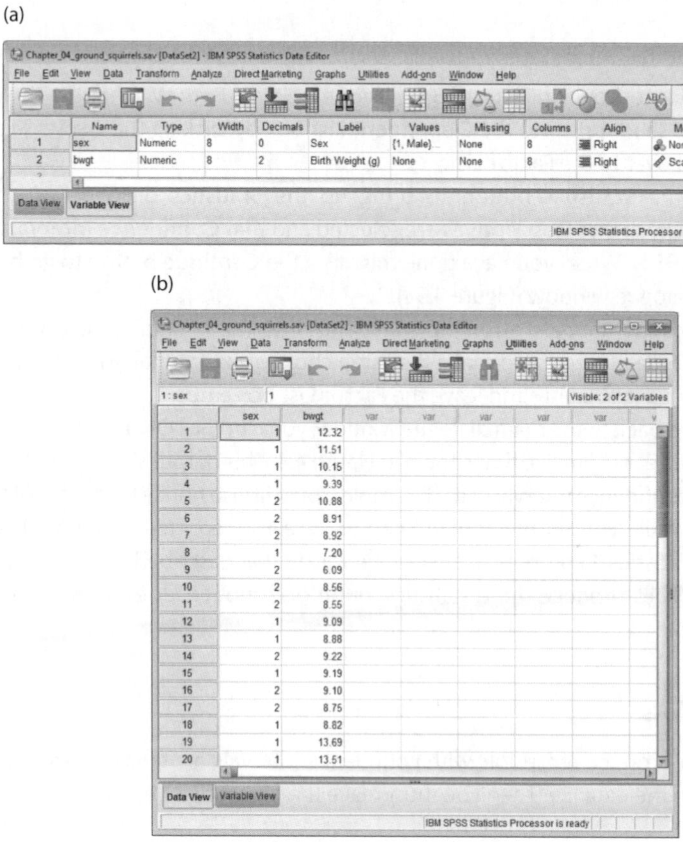

Figure 4.2 Example data in SPSS: ground squirrel sex and birth weight. (a) Variable View. (b) Data View.

(a) (b)

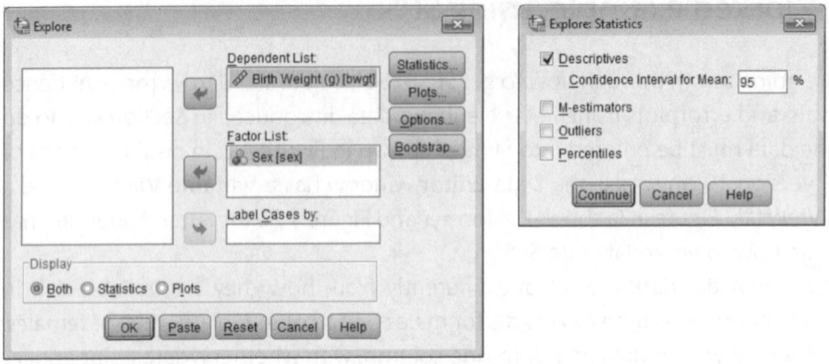

Figure 4.3 Producing standard error and confidence intervals using SPSS. (a) Main dialogue window. (b) Statistics dialogue window.

4.4.1 **Standard error and confidence intervals**

To get SPSS to produce standard errors and confidence intervals you must first open the data file. Then you must make the following selections:

Analyze
→ Descriptive Statistics
→ Explore...

A window like that shown in Figure 4.3a will appear. Select *Birth Weight* from the list on the left and send it over to the **Dependent List** box by clicking the arrow. Select *Sex* and send it over to the **Factor List**.

Next click the **Statistics** button to bring up the **Statistics** dialogue window (Figure 4.3b). Check that **Descriptives** are selected and that **Confidence Interval for Mean** is set at **95%**. When you have done this, click the **Continue** button to go back to the main dialogue window (Figure 4.3a).

If you wanted the standard error and confidence intervals of data in a just a single column of SPSS data you would make the same choices, except that you would use the **Dependent List** box only and leave the **Factor List** box empty.

Once you are back to the main dialogue window, you can click **OK** and, within moments, you will get output like that shown in Figure 4.4. This output shows the standard error and confidence interval of each sample. You will also notice that some of the descriptive statistics you learnt about in Chapter 3 are also reported. Using **Analyze/Descriptive Statistics/Explore** is an alternative way to that described in Section 3.6.1 for getting SPSS to produce these statistics, but it does not cover everything mentioned in Chapter 3.

4.4.2 **Errorplots**

To get SPSS to produce errorplots with confidence intervals or standard errors you must first open the data file. Then you must make the following selections:

Graphs
→ Chart Builder...

Case Processing Summary

		Cases					
		Valid		Missing		Total	
	Sex	N	Percent	N	Percent	N	Percent
Birth Weight (g)	Male	14	100.0%	0	.0%	14	100.0%
	Female	11	100.0%	0	.0%	11	100.0%

Descriptives

	Sex			Statistic	Std. Error
Birth Weight (g)	Male	Mean		11.0707	.63468
		95% Confidence Interval for Mean	Lower Bound	9.6996	
			Upper Bound	12.4418	
		5% Trimmed Mean		11.0863	
		Median		10.8300	
		Variance		5.639	
		Std. Deviation		2.37474	
		Minimum		7.20	
		Maximum		14.66	
		Range		7.46	
		Interquartile Range		4.40	
		Skewness		.020	.597
		Kurtosis		−1.493	1.154
	Female	Mean		9.4855	.60686
		95% Confidence Interval for Mean	Lower Bound	8.1333	
			Upper Bound	10.8376	
		5% Trimmed Mean		9.4233	
		Median		8.9200	
		Variance		4.051	
		Std. Deviation		2.01274	
		Minimum		6.09	
		Maximum		14.00	
		Range		7.91	
		Interquartile Range		2.32	
		Skewness		.895	.661
		Kurtosis		2.161	1.279

Annotations: Mean ($\bar{y}$); Standard error of the mean ($S_{\bar{y}}$); Confidence interval

Figure 4.4 SPSS output for standard error and confidence intervals with key information annotated.

This will bring up the **Chart Builder** dialogue window (Figure 4.5a). (If a window called **Element Properties** is in the way, move it to one side or close it.) On the gallery tab in the bottom half of the window select **Bar** from the **Choose from** list, then select the simple errorplot (bottom row third picture from the left) and drag it to the **Chart preview** area. Now select the variable called *Sex* from the **Variables** box on the left and drag it to the area marked **x-axis?** in the **Chart preview** area. You might have

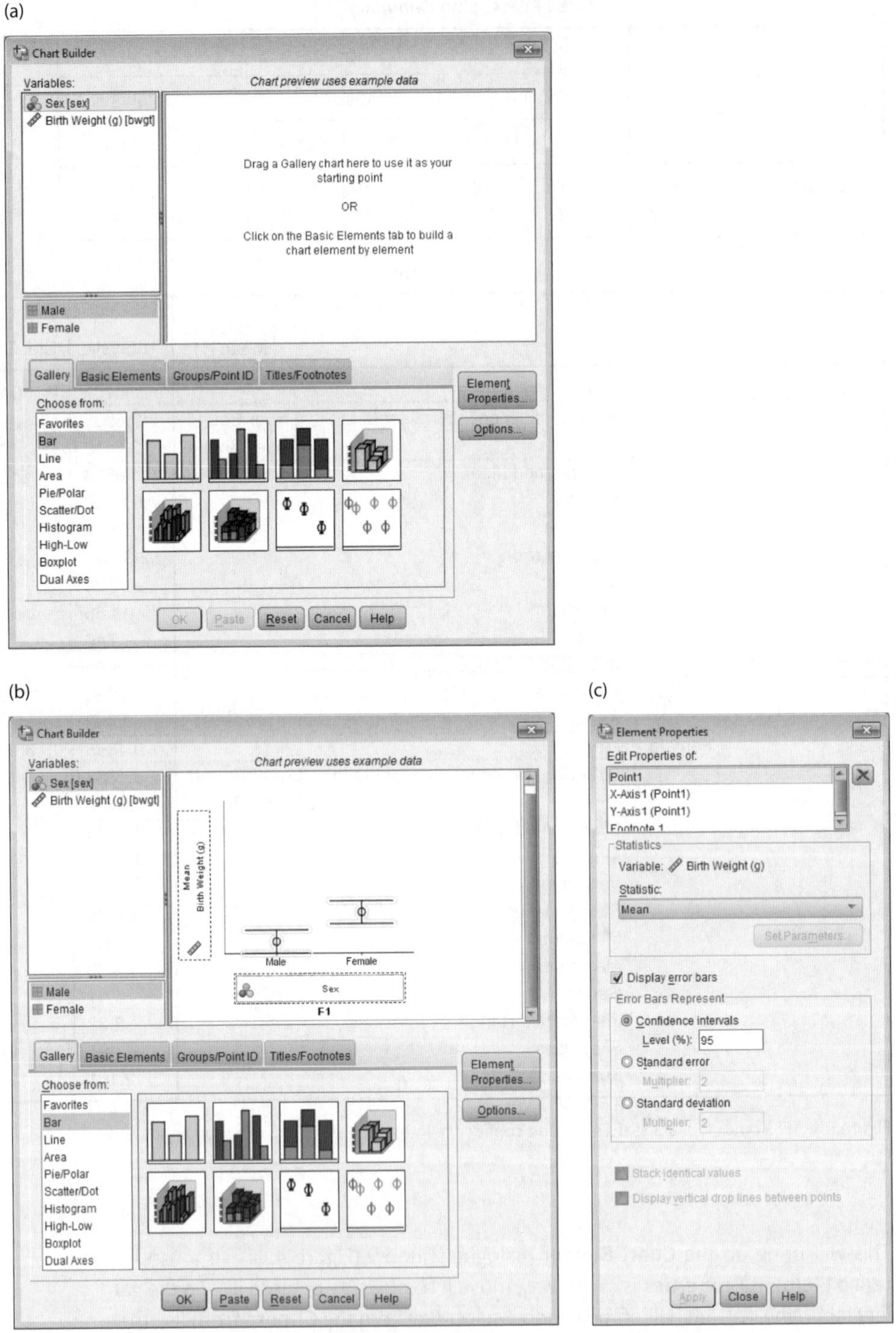

Figure 4.5 Producing an errorplot using SPSS. (a) Chart Builder dialogue window. (b) Chart Builder dialogue window with simple errorplot and variables selected. (c) Element Properties window.

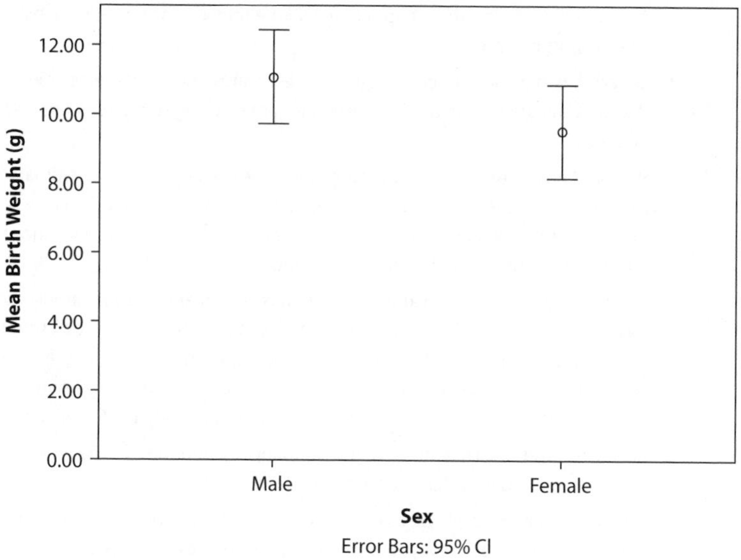

Figure 4.6 SPSS output for errorplots.

to wiggle it around a bit until it 'connects', which is indicated by the blue rectangle going black. Then select the variable called *Birth Weight* from the **Variables** box on the left and drag it to the area marked **y-axis?** in the **Chart preview...** box. Again some wiggling might be necessary. At this point what you are looking at should look like Figure 4.5b. Now move to the **Element Properties** window (Figure 4.5c). If necessary open it by using the button of this name. Check that **mean** is selected from the drop down menu under **statistics** and that **confidence intervals** at **95%** are selected under **Error Bars Represent**. (If you want to go with **Standard error** instead, you need to select this option and then indicate how many standard errors above and below the mean you want plotted in the **Multiplier** box.) Click **Apply** then return to the **Chart Builder** dialogue window and press **OK** and the output shown in Figure 4.6 will appear.

Checklist of key points

- Inferential statistics are used to infer information about populations from samples. They are used in two contexts:
  1. Estimation (estimating parameters from statistics).
  2. Null hypothesis significance testing (procedures used to decide if chance alone can account for apparent patterns in data).
- Since we only infer information about populations from samples, we have to deal with uncertainty about this information. We use the concept of probability to express how much uncertainty we are dealing with in any particular situation. To understand the material covered in this book, it is sufficient that you draw on your everyday understanding of probability in words such as chance, risk, and likelihood, all of which embody the concept of probability.

- A probability of something happening half the time can be written as a percentage (50%) or a proportion (0.5).
- Descriptive statistics can be thought of as estimates of their corresponding parameters. We can evaluate how good these estimates are by using standard errors and confidence intervals.
- Standard errors and confidence intervals can be calculated for any estimate but we focus here on calculating the standard error of the mean and the confidence interval of the mean. If the words standard error or confidence interval are written without qualification, you should assume that they refer to the mean.
- If lots of samples are taken and a histogram of the means of these samples constructed, this histogram is called a sampling distribution. When this sampling distribution is normally distributed the standard error is the standard deviation of the sampling distribution. Under such circumstances, the sample error of a single sample can be estimated from the standard deviation of that sample using the formula in Box 4.1.
- Because of the special features of the sampling distribution of a mean, the formula in Box 4.1 can be used with just about all scale-level data.
- Confidence intervals can be calculated from standard errors using the *t* statistic. The value of *t* depends on the size of your sample and the level of confidence you want to work with (95% is typical). The value of *t* can be looked up in special tables (e.g. Appendix II Table AII.2).
- Confidence intervals have an upper and a lower limit that define a range of values around the mean. Using a 95% confidence interval, if we were to take 100 samples it is likely that 95 of them would include the value of the population mean. In other words, there is a good chance that the population mean will fall within the 95% confidence intervals of any one sample.
- Error bars can be used in association with points or bars marking mean values to indicate the variability of the data. It is best to use error bars to show ranges or standard deviations, if the variability within a sample is what you are writing about. It is best to use standard error or, even better, confidence intervals if you are comparing the variability between samples.
- The more the error bars of different samples overlap, the more likely it is that the samples are from the same population.

Self-help questions

1. Why do we need inferential statistics? *Link back: 4.1.1.*
2. What are the two main ways of using inferential statistics? *Link back: 4.1.1.*
3. Write 5% probability as a proportion. *Link back: 4.1.2.*
4. Can standard errors and confidence intervals be established for measures of central tendency other than the mean? *Link back: 4.2.*
5. What does a 95% confidence interval tell you? *Link back: 4.2.2.*
6. What are the circumstances under which you can use the formula in Box 4.1 to calculate standard error? *Link back: 4.2.1.*
7. Study Figure 4.6. Do you think that the birth weights of the male and female ground squirrels studied by Peter Neuhaus are different? *Link back: 4.2.4.*

5 Overview of null hypothesis significance testing

Chapter Aims

Chapter 4 introduced inferential statistics in general, both estimation and null hypothesis significance testing (NHST), but focused mostly on estimation. In particular, it led you through the procedures for estimating the mean of a population from a sample, using techniques applicable to just about all situations involving raw data measured at the scale level. Estimation for other descriptive statistics and levels of measurement was mentioned briefly.

The aim of this chapter is to give you a more detailed overview of null hypothesis significance testing procedures. It is designed to give you a foundation on which to build when you read subsequent chapters considering specific tests in some detail. This chapter emphasizes commonalities between tests and, by doing this, reduces the amount of learning you will need to do.

Six areas are covered:

- The four steps of any test.
- Error (types I and II) and power.
- Parametric versus nonparametric tests.
- Test tails (one or two)
- Effect sizes and their confidence intervals.
- Limitations, extensions and alternatives.

At the very least you should come away knowing that:

- There are four stages to any test and roughly what they involve.
- Tests are either parametric or nonparametric and that to perform parametric tests special criteria must be met.

I recommend that you return to this chapter again after studying some of the specific tests in later chapters.

5.1 Four steps of null hypothesis significance testing

Null hypothesis significance tests (or just **NHST** or **stats tests** for short) are procedures used to make decisions about whether chance alone can account for apparent patterns in our data.

In the previous chapter we reviewed the reason why we use inferential statistics and looked particularly at their use in estimation. In this chapter we turn our attention to their use in **null hypothesis significance testing**. In brief, **null hypothesis significance tests** (or just **NHST** or **stats tests** for short) are procedures used to make decisions about whether chance alone can account for apparent patterns in our data. If this doesn't ring any bells for you then return to and re-read Section 4.1.1.

In this section I present four steps for you to follow whenever you conduct a statistical test. These are:

1. Construct a (statistical) null hypothesis.
2. Choose a critical significance level.
3. Calculate a statistic.
4. Reject or accept the (statistical) null hypothesis.

In doing so I'm following a particular philosophical approach called **frequentist**. It is the most widely used approach to statistical decision making. The founders of the frequentist school of thought, in the 1930s, were Ronald Fisher, Jerzy Neyman, and Egon Pearson. It is sometimes called the **Neyman–Pearson** approach.

5.1.1 Construct a (statistical) null hypothesis (H_0)

For every hypothesis there is a **null hypothesis (H_0)**. A null hypothesis is typically a statement embodying the idea that there is nothing going on. Null hypothesis significance tests are built around rejecting or accepting **statistical null hypotheses**. A statistical null hypothesis will say that there is no pattern in the data; for example, that there is no difference between samples or no relationship between variables.

Statistical hypotheses and **statistical null hypotheses** are specific statements about patterns, or a lack of patterns, in data. They are not synonymous with **research hypotheses** (Chapter 1) and their null forms. Rather, statistical hypotheses and null hypotheses parallel specific predictions generated by research hypotheses.

The first thing you need to do is to construct what is called a statistical **null hypothesis**. Before explaining what this is, we need to review another sort of hypothesis, which we met in Chapter 1.

The type of hypothesis we talked about in Chapter 1 in connection with the hypothetico-deductive approach should really be referred to as a **research hypothesis** to distinguish it from a **statistical hypothesis**, which we will consider here. Normally, a research hypothesis is a more general statement than a statistical hypothesis. A statistical hypothesis states specific relationships between variables and therefore parallels a prediction generated by a research hypothesis. The more specific the prediction the more closely it will parallel a statistical hypothesis.

Both sorts of hypothesis can be expressed in their null form, giving a null research hypothesis and a null statistical hypothesis respectively. A null hypothesis expresses the idea that nothing is happening. A statistical null hypothesis is a statement of the scenario in the top half of Figure 5.1 when both samples are drawn from the same population—any apparent difference being due to sample error alone.

People are rarely explicit about the type of hypothesis or null hypothesis they are talking about. You will have to interpret this from the context. That is why I have put the word statistical in brackets in the title of this section, and when I have referred to null hypothesis significance testing procedures in the preceding section. From now on, unless I state otherwise, when I say hypothesis I will mean statistical hypothesis and when I say null hypothesis I will mean statistical null hypothesis.

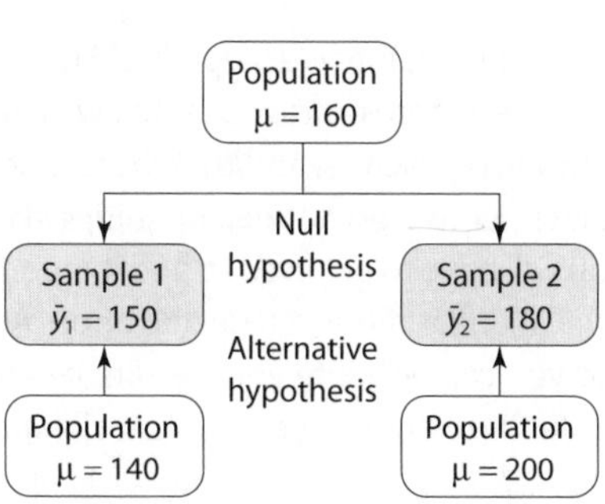

Figure 5.1 Dilemma caused by sample error revisited. When two samples have a different mean is this due to sample error alone (top half of the figure), or to sample error plus something of biological interest (bottom half of the figure)? The top half of the figure represents the null hypothesis. As for Figure 2.1, $\bar{y}_1$ = sample mean (sample 1); $\bar{y}_2$ = sample mean (sample 2), μ = population mean.

The symbol used for a null hypothesis is H_0. The symbol for the **statistical alternative hypothesis** is H_1. Here is an example of a null hypothesis and its corresponding alternative statistical hypothesis.

A **statistical alternative hypothesis (H_1)** is a reverse statement of a statistical null hypothesis. For example, if a null hypothesis is that there is no difference between two samples, then the alternative will be that there is a difference.

> H_0: lipid levels in the mantles of pimpleback mussels exposed to air and remaining immersed in water are the same.
>
> H_1: lipid levels in the mantles of pimpleback mussels exposed to air and remaining immersed in water are different.

At first, most biologists would find building their hypothesis tests on a null hypothesis counterintuitive. The null situation is usually not what excites us about the natural world. In most situations thinking that something isn't happening does not simulate our curiosity. For example, the idea that a vaccine isn't reducing disease level doesn't get our juices going. It is more likely that the alternative hypothesis reflects our thinking. Nevertheless, it is the null hypothesis upon which we build our test, because the maths work that way. This may feel strange at first but with time you will get used to this approach.

In review, the first step in null hypothesis significance testing is to state your question or prediction in terms of a null hypothesis: that nothing is happening.

5.1.2 Decide on a critical significance level (α)

The second step you need to take is to decide on a **critical significance level** (otherwise denoted by the Greek letter alpha, α). This sets the level of uncertainty we are prepared to accept when doing null hypothesis significance tests. As discussed in Section 4.1.2, complete certainty in science is not an option. We have to live with uncertainty. What a critical significance level does is give us an explicit benchmark with respect to this uncertainty.

A **critical significance level (α)** sets the decision point determining whether a null hypothesis is accepted or rejected. It is expressed as a probability. A 5% (0.05) critical significance level is often used.

It is very common for biologists to use 5% as their critical significance level. Expressed as a proportion this would be 0.05. In some areas of the biosciences, for example biomedicine, levels of 1% (or 0.01) are more common. Nevertheless, whatever level it is, the person conducting the test has decided upon it.

It might not feel right that uncertainty is acceptable! However, it is and you have to get used to it. It may seem even more outrageous that we can decide the level of uncertainty we want to work with ourselves, and that there is no predefined or set level. You might be thinking that scientists do not decide things like that. Well, yes they do! Again, it is something you will have to get used to and it is perfectly fine as long as you are explicit about the level of uncertainty you decide upon. As long as you know what benchmark is being used you can evaluate your results, or those of other scientists.

In review, the second step is to choose your critical significance level, for example 5% (or 0.05).

5.1.3 **Calculate your statistic**

Once you have constructed a null hypothesis and chosen a critical significance level, you need to calculate an appropriate statistic. Different statistical tests use different statistics. For example the *t*-test uses the *t* statistic while the Mann–Whitney U test uses the U statistic. We will meet these and other tests in subsequent chapters, where I will lead you through the calculation by hand and by computer.

In association with calculating a statistic you will need to find a bit of information related to how much data you are working with, for example the sample size or **degrees of freedom**. We met degrees of freedom when using the *t* statistic to calculate confidence intervals in Chapter 4. Again, the details of this will be covered with specific tests in subsequent chapters.

Degrees of freedom is a number related to sample size which often has to be calculated along with the statistic when conducting a hypothesis test.

In review, the third step is to calculate the statistic and degrees of freedom if appropriate.

5.1.4 **Reject or accept the null hypothesis**

Finally, you have to use your critical significance level from step 2 and calculations from step 3 to decide whether to reject or accept the null hypothesis that you created in step 1. If you accept the null hypothesis you are accepting that chance alone explains any apparent patterns in your data. Thus you are saying you have a non-significant result. Conversely, if you reject the null hypothesis you are saying that you are rejecting the proposal that the results are due to chance alone. Your results in this case are significant. This logic can be quite mind-blowing at first. Yet again you will just have to stick with it and in time it will become second nature to you.

The procedure for rejecting or accepting a null hypothesis is different when you do this by hand compared with when you do it using a statistical computing package such as SPSS. When you do the calculations by hand you use **critical values** that you look up in **critical-value tables**. When you do calculations using SPSS, you use the ***P* value** on the output.

Critical values are values of statistics corresponding to a specific critical significance level and sample size(s) or degrees of freedom. They can be looked up in **critical-value tables**.

A ***P* value** (or **significance level**) reflects the probability of getting a value for the statistic equal to or more extreme than the one calculated in a hypothesis test, if the null hypothesis is true.

Step 4: using critical-value tables

When deciding to reject or accept a null hypothesis when you are calculating by hand, you have to compare your calculated statistic with a critical value for that statistic, which you look up in a table (see examples in Appendix II). We have already gone through a procedure like this when calculating confidence intervals (Section 4.2.2). The table gives different values for different critical significance levels, so you need to have done step 2 first. You will also need to know your sample sizes or degrees of freedom to look values up in these tables, which you typically calculate along with the statistic in step 3.

Critical values are denoted using a subscript, for example $t_{critical}$. The rule is usually that if the value you calculate is greater that the critical value, then you reject. For example, you reject your null hypothesis if your t value is bigger than $t_{critical}$ for the t-test. However, annoyingly, this rule is sometimes the reverse, so you need to double check them by looking them up. In most cases they will be in the legend of the relevant table of statistical values.

Step 4: using *P* values on computer output

When deciding to reject or accept a null hypothesis when you are using a statistical computing package, you have to find the ***P* value** on the computer printout and compare it directly with your critical significance level. A P value reflects the probability of getting a value for the statistic equal to or more extreme than the one calculated in a null hypothesis significance test, if the null hypothesis is true. P is often called the **significance level**.

What is nice about this route, other than that the computer does all the calculations for the statistics, is that the rule for rejecting or accepting is always the same. This rule, which is well worth committing to memory, is summarized in Box 5.1. As a supplement to this basic rule you can interpret P values in a more graded fashion as outlined in Table 5.1.

Box 5.1 Decision on rejecting or accepting the null hypothesis using P, where α is the critical significance level

If $P \leq \alpha \rightarrow$ reject the null hypothesis.

If $P > \alpha \rightarrow$ accept the null hypothesis.

One thing to note about P on a computer printout is that you will sometimes get a value of 0.000. In this case P has been rounded up to three decimal places and what it actually means is that P is less than 0.0005. People will often prefer to write $P < 0.0005$ or $P < 0.001$, rather than $P = 0.000$, when reporting such a result.

In review, the fourth step is to reject or accept your null hypothesis. The way you go about this depends on whether you are using a statistical computer package or

Range of *P* values	Interpretation
>0.05	Non-significant
0.10–0.05	Suggestive
0.05–0.01	Significant
0.01–0.001	Highly significant
<0.001	Extremely significant

Table 5.1 Graded interpretation of P values. Rather than using 0.05 as an absolute significance/non-significance, as we will do in the worked examples in this book, you can use a graded system.

not. In all cases rejecting or accepting your null hypothesis means the same thing. If you reject your null hypothesis, your result is significant and you consider that the pattern in your data cannot be accounted for by chance alone. If you accept your null hypothesis, your result is non-significant and you consider that any apparent pattern in your data is a product of chance. This is summarized in Box 5.2.

Box 5.2 What your decision on the null hypothesis means

Reject H_0: reject pattern in data as being a product of chance → significant result.
Accept H_0: accept pattern in data as being a product of chance → non-significant result.

Returning to our pimpleback mussel example, if we reject the null hypothesis (H_0), that lipid levels in the mantles of pimpleback mussels exposed to air and remaining immersed in water are the same, then we accept the alternative hypothesis (H_1), that lipid levels in the mantles of pimpleback mussels exposed to air and remaining immersed in water are different.

5.2 Error and power

It is inevitable that you will sometimes make errors. I am not talking about errors of computation, although you may make those from time to time. I am talking about a type of error that you cannot help making, which is integral to the procedure itself. Given that we are working with probabilities and not certainities, it is possible that you will reject a null hypothesis that is true. This is known as making a **type I error**, and the probability of making a type I error is set by the critical significance level (α). If you always use a critical significance level of 5% throughout your career, the chances are that you will make a type I error once for every 20 stats tests you carry out.

A **type I error** is rejecting a true null hypothesis. The critical significance level (α) sets the probability of making a type I error. A **type II error** is accepting a false null hypothesis. The probability (β) of making a type II error is harder to assess.

You make a **type II error** if you accept a false null hypothesis. In most circumstances the probability of making a type II error (β) cannot easily be quantified. It is actually for this reason that null hypothesis significance testing is built on the null hypothesis and not the alternative. You might have noticed that the null hypothesis is always very specific, whereas the alternative hypothesis is less so. The null hypothesis about differences, for example, is specifically about the one case when there is no difference, but the alternative covers the range of possibilities if there is a difference.

The **power of a test** ($1 - \beta$) is the probability of rejecting a false null hypothesis. In other words it's the probability of not making an error.

Type I and II errors are about the probabilities of getting it wrong. But what about the probability of getting it right? The **power of a test** is the probability of rejecting a false null hypothesis. In other words, it is the probability of not making an error and of getting a significant result when there really is a significant difference. The power of a test is 1 minus the probability of making a type II error ($1 - \beta$). Since the probability of making a type II error is generally not quick and easy to determine, neither is the power of a test. Without going into further detail here, it is worth appreciating that the power of a test increases with:

- Increasing sample size (Section 2.1.1);
- Increasing effect size (Section 5.5), that is the effect of the source of variation which we are investigating (Section 2.2);
- Decreasing background variation, that is the effect of variation generated by other sources (Section 2.2).

Power analyses, mentioned in Section 2.1.1, can be used to determine the sample size you need to get reasonable power. The understanding and reporting of the power of a test and the use of power analysis in planning research is very good practice. Think about what a waste of time it would be to carry out an experiment on which you could only conduct a test with zero power! Nonetheless, these topics are beyond our remit here and you will need to consult one of the books recommended in Selected further reading at the end of this volume to learn more.

5.2.1 **Implications for decision making**

Accepting a null hypothesis does not mean it is definitely true. All you have done is decided not to reject it. This might sound a bit like philosophical mind bending, but it is an important distinction. The existence of type I errors means that if we accept a null hypothesis we are merely deciding to proceed on the basis that any apparent pattern in the data is due to chance alone. If you are told, for example, that lipid levels in the mantles of pimpleback mussels exposed to air and remaining immersed in water are different, according to this process then you need to have in the back of your mind, it's not impossible that they are not! When you conduct a statistical analysis you follow a procedure that enables you to reach an objective and consistent decision about a particular set of data. You, and the scientists to whom you communicate your results, should understand the context in which you have made your decisions.

5.3 Parametric and nonparametric

Null hypothesis significance testing procedures fall into two categories: **parametric tests**, such as the *t*-test, and **nonparametric tests**, such as the chi-square test. The first section below compares these two categories. The second section looks at how you can check for the special criteria that apply to parametric tests.

Parametric tests are null hypothesis significance testing techniques for which data must meet special criteria. In particular they assume that the dependent variable can be described by a specified distribution. For the parametric techniques in this book this is the normal distribution.

Nonparametric tests are null hypothesis significance testing techniques that require fewer assumptions to be made about data and do not rely on the normal distribution.

5.3.1 **Comparison of parametric and nonparametric**

Parametric tests are null hypothesis significance testing techniques for which data must meet special criteria. In particular they assume that the dependent variable

Table 5.2 A comparison of parametric tests using the normal distribution and nonparametric tests.

	Parametric tests (assuming the normal distribution)	**Nonparametric tests**
Example test	*t*-Test, paired *t*-test, one-way Anova.	Chi-square, Mann–Whitney U test, Kruskal–Wallis Anova.
Level of measurement of data	Can only be used with scale level of measurement.	Can be used with all levels of measurement.
Criteria	Requirements specific to each test to do with: (1) the normal distribution of data; (2) similarity of sample variances.	Distribution-free.
Sample sizes	Need larger sample sizes to check criteria.	Useful when sample sizes are too small to check criteria.

conforms to a specific frequency distribution (Section 3.3). The normal distribution is the distribution which most often applies in parametric tests, but some parametric procedures use distributions other than normal. The parametric tests presented in this book are all ones that use the normal distribution. **Nonparametric tests** are hypothesis significance testing techniques that require fewer assumptions to be made about data, and do not rely on any distribution, normal or otherwise. Take a look at Table 5.2., which compares parametric tests using the normal distribution to nonparametric tests.

For each of the basic parametric tests, there is a nonparametric test that you can use instead, if the requirements of the parametric test are not met. You might ask, why bother learning about parametric tests? Well, parametric tests are more powerful at rejecting false null hypotheses and are therefore better at finding significance. Furthermore, the more sophisticated parametric tests do not have a nonparametric equivalent; there are things that you can do with parametric statistics that you cannot do with nonparametric statistics! You need, therefore, to know about both.

5.3.2 **Checking criteria for parametric tests using the normal distribution**

There are certain criteria that you must be satisfied that your data fulfil before carrying out a parametric test. If you do not then your results will be unreliable. The details of these parametric criteria vary between tests, but they all involve something about the distribution of the data and often require the variances of samples to be similar. The distribution involved is typically the normal distribution (Section 3.3.3) and that's the only case we consider here.

If you have worked through Chapter 3, you should already have acquired the basic skills for assessing whether your scale data meet parametric criteria of normality and variance similarity. You can construct a frequency distribution graph to assess for normality and you can calculate standard deviation to assess the similarity of variances (remembering that standard deviation is the square root of the variance) according to the following rules of thumb:

- Assessing for normality: if your frequency-distribution graph is roughly symmetrical and not obviously skewed, then you can assume that your data approximate the normal distribution.
- Assessing for similarity of variances: if the standard deviation of your most variable sample is less than 10 times the standard deviation of your least variable sample, then you can assume that your variances are sufficiently similar.

These are not the most-sophisticated tools for checking parametric requirements. Nonetheless, you will be doing pretty well to carry out any checks at all, since it is common for scientists to just assume their data meet the requirements. As you progress in your career you may wish to learn about the more sophisticated techniques but, if you are just starting out, it is beyond the call of duty!

5.3.3 **Choosing between parametric and nonparametric**

When deciding whether to use a parametric or a nonparametric test suitable for the same data, the first thing to do is to check that your data are at scale level. If they are at the ordinal or nominal levels, you should forget about pursuing a parametric test that uses the normal distribution there and then, because the parametric criteria definitely won't be fulfilled. Next you need to check that your data conform to the parametric criteria for the parametric test under consideration. If they do not, then go with the nonparametric alternative.

Sometimes, for example if sample sizes are small, it will not be possible to check your data. In this case, if you have no reason to believe that your data do not fulfil the parametric criteria, then you might decide to just assume that they do. This is not the most cautious approach but it is generally a reasonable one. The cautious approach would be to do the nonparametric alternative, should there be one.

5.3.4 **Transformation**

It may be that you really want to use parametric statistics; perhaps only a more sophisticated test will satisfy your needs. If your data are at the scale level and okay for parametric analyses, except that they are not normally distributed and/or the variances are dissimilar, all is not lost. You may be able to **transform** your data. Initially this may seem a bit like fiddling your data to fit, but because the same procedure is carried out on every data item, it is valid mathematically. The type of transformation that will make your data suitable for parametric testing will depend on the distribution of your data. For example, something called an arcsine transformation often works on percentage data. We are not going to consider transformations in any detail in this book. It suffices at this stage just to plant this option in your consciousness and to give you the term transformation to look up in the index of another textbook (see Selected further reading).

Transformation is a way of converting scale-level data that do not fulfil the parametric criteria for a test so that they do meet the criteria.

5.4 One- and two-tailed tests

A **two-tailed test** is used when the alternative hypothesis is **nondirectional**, stating, for example, that populations are different. A **one-tailed test** is used when the alternative hypothesis is **directional**, stating, for example, that one population has greater values than another.

A **two-tailed test** is used when the alternative hypothesis is **nondirectional**, stating, for example, that populations are different. A **one-tailed test** is used when the alternative hypothesis is **directional**, stating for example, that one population has greater values than another. Let us look at an example of a two-tailed situation. In Section 5.1.1 on constructing a null hypothesis, I gave the following example of a null hypothesis and its alternative:

H_0: lipid levels in the mantles of pimpleback mussels exposed to air and remaining immersed in water are the same.

H_1: lipid levels in the mantles of pimpleback mussels exposed to air and remaining immersed in water are different.

Notice that the alternative hypothesis says that the lipid levels are different but it does not specify in which direction. It does not specify whether lipid levels in the exposed group are higher or lower than the control group. This is how alternative hypotheses are phrased for two-tailed tests; they are **nondirectional**. Conversely, the alternative hypothesis for a one-tailed test is **directional**.

It is rather unfortunate that you will usually think in directional terms in your question or prediction, and yet you will almost always be working with the nondirectional two-tailed situation from the statistical perspective. This is unfortunate because it is rather confusing! For example, Greseth *et al.* (2003) were interested in elevated lipid levels in their pimpleback mussels. The word elevated indicates directionality, but from the statistical point of view the system is not constrained in this way—levels could be lower in exposed mussels.

You need an exceptional reason for doing the one-tailed version of a test when you have the option of doing a two-tailed test. Just being interested in one direction is not enough. The system you are working on must be constrained and only able to vary in a particular direction, or you must have absolutely no interest in variation in one direction. An example of the latter would be if you were testing a cancer drug and were absolutely sure the drug would not decrease survival, only make no difference or improve survival.

As a rule of thumb, I recommend avoiding one-tailed tests. You might lose a bit of power (Section 5.2), but you can never be wrong using the two-tailed version of a test when you could have used the one-tailed version. However, if you use a one-tailed version inappropriately, that makes your result less reliable. Remember that you can refer to your descriptive statistics to substantiate the direction of any significant results that you find.

Although my advice is not to use a one-tailed test on your own data if you have the choice, there are three reasons for being familiar with the idea of one- and two-tailed tests. Firstly, you need to understand the concept to be able to evaluate the use of the one-tailed test by other scientists. Secondly, when asking a computer to perform a test, you may well need to tell it whether you want a two-tailed or one-tailed option (although two-tailed is typically the default). Thirdly, in tables of critical values of statistics, or on computer printouts, you may be offered a choice of values depending on whether you want to use one or two tails. You need not be thrown by this choice and, as I've already said, you should go for the two-tailed option.

5.5 Effect sizes and their confidence intervals

An **effect size** is a measure of the strength of the pattern in your data. To be useful in ways outlined in Section 5.5.1, they need to be calculated along with their 95% confidence intervals and you can find examples of this in Chapters 7, 8, and 13.

An **effect size** is a measure of the strength of the pattern in your data. An unstandardized effect size (e.g. the difference between sample means) can be converted to a standardized effect size (e.g. Cohen's *d*) using the standard deviation.

In spite of the usefulness of effect size measures, statistical packages do not currently tend to produce useful effect size estimates, and especially effect size estimates with their corresponding 95% confidence intervals in their outputs by default, or even at all.

5.5.1 **Uses of effect size**

Effect size is useful in three main ways. One way is through the assessment of biological significance. If you get a significant result it does not necessarily mean you have found something biologically meaningful. For example, results of a statistical test might indicate a significant difference in the levels of cholesterol between a group of people given a new drug and a control group, but if the amount by which it is reduced would have any impact on patient health then this is only of theoretical rather than practical interest. You could assess this by looking at the 95% confidence intervals of the effect size in relation to the size of effect needed to produce health benefits.

A second way in which effect size is useful is through the assessment of non-significant results. In Section 5.2, we considered that getting a non-significant result might be a consequence of a type II error (accepting a false null hypothesis).This will be related to the power of the test, the less powerful it is the more likely you are to make a type II error; but calculating the power of your test post hoc won't help you assess the cause of your non-significance. But you can assess this by looking at the confidence intervals of the effect size; if they are quite narrow then you can have confidence in your interpretation, but if they are wide then about all you can conclude is that you need to collect more data.

A third way in which effect size is useful is through use in meta-analyses. Effect sizes from different studies can be compared and combined to give a broader picture of what is going on.

5.5.2 **Ways of measuring effect size**

There are many different ways to measure effect size. In most cases a test statistic and the measure of effect size for that test are not the same, but the Pearson correlation coefficient (*r*) is both a test statistic and an effect size measure. The Pearson correlation coefficient measures the strength of a relationship between two variables and it appears in Chapter 11. A measure that is just an effect size and not a test statistic is the difference between group means, which you will find in Chapters 7 and 8. Two other related effect size measure are odds ratio, which appears in Chapter 13, and relative risk (Box 5.3).

Box 5.3 Odds ratios and relative risk effect sizes

The table below has data from the paper by Doll and Hill (1950) on smoking and carcinoma. They recorded the smoking habitat (smoker or non-smoker) and the type of cancer they had (long or other) of 1298 cancer patients. This is the sort of data set on which you could conduct a 2 × 2 chi-square test (Chapter 6).

	Smoker	Non-Smoker
Number of patients with lung cancer	647	2
Number of patients with other cancers	622	27
Total	**1269**	**29**
Risk of getting lung cancer	0.510	0.069
Odds of getting lung cancer	1.040	0.074

Relative risk (or risk ratio) compares probabilities:

For smokers the probability of having lung cancer is 647/1269 = 0.510

For non-smokers the probability of having lung cancer is 2/29 = 0.069

The relative risk of smokers having lung cancer is 0.510/0.069 = 7.391.

The risk of having lung cancer is over seven times greater for smokers than non-smokers.

Odds ratio compares odds:

For smokers the odds of having lung cancer are 647/622 = 1.040

For non-smokers the odds of having lung cancer are 2/27 = 0.074

The odds ratio of smokers having lung cancer is 1.040/0.074 = 14.054.

The odds of having lung cancer is over 14 times greater for smokers than non-smokers.

Measures of effect size come in unstandardized and standardized forms. For example, the difference between group means in Chapters 7 and 8. These could be standardized by dividing by the average standard deviation of the groups to produce an effect size known as Cohen's *d*. Standardized measures of effect size are good if you want to make comparisons across different studies and conduct meta-analyses (Section 5.6.2)

5.6 Limitations, extensions, and alternatives to null hypothesis significance testing (NHST)

5.6.1 Limitations of NHST

There are a number of criticisms that can be made against NHST, and I give three examples along with my responses in Box 5.4. There are even those who argue that we should abandon NHST completely, but this is a rather extreme position! Criticisms of NHST tend to be true but do not in themselves justify the abandoning of NHST.

It is a matter of appreciating that you are taking a particular philosophical approach and understanding this approach and its limitations, which is something you need to do for any method. Moreover, even if all biologists stopped using NHST today, you would still need to learn about it because there is more than half a century of biological literature out there, which you need to be able to understand and evaluate critically, which relies heavily on NHST. So no excuses, you still need to learn about NHST!

Box 5.4 Example criticism of NHST and responses

CRITICISM: The critical significance level is arbitrary.

RESPONSE: Critical significance levels are defined by scientists and traditionally set at 0.05 or lower (Section 5.1.2). As long as you understand this and report statistical results in full, including actual *P* values, so that others are able to make their own interpretations, then it is okay.

CRITICISM: Accepting the null hypothesis means you are accepting that there is not enough evidence to reject it, rather than actually establishing evidence in its support.

RESPONSE: Perhaps the criticism should be of the use of the term 'accept' as short for 'not enough evidence to reject', but what is important is your interpretation of non-significant results. An understanding of the underlying philosophy (Section 5.1), error and power (Section 5.2), and the use of effect size and their 95% confidence intervals is recommended (Section 5.5.1).

CRITICISM: A statistically significant result is not necessarily biologically interesting.

RESPONSE: As long as you understand that finding statistical significance just tells you that you have a non-random pattern, but that this could be of no consequence biologically, then all is well. You can look at effect sizes and their 95% confidence intervals to help you to interpret the biological significance of your results (Section 5.5.1).

5.6.2 Extensions and alternatives to NHST

The statistical tools available to research biologists are increasing at an exciting, or depending on your perspective, alarming rate. The computing power that became available at the end of the twentieth century made the unthinkable thinkable in terms of data analyses, to paraphrase the famous statistician Bradley Effron. This has opened up a world of new possibilities that are still emerging. In this section I am going to mention some alternatives and extensions to NHST which come up elsewhere in this book, then I am going to give you a flavour of what else is out there. As time goes on I suspect that more and more of this will become central to the undergraduate curriculum, but for now this is largely the realm of higher degrees and research.

The use of effect sizes and their confidence intervals was introduced in Section 5.5 and can radically enhance, and help to address some of the criticisms made of NHST (Section 5.3.1). Effect size is not part of traditional NHST but is an extension based on the same underlying maths. Similarly GLZM, covered in Chapters 12 and 13, introduces a modelling approach which is related mathematically to NHST, but offers a different philosophical approach, especially when part of a multi-modelling and model selection process (Section 12.6). Meta-analysis has been mentioned in Section 5.5. This involves combining results from lots of different studies on the same topic and is facilitated by the reporting effect size.

The intriguingly named techniques, bootstrapping and jackknifing, came up in Section 4.2.1. These, and the related resampling Monte Carlo and randomization methods, can enhance NHST and be usefully deployed in modelling techniques. An important aspect of these techniques is that they can liberate you from needing to make assumptions about the distribution of your data and thus provide an alternative to nonparametric statistics.

Multivariate methods, such as principal components analyses and factor analyses, have been around for a while. They are not currently as widely taught as NHST, although my colleague Philip Pugh has been successfully teaching these methods to undergraduates for over a decade, but they are becoming increasingly accessible with the availability of greater computing power. They are also gaining importance because they are well suited to handling the massive data sets (otherwise known as 'big data') that are increasingly becoming available these days. Multivariate methods can play an important role in generating research hypotheses and they complement NHST and modelling techniques.

In terms of more fundamental alternatives, Bayesian inference, as an alternative to frequentist inference (Section 5.1), is gaining ground and becoming well established in a number of areas of research in biology. Thomas Bayes was an English mathematician and Presbyterian minister who lived in the eighteenth century, but it wasn't until recently, with the assistance of late-twentieth-century computing power, that his ideas could be widely applied. One of the strengths of Bayesian statistics is that it allows you to take into account existing knowledge. This is definitely something to get into if you pursue a career in research.

Checklist of key points

- There are different approaches to using inferential statistics for making decisions about data. In this book we follow the most widely used of these: The Neyman–Pearson frequentist approach.
- The four steps of a statistical test are:
  1. Construct a null hypothesis (H_0).
  2. Choose a critical significance level (α).
  3. Calculate a statistic.
  4. Reject or accept the null hypothesis.
- There are two ways of doing step 4:
  - Comparing the calculated value of the statistic with a critical value looked up in a table (the rule varies between statistics).
  - Comparing P with α (according to the rule in Box 5.1).
- Rejecting a null hypothesis means finding a significant result. Accepting a null hypothesis is a non-significant result.

- Type I and II errors are about getting it wrong.
  - Type I error is rejecting a true null hypothesis. The probability of committing a type I error is α.
  - Type II error is accepting a false null hypothesis. The probability of committing a type II error is β.
- Accepting a null hypothesis does not mean that the null hypothesis is definitely true. You have just decided not to reject it.
- Power is about getting it right. The power of a test is the probability of rejecting a false null hypothesis ($1 - \beta$). Power increases with increasing sample size and effect, and decreasing background variation.
- There are two basic categories of test: parametric and nonparametric. To perform a parametric test your data must be scale level and must fulfil the parametric criteria of the test. The parametric criteria are test-specific, but typically have something to do with the data being normally distributed (which you can check by drawing a frequency-distribution graph) and variances being similar (which you can check by seeing if sample standard deviations are within a factor of 10).
- Tests can be one-tailed (where the null hypothesis is directional) or two-tailed (where the null hypothesis is non-directional). It is safest to always go for a two-tailed test when you have the option.
- Biological significance and statistical significance are not necessarily the same thing. Consider the size of the effect as well as its statistical significance.
- Effect size is a measure of the strength of a pattern and is usually different to the test statistic and comes in standardized and unstandardized forms. In association with their 95% confidence intervals, effect sizes can be used to assess biological significance and non-significance. Effect size also facilitates meta-analyses.
- Null hypothesis significance testing has its limitations, but this can largely be mitigated by understanding what it actually does and also by using effect size and their confidence intervals.

Self-help questions

Questions 3–6 are multiple choice (only one correct answer).

1. What is the name of the approach to statistical decision making followed in this book? *Link back: Section 5.1.*
2. How many steps are there in a hypothesis test and what are they? *Link back: Section 5.1.*
3. The null hypothesis: *Link back: Section 5.1.1.*
   (a) Predicts, for example, that an experimental treatment will have no effect.
   (b) Is denoted by the symbol H_0.
   (c) Predicts, for example, that two variables are not correlated.
   (d) All of the above.
4. Which of the following statements is true about critical significance levels? *Link back: Section 5.1.2.*
   (a) All biologists must use a critical significance level of 0.05.
   (b) A researcher decides on his/her own critical significance level.
   (c) Critical significance levels are calculated using a *t*-test.
   (d) All of the above.
5. A researcher risks a type I error: *Link back: Section 5.2.*
   (a) Anytime H_0 is rejected.
   (b) Anytime H_0 is accepted.

(c) Only when he or she chooses a critical significance level of 0.05.
(d) All of the above.

Link back: Section 5.2.

6. The power of a test increases with:
   (a) Decreasing sample size.
   (b) Decreasing effect size.
   (c) Decreasing background variation.
   (d) All of the above.

Link back: Section 5.1.4

7. State true or false for each of the following statements.
   (a) If the significance level P is less than the critical significance level α, we reject H_0.
   (b) If H_0 is rejected the result is statistically non-significant.
   (c) If H_0 is rejected when $\alpha = 0.05$, then it would definitely be rejected for $\alpha = 0.01$.
   (d) Changing the critical significance level from 0.01 to 0.05 increases the risk of a type I error.
   (e) If your value for P is 0.04 and your significance level is 0.05, you accept H_0.
   (f) If you accept your null hypothesis you are, for example, accepting that there is no difference between your samples.

Link back: Section 5.2.

8. The critical significance level (α) is the probability of committing which type of error?

Link back: Section 5.3.

9. What do you have to consider when deciding whether to do a parametric test?

Link back: Section 5.5.1.

10. List three reasons for calculating and reporting effect sizes and the 95% confidence intervals.

6 Tests on frequencies

Chapter Aims

Chapter 5 will have given you an overview of null hypothesis significance testing, including an introduction to four steps which apply to any test. This chapter is the first of five chapters dealing with specific named null hypothesis significance tests. In this chapter the focus is tests dealing with frequency data. In particular, it considers chi-square tests. You will be led through some general points about chi-square tests, then shown when and how to perform the two main types of chi-square test (one-way and two-way). The four steps of null hypothesis significance testing introduced in Chapter 5 are applied to one-way and two-way chi-square tests. This is first done in a general context and then using example data worked through by hand and using SPSS. An example of the use of each test in the literature is also presented.

6.1 Introduction to chi-square tests

This chapter is all about a group of nonparametric inferential statistical tests known as chi-square tests. Chi-square tests are used to assess whether differences between observed and expected frequencies are due to chance alone, or to something more than sample error. **Expected frequencies** are those you would expect according to the null hypothesis. In this introductory section, I am going to go over the following five general issues relating to chi-square tests.

A **frequency** is the number of times a particular value occurs or the number of times values in a stated range occur. An **expected frequency** is the number of times you expect a particular value to occur, or the number of times you expect values in a stated range to occur, based on the null hypothesis being tested.

- The type of data on which you can conduct chi-square tests.
- The types of chi-square test available.
- Circumstances requiring caution when using a chi-square test.
- Alternatives to chi-square tests.
- The name and symbol for the chi-square statistic.

6.1.1 Only use frequency data

Chi-square tests can only be conducted on **frequencies**. It's amazing how often even quite experienced researchers mess up on this. Before embarking on a chi-square test, be sure that you are dealing with frequencies, not percentages, descriptive statistics of scale data, or any other data that are not in the form of frequencies.

We have already worked with frequencies in Section 3.3. Frequencies are counts of items in different categories. For nominal, ordinal, and discrete levels of measurement the data themselves tend to suggest the categories you should use. For the continuous scale level of measurement, values have to be grouped into categories. The frequencies you met in Section 3.3 involved just one set of categories based on one variable. However, characteristics from more than one variable can be combined into a single set of categories. This is illustrated by the pea-seed example data that we are going to use later in this chapter (Section 6.2.1).

It is important to be clear that observations can be made using any level of measurement (nominal, ordinal, discrete scale, or continuous scale), but it is the frequency distributions derived from these data that are used in chi-square tests. Frequency distributions of observations made can be generated directly, by counting the number of times in each category, or indirectly, by recording the observations and then tallying them up.

Let me explain this using an example of a simple type of frequency distribution that you met in Section 3.3. Say you had a sample of 20 crocus plants with flower colours of yellow, purple, or white. You could generate a frequency distribution of flower colour either of two ways. One way would be to make records of your observations (take a look at Table 6.1a to see what this would look like) and then tally up the records for each colour to produce a frequency distribution like the one shown in Table 6.1b. The other way would be to count the numbers of plants of each

(a)

Crocus plant ID	Flower colour
A	White
B	Yellow
C	Yellow
D	Purple
E	Yellow
F	Yellow
G	Yellow
H	Yellow
I	Purple
J	White
K	Purple
L	Yellow
M	Purple
N	Purple
O	White
P	White
Q	Yellow
R	White
S	White
T	Purple

(b)

Flower-colour category	Tally	Frequency
Yellow	IIIIIIII	8
Purple	IIIIII	6
White	IIIIII	6
Total		**20**

Table 6.1 (a) Records of the flower colour of 20 crocus plants. (b) Frequency distribution of the flower colour of 20 crocus plants.

colour to produce a frequency distribution, which would lead you to something like Table 6.1b directly.

In the second of these methods, because records of the colour of each flower are not written down, it is easy to miss that measurements of flower colour using a nominal scale are still being made.

6.1.2 **Types of chi-square test**

In the simple frequency distributions reviewed so far, items have been counted into one set of categories. A more complex arrangement is to count your items against two sets of categories and produce a contingency table frequency distribution. Corresponding to these two arrangements are two types of chi-square test:

1. **One-way classification chi-square test**. This is for situations where frequencies are assigned according to a single set of categories. You can see this illustrated in Table 6.1b and Table 6.2b, which I will return to in more detail in Section 6.2.1. Other examples can be found in Section 3.3. You can shorten the name to **one-way chi-square** and you may also see it referred to as a **goodness-of-fit chi-square test**. As will be explained later, a **test of homogeneity** is a particular type of one-way chi-square test.
2. **Two-way classification chi-square test**. This is for situations where frequencies are assigned according to two sets of categories. The two-dimensional frequency table produced is termed an **R × C contingency table**, where R is the number of rows and C is the number of columns. Table 6.3b, which is featured in detail in Section 6.2.2, is a 2 × 4 contingency table. You can shorten the name to **two-way chi-square**, and you may also see a test of this type referred to as a **contingency table test**, a **test of association**, or a **test of independence**.

All chi-square tests are in essence goodness-of-fit tests, because they compare observed and expected frequencies. As we shall see, for single-classification tests the expected frequencies are based on some theoretical distribution across the single set of categories. In a two-way classification, the expected frequencies are calculated assuming that there is no association between the two sets of categories. Nevertheless, the term goodness-of-fit is often used to refer just to one-way classification.

6.1.3 **When to use chi-square with caution**

Chi-square tests become unreliable if some expected values are small. As a rule of thumb, if any expected value is 0 or more than one-fifth of the expected values are less than five, the chi-square test is not dependable. You should note that this applies to expected values. It does not matter what the observed values are. One way of overcoming this problem can be to merge cells in the frequency table and combine their totals.

Another circumstance under which chi-square tests become unreliable is when they have only one **degree of freedom**, especially if the grand total of the frequencies is small, say less than 20. There are a couple of quick fixes for this; one is called **Yates' correction** (for continuity) and the other is called **Williams' correction**. Both of these

A **one-way classification chi-square test** or **one-way chi-square**, for short, is often referred to as a **goodness-of-fit-chi-square test**. It compares frequencies assigned according to a single set of categories in a simple frequency distribution to those expected based on some theoretical consideration. When all the expected values are the same this type of test can be called a **test of homogeneity**.

A **two-way classification chi-square test** or **two-way chi-square**, for short, is also known as a **test of association** or a **test of independence**. It compares frequencies assigned according to two sets of categories in an **R × C contingency table** frequency distribution to those expected based on no association between the two sets of categories. In a **contingency table**, R is the number of rows (that is, number of categories in one set) and C is the number of columns (that is, categories in the other set).

Degrees of freedom is a number related to sample size which often has to be calculated along with the statistic when conducting a null hypothesis significance test.

Yates' correction (for continuity) and **Williams' correction** are adjustments to the chi-square formula that you should consider using if you reject a null hypothesis based on a chi-square value with only one degree of freedom, especially if the grand total of all frequencies is less than 20.

corrections reduce the value of chi-square that you calculate and make it less likely that you will reject your null hypothesis when it is true (that is, less likely to commit a type I error). If a chi-square includes a correction then a subscript 'adj' is typically used (X^2_{adj}).

If you conclude to accept your null hypothesis without a correction factor you will find yourself accepting it with a correction factor. In practice, the only time it may make a difference to your conclusion is if you have rejected a null hypothesis but it is a borderline decision. I am not going to explain how to calculate adjusted chi-square in this book. You will need to look elsewhere, for example Siegel and Castellan (1988) or Sokal and Rohlf (2012), if you are in this situation and are feeling particularly keen. My main priority here is to make sure that you know what an author is talking about if they mention having used Yates' or Williams' corrections.

6.1.4 **Alternatives to chi-square tests**

Chi-square tests are all nonparametric. Unlike the tests presented in subsequent chapters there are no parametric alternatives to consider. There are other tests that can be done on frequency data, including G tests (one-way or two-way classification), the Kolmogorov–Smirnov test (one-way classification), and the Fisher's exact test (two-way classification). Data suitable for analyses using chi-square tests can also potentially be analysed using a loglinear model with GLZM (see Chapter 13). These are better alternatives to the chi-square under particular circumstances, which you could look up in, for example, Dobson (2001), Siegel and Castellan (1988), or Sokal and Rohlf (2012). G tests especially are generally easier and better than their chi-square counterparts. I have chosen to focus on chi-square tests in this book because they are, and have been, used so widely. You can use chi-square tests in almost all the circumstances that you can use these alternatives. In other words, chi-square tests are the best tests of frequencies to get under your belt first, because they are traditionally the most widely used and they will see you through the majority of situations satisfactorily.

Chi-square tests are not designed for dealing with related data. Each item sampled must contribute to only one category or cell in the frequency table. If you find yourself in a situation where you need a test of frequencies for repeated or matched measures, look up the McNemar test in a book like Siegel and Castellan (1988).

6.1.5 **Names and symbols for the chi-square statistic**

Finally, there is something that you need to know in order to understand your SPSS output for a two-way chi-square test. The statistic used in a chi-square test is the chi-square statistic. Strictly speaking, the chi-square tests don't actually use the real chi-square statistic, but an approximation to it. The G test also uses a statistic that is an approximation of a chi-square. Some people, including me in this book, use the symbol $\mathbf{X}^2$ for an approximated chi-square and χ^2 for a real one. To distinguish an approximated chi-square used in the chi-square test from the approximated chi-square used in the G test, the former is often called the **Pearson chi-square** and

χ^2 is the symbol for a real chi-square. Statistical tests often use approximations that should be distinguished by use of the symbol $\mathbf{X}^2$, instead of χ^2. Chi-square tests use an approximation often referred to as a **Pearson chi-square**.

the latter the likelihood ratio chi-square. The real chi-square crops up in statistical procedures not considered by this book. You shouldn't get hung up on this; the approximations do their job just fine and all that we need to worry about is being able to spot the chi-square value that comes from a traditional Pearson chi-square test.

6.2 Example data

In Sections 6.3 and 6.4 I will be leading you through the recipes for conducting one-way and two-way chi-square. This is best done using real data and in this section I am going to introduce you to the data sets we will be using. I will give some brief background to the data, highlight the features of the data that make it suited to one-way or two-way chi-square testing, and remind you why we need to test the data statistically.

6.2.1 **One-way classification: Mendel's peas**

Over 100 years before biologists regularly started using inferential statistics, the Austrian Augustinian monk Gregor Mendel collected data ideally suited to exploration using chi-square analyses. He collected frequency data on physical characteristics of pea plants, such as seed colour, in order to investigate how these characteristics were inherited. We are going to use the results of just one of his experiments to demonstrate the use of chi-square. In this experiment Gregor focused on two characteristics, or variables, of the pea plant: seed form (round or wrinkled) and seed colour (yellow or green). Combining these two variables he got a single set of four categories describing seed form and colour together:

1. Round in form and yellow in colour (Round Yellow).
2. Round in form and green in colour (Round Green).
3. Wrinkled in form and yellow in colour (Wrinkled Yellow).
4. Wrinkled in form and green in colour (Wrinkled Green).

In the experiment, Mendel artificially fertilized plants grown from round, yellow seeds with the pollen of plants grown from wrinkled, green seeds. The seeds produced, that is the offspring of this union, included representatives from all four categories of seed form and colour. Look at Table 6.2a. If Mendel had made records of each seed, this is how his data might have appeared. You should notice that the level at which seed colour and seed form are measured is nominal. Mendel would then have tallied the number of seeds up in each combined form and colour category as has been done in Table 6.2a and produced the frequency distribution shown in Table 6.2b. Alternatively, he could have skipped Table 6.2a and just counted the number of seeds of each type and produced Table 6.2b directly.

The numbers in Table 6.2b come from Gregor Mendel's famous paper to the Brünn (now Brno) Natural Science Society in the Czech Republic, in 1865, published a year later in the society's journal. The numbers represent the number of seeds falling into

Table 6.2 Example data: Mendel's peas. (a) Records of seed colour and form from 10 plants. This is how Mendel's records could have looked for 10 of the 110 seeds in our example data set. (b) Frequency distribution of seed colour and form from all 110 plants.

(a)

Seed number	Seed form	Seed colour	Category of seed form and colour
1	Round	Yellow	Round Yellow
2	Round	Green	Round Green
3	Round	Green	Round Green
4	Wrinkled	Green	Wrinkled Green
5	Wrinkled	Yellow	Wrinkled Yellow
6	Wrinkled	Yellow	Wrinkled Yellow
7	Round	Green	Round Green
8	Wrinkled	Yellow	Wrinkled Yellow
9	Round	Yellow	Round Yellow
10	Round	Yellow	Round Yellow

(b)

Category of seed form and colour	Number of seeds
Round Yellow	31
Round Green	26
Wrinkled Yellow	27
Wrinkled Green	26
Total	**110**

Source of data: http://www.mendelweb.org/Mendel.html

each category and are therefore frequencies. We can use a chi-square test to assess if these observed frequencies are different from those expected according to various so-called inheritance ratios. Knowing if the frequencies of different characteristics conform to a particular inheritance ratio is useful, because it can indicate the numbers and types of chromosomes and genes involved in the inheritance of a particular characteristic. If you would like to learn more about this you should consult an introductory genetics textbook, such as Thomas (2003).

We would not expect the observed frequencies in our sample to fit a particular ratio exactly, even if the sample came from a statistical population in which the categories did in fact occur in this ratio. Why? Remember sample error (Section 2.1.2)? That's why. A chi-square test is a procedure for allowing us to assess whether the difference between our observed and expected frequencies is acceptable, according to the level of certainty we set with our critical significance level. Gregor Mendel lived at a time when this was not yet appreciated. Fortunately, Mendel's data have since been subject to such analyses and his conclusions have been supported.

The frequencies in Table 6.2b are assigned according to a single set of categories and thus the frequency distribution produced is a simple, one-way organization. This means a one-way chi-square type of analysis is appropriate. A further characteristic of the data in Table 6.2b that makes them suited to investigation with a chi-square test is that each seed contributes to one category only; the data are unrelated.

6.2.2 **Two-way classification: Mikumi's elephants**

Elephants live in groups of different types. Female elephants live in family groups with related females and young offspring, either without a male in attendance (Family Group) or with one (Family with Bull). Male elephants can live alone (Solitary Bull) or in bachelor groups (Bull Group) or, as already mentioned, with a female group (Family with Bull). As part of an extensive study of the largest remaining population of African elephants, researchers in Mikumi National Park, in Tanzania, wanted to know if the relative numbers of different types of group vary between seasons. Looking for an association between group type and habitat in elephants is of applied as well as theoretical interest. For example, an understanding of habitat use in relation to social behaviour is useful to wildlife managers directing anti-poaching and crop-protection efforts.

To begin to answer this question, April Ereckson made a series of observations of elephant groups in the dry and wet seasons of 1998/1999 (Ereckson, 2001). Her records would have looked something like Table 6.3a. From these records, April tallied up the number of groups of each type seen in each season, producing a table like Table 6.3b. The data in Table 6.3b are frequencies classified according to two sets of categories. Table 6.3b has two categories in one set and four in the other, producing a 2×4 contingency table. The question of interest is, is group type associated with season? The situation lends itself perfectly to analysis using a two-way chi-square procedure.

If April had combined group type and season into a single set of categories, as for our previous example with Mendel's pea seeds, she could have tallied the number of sightings up according to a single set of categories based on these characteristics

(a)

Sighting record number	Group type	Season
1	Bull	Dry
2	Family	Dry
3	Family	Dry
4	Solitary Bull	Dry
5	Family	Dry
6	Family with Bull	Wet
7	Solitary Bull	Wet
8	Bull	Wet
9	Family	Wet
10	Family	Wet

(b)

		First set of categories: Group type				
		Solitary bull	Bull group	Family group	Family with bull	Total
Second set of categories: Season	Dry	43	4	196	7	250
	Wet	92	17	195	8	312
	Total	135	21	391	15	562

Source of data: courtesy of April Ereckson.

Table 6.3 Example data: Mikumi's elephants. (a) Records of elephant group type and season for sightings of elephant groups in Mikumi National Park 1998/1999, for 10 observations. This is how April Ereckson's records could have looked for 10 of the 562 elephant group sightings in Mikumi National Park. (b) Frequency distribution of elephant group sightings by group type and season in Mikumi National Park 1998/1999 for all 562 sightings.

combined. If she had done this she would have ended up with the same numbers. However, as you will see in Section 6.4, it is important to lay the observed frequencies out in a contingency matrix, like that illustrated in Table 6.3b, when using a two-way chi-square, in order to appreciate how the expected values and the degrees of freedom are calculated. The key point is that April was looking for an association, not just a match, between her observed frequencies and a distribution based on a predefined ratio.

Sampling error means we are unlikely to find that the observed frequencies match those expected exactly, even if the two sets of categories are unassociated. The two-way chi-square test gives us an objective procedure for deciding whether any difference is due to chance alone or chance plus something of potential biological interest.

April's data are suited to a two-way chi-square because the data are frequencies classified according to two sets of categories and the question is one of association between the two sets of categories. But are the data unrelated? April decided that it was safe to assume that they were not. Certainly only one particular group sighting contributed to one cell in the frequency matrix and, because she recognized many of the animals, she could mostly avoid recording the same group inadvertently more than once within the same season. Any transgression of the requirements of the data to be unrelated for chi-square analyses were minor and it is reasonable to go ahead with a chi-square analysis under such circumstances.

6.3 One-way classification chi-square test

This section focuses on one-way chi-square. It covers:

- When to use and when not to use a one-way chi-square test.
- The four steps as applied to a one-way chi-square.
- Worked examples by hand.
- Worked examples using SPSS.
- An example of the use of a one-way chi-square test from the literature.

The data used in the worked examples come from Gregor Mendel's work and were introduced in Section 6.2.1.

6.3.1 When to use

One-way classification tests are used for analysing frequencies classified according to one set of categories. These categories can be generated from a single variable (e.g. like those in Table 6.1b), or from more than one variable (e.g. like those in Table 6.2b). Any of the example frequency distributions in Section 3.3 (all one set of categories based on one variable; Tables 3.1b and 3.2b) are potentially suitable for testing using a one-way chi-square test. I say potentially because, as noted in Section 6.1.3, chi-square tests must be viewed cautiously if there are small expected values and/or only one degree of freedom.

One-way classification tests compare your observed frequency distribution to the frequency distribution you expect based on some theoretical consideration. Your expectation might be that the frequencies in all categories should be the same. In this case the test is known as a test of homogeneity. Another common use of this type of test is to generate expected frequencies based on standard Mendelian inheritance ratios.

Box 6.1 summarizes when to use and when not to use one-way chi-square tests. It covers points raised in this section, as well as those raised in Section 6.1 applying to all chi-square tests.

Box 6.1 When to use a one-way classification chi-square test

Use this test when:

- Dealing with the frequencies classified according to one set of categories.
- Each item sampled contributes to only one frequency category (that is, the categories are mutually exclusive).
- Looking for a difference between an observed frequency distribution of a single sample and an expected frequency distribution based on theoretical reckoning.

Do not use this test when:

- Using descriptive statistics, percentages, or anything other than frequencies.
- Expected values include small values. As a rule of thumb, if any expected value is 0 or more than one-fifth of the expected values are less than 5, the chi-square test becomes unreliable. (Note, that means expected values; it does not matter what the observed values are.)

6.3.2 **Four steps**

Here are the four steps of a null hypothesis significance test, outlined in Chapter 5, specifically applied to a one-way classification chi-square test.

STEP 1 State the null hypothesis (H_0)

In the case of a one-way classification chi-square, the null hypothesis takes the general form:

> H_0: There is no difference between the observed frequency distribution and that expected based on...

STEP 2 Choose a critical significance level (α)

Typically this is 5% (0.05).

STEP 3 Calculate the test statistic

For chi-square tests the statistic is X^2. The formula is given in Box 6.2. The degrees of freedom equal the number of categories minus 1.

Box 6.2 The formula for chi-square in a one-way classification chi-square test

$$X^2 = \sum \frac{(O-E)^2}{E}$$

Where:
O = observed frequency,
E = expected frequency.

Degrees of freedom = number of categories − 1

STEP 4: Reject or accept your null hypothesis
How you do this will depend on whether you are using critical values for your statistic or *P* values generated by a statistical software package on a computer.

Using critical-value tables

First you need to look up the critical value of chi-square, given the critical significance level you have chosen in step 2 and the degrees of freedom calculated in step 3, in a chi-square critical-values table (Appendix II, Table AII.1). Then you need to compare your value for chi-square with this critical value. Finally, you should decide to reject or accept your null hypothesis according to the rule given in Box 6.3.

Box 6.3 Decision using critical values for a one-way chi-square test

If $X^2 \geq \chi^2_{critical} \rightarrow$ reject $H_0 \rightarrow$ significant result.
If $X^2 < \chi^2_{critical} \rightarrow$ accept $H_0 \rightarrow$ non-significant result.

Using *P* values on computer output

Find *P* on the computer output and make your decision according to the following rule given in Box 6.4.

Box 6.4 Decision using *P* values for a one-way classification chi-square test

If $P \leq \alpha \rightarrow$ reject $H_0 \rightarrow$ significant result.
If $P > \alpha \rightarrow$ accept $H_0 \rightarrow$ non-significant result.

6.3.3 **Worked examples: by hand**

We are going to run through the four steps of a one-way chi-square test using the data introduced in Section 6.2.1. First we will test the observed frequencies against those expected based on a 1:1:1:1 ratio (a test of homogeneity), and then those expected based on a 9:3:3:1 ratio. These are both Mendelian ratios, which can be used to explore the genetic inheritance of characteristics. In practice, it is not common to check observed frequencies against two ratios like this, but I am going to use the same data set in both cases to emphasize that these are variations of the same test.

With expected according to a 1:1:1:1 Mendelian ratio (test of homogeneity)

STEP 1 State the null hypothesis (H_0)

H_0: There is no difference between the observed frequency distribution and the expected frequency distribution based on a 1:1:1:1 ratio.

STEP 2 Choose a critical significance level (α)
We will use 5% (0.05).

STEP 3 Calculate the test statistic

The formula is given in Box 6.2 (Section 6.3.2). The easiest way to apply the data to this formula is to use a table like the one shown in Table 6.4. The observed values come from Table 6.2b. In the case of a test of homogeneity the expected values are in a 1:1:1:1 ratio and are found by dividing the total of the frequencies by the number of categories, in this case 110 divided by 4, which gives 27.5 for each category.

Our chi-square statistic is the sum of the final column, which is 0.62 calculated to two decimal places. Since there are four categories there are three degrees of freedom. In summary, the key information is: $X_3^2 = 0.62$, $N = 110$.

STEP 4 Reject or accept your null hypothesis using critical-value tables

The critical value of chi-square at a 5% critical significance level with three degrees of freedom is 7.8 (from Appendix II, Table AII.1). Since the critical value 7.8 is larger than the value of 0.62 that we calculated, we must accept the null hypothesis according to Box 6.3 (Section 6.3.2). In short:

$$X^2\,(0.62) \le \chi^2_{\text{critical}}\,(7.8) \rightarrow \text{accept } H_0 \rightarrow \text{non-significant result.}$$

If we accept the null hypothesis, we must accept that there is no difference between the observed frequency distribution and the expected frequency distribution

Category	Observed frequency (*O*)	Expected frequency (*E*)	*O* – *E*	$(O-E)^2$	$(O-E)^2/E$
Round Yellow	31	27.50	3.50	12.25	0.45
Round Green	26	27.50	−1.50	2.25	0.08
Wrinkled Yellow	27	27.50	−0.50	0.25	0.01
Wrinkled Green	26	27.50	−1.50	2.25	0.08
Total	**110**	**110.00**	**0.00**	**17.00**	**0.62**

Table 6.4 Table for calculating the chi-square statistic for a 1:1:1:1 expected ratio.

based on a 1:1:1:1 ratio. In other words, the frequencies of the four categories are in a 1:1:1:1 ratio (one-way classification chi-square: $X^2_3 = 0.618$, $N = 110$, $P > 0.05$). You should return to your basic genetics text to find out the implications of this result (Section 6.2.1).

With expected according to a 9:3:3:1 Mendelian ratio

STEP 1 State the null hypothesis (H_0)

H_0: There is no difference between the observed frequency distribution and the expected frequency distribution based on a 9:3:3:1 ratio.

STEP 2 Choose a critical significance level (α)
We will use 5% (0.05).

STEP 3 Calculate the test statistic

The formula is given in Box 6.2 (Section 6.3.2). The easiest way to apply it is to use a table like the one shown in Table 6.5. The observed values come from Table 6.2b. The expected frequencies are calculated by apportioning the total of 110 according to a 9:3:3:1 ratio. For example, nine-sixteenths (9/16ths) of 110 seeds are expected in the round, yellow category. This is calculated by multiplying 9/16 by 110, which gives 61.88. The number 16 at the bottom of the 9/16 fraction comes from summing the ratio: 9 plus 3 plus 3 plus 1 equals 16. The 9 at the top of the 9/16 fraction is the 9 corresponding to the round, yellow category in the 9:3:3:1 ratio.

Our chi-square statistic is the sum of the final column, which is 71.98. Since there are four categories there are three degrees of freedom. The key information is therefore: $X^2_3 = 71.98$, $N = 110$.

STEP 4 Reject or accept your null hypothesis using critical value tables

The critical value of chi-square at a 5% critical significance level with three degrees of freedom is 7.8 (from Appendix II, Table AII.1). Since the critical value 7.8 is smaller than the value of 71.98 that we calculated, we must reject the null hypothesis according to the rule outlined in Box 6.4 (Section 6.3.2). In short:

$$X^2\,(71.98) > \chi^2_{\text{critical}}\,(7.8) \rightarrow \text{reject } H_0 \rightarrow \text{significant result.}$$

Table 6.5 Table for calculating the chi-square statistic for a 9:3:3:1 ratio.

Category	**Observed frequency (*O*)**	**Expected frequency (*E*)**	***O* – *E***	**(*O* – *E*)²**	**(*O* – *E*)²/*E***
Round Yellow	31	61.88	–30.88	953.27	15.41
Round Green	26	20.63	5.38	28.89	1.40
Wrinkled Yellow	27	20.63	6.38	40.64	1.97
Wrinkled Green	26	6.88	19.13	365.77	53.20
Total	**110**	**110.00**	**0.00**	**1388.56**	**71.98**

If we reject the null hypothesis, we must reject that there is no difference between the observed frequency distribution and the expected frequency distribution based on a 9:3:3:1 ratio. In other words, the frequencies in the four categories are not in a 9:3:3:1 ratio (one-way classification chi-square: $X^2_3 = 71.98$, $N = 110$, $P > 0.05$). You should return to your basic genetics text to find out the implications of this result (Section 6.2.1).

6.3.4 **Worked example: using SPSS**

online resource centre

We are going to run through the four steps of a one-way chi-square test using the data on Mendel's peas (Section 6.2.1) again, but this time using the SPSS computing package. As before, we are first going to test the observed frequencies against those expected based on a 1:1:1:1 ratio (a test of homogeneity), and then those expected based on a 9:3:3:1 ratio.

Figure 6.1 shows how the example data need to be entered into SPSS. The column we will be using is the category column, which has been created by combining the information in the first and second columns. In newer versions of SPSS (version 10 onwards) the **Data Editor** has a **Variable View** and a **Data View** tab. Figure 6.1a shows the former and Figure 6.1b shows the latter. Appendix I has notes on how to enter data into SPSS.

Notice how Figure 6.1b resembles Table 6.2a. The data are in the form of raw measures of the items sampled. SPSS calculates the frequencies as part of the procedure. If your data is already in the form of a frequency distribution, like Table 6.2b, your best bet is to use the 'by-hand' route (Section 6.3.3).

With expected according to a 1:1:1:1 Mendelian ratio (test of homogeneity)

STEP 1 State the null hypothesis (H_0)

H_0: There is no difference between the observed frequency distribution and the expected frequency distribution based on a 1:1:1:1 ratio.

STEP 2 Choose a critical significance level (α)
We will use 5% (0.05).

STEP 3 Calculate the test statistic

To get SPSS to conduct a one-way chi-square test on your data you must first open the data file. Then you must make the following selection:

Analyze
→ Nonparametric Tests
→ Legacy Dialogs
→ Chi-Square...

(a)

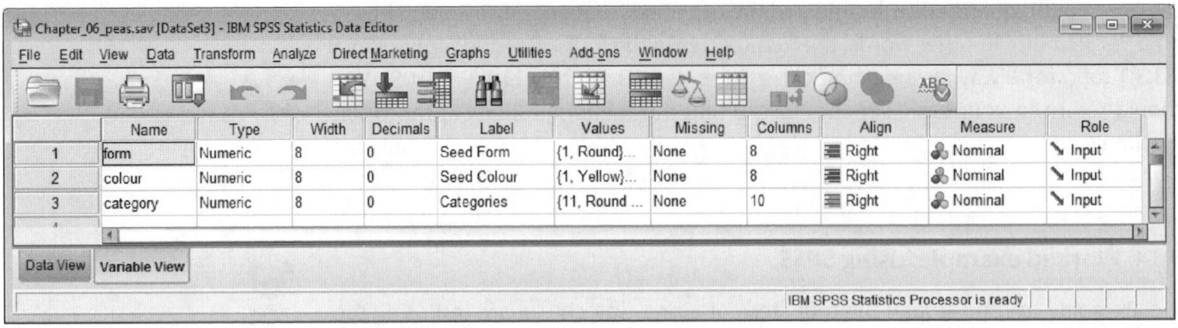

(b)

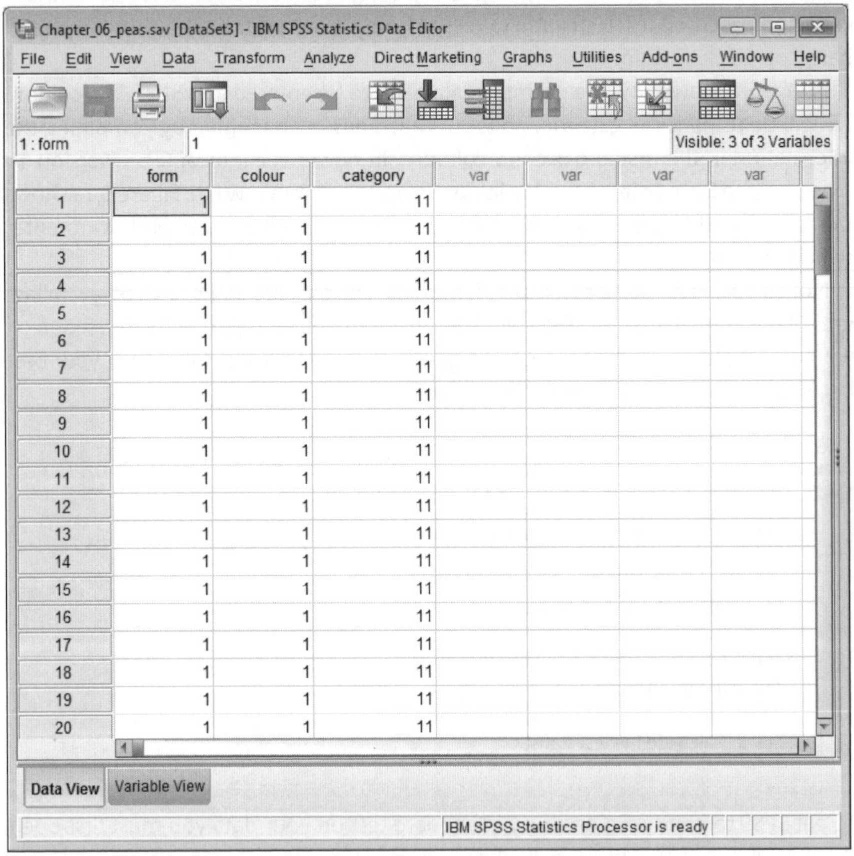

Figure 6.1 Example data in SPSS, Mendel's peas. (a) Variable View. (b) Data View.

A window like that shown in Figure 6.2 will appear. You need to select the variable from the list on the left, which contains the data you want to analyse, and send it to the **Test Variable List** box. In our case this is *Categories*. Under **Expected Range** check that the option **Get from data** is selected. Under **Expected Values** the **All categories equal** is the option that should be selected for a test of homogeneity.

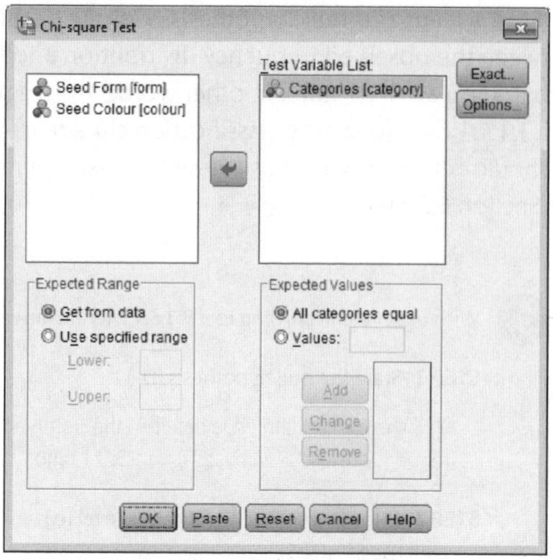

Figure 6.2 Conducting a one-way classification chi-square test of homogeneity using SPSS. Chi-Square Test dialogue window.

Chi-Square Test

Test Statistics

	Categories
Chi-Square	.618[a]
df	3
Asymp. Sig.	.892

Statistic (X^2) ← .618

Degrees of freedom ← 3

P ← .892

[a]0 cells (.0%) have expected frequencies less than 5. The minimum expected cell frequency is 27.5.

This footnote is useful for determining whether you have a problem with small expected values.

Frequencies

Categories

	Observed N	Expected N	Residual
Round Yellow	31	27.5	3.5
Round Green	26	27.5	−1.5
Wrinkled Yellow	27	27.5	−.5
Wrinkled Green	26	27.5	−1.5
Total	110		

Total sample size (N) ← 110

Figure 6.3 SPSS output for a one-way classification chi-square test of homogeneity with key information annotated.

Once you have done all this you should click **OK** and, as if by magic, you will get output looking like Figure 6.3. The key elements of this output have been annotated and are: $X^2 = 0.618$, df = 3, N = 110, $P = 0.892$.

STEP 4 Reject or accept your null hypothesis using *P* values on the computer output

Since the critical significance level 0.05 is smaller than 0.89, the value of *P*, we must accept the null hypothesis according to Box 6.4 (Section 6.3.2). In short:

$P(0.89) > \alpha(0.05) \rightarrow$ accept $H_0 \rightarrow$ non-significant result.

If we accept the null hypothesis, we must accept that there is no difference between the observed frequency distribution and the expected frequency distribution based on a 1:1:1:1 ratio. In other words, the frequencies of the four categories are in a 1:1:1:1 ratio (one-way classification chi-square: $X^2_3 = 0.618$, $N = 110$, $P = 0.892$). You should return to your basic genetics text to find out the implications of this result (Section 6.2.1).

With expected according to a 9:3:3:1 Mendelian ratio

STEP 1 State the null hypothesis (H_0)

H_0 = There is no difference between the observed frequency distribution and the expected frequency distribution based on a 9:3:3:1 ratio.

STEP 2 Choose a critical significance level (α)
We will use 5% (0.05).

STEP 3 Calculate the test statistic

To get SPSS to conduct a one-way chi-square test on your data you must first open the data file. Then you must make the following selection:

Analyze
→ Nonparametric Tests
→ Legacy Dialogs
→ Chi-Square...

A window like that shown in Figure 6.4 will appear. You need to select the variable from the list on the left, which contains the data you want to analyse, and send it to

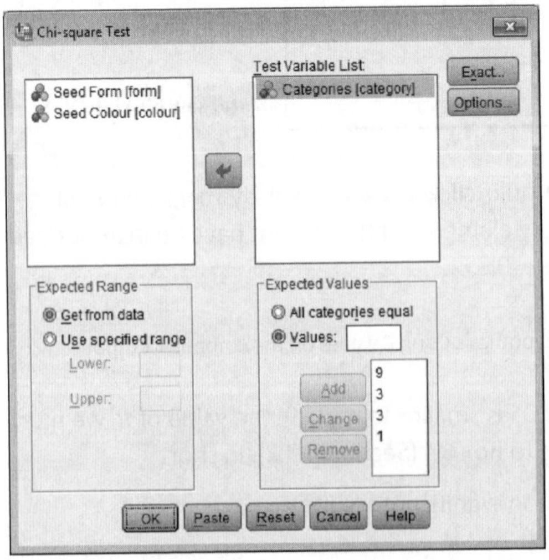

Figure 6.4 Conducting a one-way classification chi-square test for unequal expected frequencies using SPSS. Chi-Square Test dialogue window.

Chi-Square Test

Test Statistics

	Categories	
Chi-Square	71.980[a]	← Statistic (X^2)
df	3	← Degrees of freedom
Asymp. Sig.	.000	← *P*

[a]0 cells (.0%) have expected frequencies less than 5. The minimum expected cell frequency is 6.9.

This footnote is useful for determining whether you have a problem with small expected values.

Frequencies

Categories

	Observed N	Expected N	Residual
Round Yellow	31	61.9	–30.9
Round Green	26	20.6	5.4
Wrinkled Yellow	27	20.6	6.4
Wrinkled Green	26	6.9	19.1
Total	110 ← Total sample size (N)		

Figure 6.5 SPSS output for a one-way classification chi-square test for unequal expected frequencies, with key information annotated.

the **Test Variable List** box. In our case this is *Categories*. Under **Expected Range** check that the option **Get from data** is selected.

The next bit is where the procedure differs from the test of homogeneity and is a bit tricky. You have to tell SPSS the expected ratio. SPSS then calculates the expected values from the expected ratio. To tell SPSS what the expected ratio is you first have to select **Values** under **Expected Values**. Then you type 9 in the adjacent box then click **Add**. Repeat this procedure for the numbers 3, 3, and 1. The order of the values is important. The 9 corresponds to the expected ratio for the category with the lowest-number code. In our example this is *Round Yellow*, which has the code 11. (Switch **Value Label** off under **View** to see the number codes rather than the word labels in your SPSS data file.) The first 3 corresponds to *Round Green* (code 12), the second 3 to *Wrinkled Yellow* (code 21), and the final entry, 1, corresponds to *Wrinkled Green* (code 22). The general rule is that you have to add the ratio values in ascending order of the category-number codes, with the first value in the list corresponding to the category with the lowest-number code.

In case you are wondering, if you put the ratio 1:1:1:1 in the **Expected Values** range it would be another way of doing the test of homogeneity.

Once you have done all this you should click **OK** and, in no time at all, you will get output looking like Figure 6.5. The key elements of this output have been annotated and in summary are: $X_3^2 = 71.98$, $N = 110$, $P < 0.001$.

STEP 4 Reject or accept your null hypothesis using *P* values on the computer output

Since the critical significance level 0.05 is larger than 0.000, the value of *P*, we must reject the null hypothesis according to Box 6.4 (Section 6.3.2). In short:

$P(<0.000) < \alpha\ (0.05) \rightarrow$ reject $H_0 \rightarrow$ significant result.

If we reject the null hypothesis, we must reject that there is no difference between the observed frequency distribution and the expected frequency distribution based on a 9:3:3:1 ratio. In other words, the frequencies of the four categories are not in a 9:3:3:1 ratio (one-way classification chi-square: $X_3^2 = 71.98$, $N = 110$, $P = 0.000$). You should return to your basic genetics text to find out the implications of this result (Section 6.2.1).

6.3.5 **Literature link: *weary* lettuces**

online resource centre

Grube *et al.* (2003) J. Exp. Bot. 54: 1259–1268 (OUP).

Grube *et al.* (2003) described a naturally occurring mutation in lettuce plants in which flowering stalks grow along the ground rather than upright. They called this mutant *weary* and used it to investigate the mechanism of gravity perception and response in plants.

The first objective of their study was to determine the inheritance of the *weary* phenotype. To this end, they bred normal, wild-type plants with *weary* mutants to produce what geneticists call an F_1 generation. They used the pollen of two wild-type cultivars, Salinas88 and 00–235, to fertilize *weary* plants. They then reversed this using pollen from *weary* plants to fertilize Salinas88 and 00–235. The technical name for such a reversal procedure is a reciprocal cross. They used a test of homogeneity (one-way classification chi-square test) to check that it made no difference which way round the pollination was done. In the Results section of the paper they report that 'Chi-square tests of homogeneity established segregation ratios in populations resulting from reciprocal crosses did not differ, providing no evidence for maternal inheritance'.

Grube and colleagues then bred this F_1 generation with itself to produce an F_2 generation. They hypothesized that one pair of genes is involved in the inheritance of this characteristic (known as monogenic inheritance). Furthermore, they postulated that the upright growth gene code is dominant to the mutant *weary* gene code. In this case the *weary* code is described as recessive and will only affect the phenotype if both genes in the pair have the *weary* code. According to Gregor Mendel's famous ratios, under such circumstances the F_2 should contain three wild-type lettuces for every *weary* lettuce. Grube and his fellow workers therefore used a one-way classification chi-square test to see how well their observed frequencies fitted a 3:1 ratio. In their own words in the Methods section, 'chi-square goodness-of-fit tests were used to examine genetic hypotheses'. In their results they reported that 'the *weary* phenotype segregated in ratios consistent with monogenic recessive inheritance (Table 1)'. A modified version of Table 1 from Grube *et al.* (2003) appears here as Table 6.6. Since the authors give us the frequency data, we have the information we need to repeat the chi-square goodness-of-fit analysis ourselves, should we so desire.

Using a 5% critical significance level with one degree of freedom, the critical value of chi-square is 3.8. Since the calculated chi-square value is less than this in both cases, the null hypotheses are accepted and there is no difference between the observed frequency and that expected based on a 3:1 ratio.

In spite of only having one degree of freedom, the authors choose not to use Yates' or Williams' corrections. This is okay. Firstly, the grand totals are reasonably large.

Cross	Number of *weary* plants in F_2	Number of wild-type plants in F_2	Expected ratio (*weary*: wild-type)	X_2[a]	P[a]
weary × Salinas88 wild-type	159	414	1:3	2.31	0.13
weary × 00-235 wild-type	159	426	1:3	1.48	0.22

[a]These are the X^2 and P values from a chi-square test of the fit of the data to expected Mendelian ratios for a single recessive gene.

Source: modified version of Table 1 from Grube *et al.*, by permission of Oxford University Press.

Table 6.6 Inheritance of *weary* lettuce mutant phenotype.

Secondly, using one of these corrections could only have made a difference to the findings if the null hypothesis had been rejected. In other words, if they had used a correction they would still have obtained a non-significant result.

6.4 Two-way classification chi-square test

This section focuses on two-way chi-square, in which frequencies are classified according to two sets of categories rather than just one. It covers:

- When to use and when not to use a two-way chi-square test.
- The four steps as applied to a two-way chi-square test.
- Worked example by hand.
- Worked example using SPSS.
- An example of the use of two-way chi-square test from the literature.

The data used in the worked examples are on elephants and were introduced in Section 6.2.2.

6.4.1 When to use

Two-way classification tests are conducted on frequencies classified according to two sets of categories. Typically such data will be presented in a contingency table (for example, Table 6.3b, introduced in Section 6.2.2). The items sampled must contribute just once each to the various combinations of categories: that is, they can count in only one cell each in the contingency table.

Two-way classification tests compare an observed frequency distribution with the frequency distribution you would expect if the two sets of categories in the contingency table were not associated. An association is when items are counted in a particular cell of a contingency table, more or less often than you would predict by chance alone. With the Mikumi elephants (Section 6.2.2), for example, if a higher proportion of groups sighted in the wet season were solitary bull, compared with the dry season, this could indicate an association between group type and season.

Box 6.5 summarizes when to use and when not to use two-way chi-square tests. It covers points raised in this section as well as those raised in Section 6.1 applying to all chi-square tests.

6.4.2 **Four steps**

Here are the four steps of a null hypothesis significance test, outlined in Chapter 5, specifically applied to a two-way chi-square test.

STEP 1 State the null hypothesis (H_0)

In the case of a two-way classification chi-square, the null hypothesis takes the general form:

H_0: There is no difference between the observed two-way frequency distribution and that expected based on no association between two sets of categories.

STEP 2 Choose a critical significance level (α)
Typically this is 5% (0.05).

Box 6.5 When to use a two-way classification chi-square test

Use this test when:

- Dealing with the frequencies classified according to two sets of categories in a contingency table.
- Each item sampled contributes to only one cell in the contingency table.
- Looking for a difference between an observed frequency distribution of a single sample and an expected frequency distribution based on no association between the two variables.

Do not use this test when:

- Using descriptive statistics, percentages or anything other than frequencies.
- Expected values include small values. As a rule of thumb, if any expected value is 0, or more than one-fifth of the expected values are less than 5, the chi-square test becomes unreliable. (Note, this means expected values: it does not matter what the observed values are.)

STEP 3 Calculate the test statistic
For chi-square tests the statistic is X^2. The formula is the same as for one-way chi-square (Box 6.6). The degrees of freedom equal the number of categories in the first set minus 1, multiplied by the number of categories in the second set minus 1. Referring to the contingency table containing the frequency data, this can be written in terms of the number of rows and columns, as shown at the bottom of Box 6.6.

Box 6.6 The formula for chi-square in a two-way classification chi-square test

$$X^2 = \sum \frac{(O-E)^2}{E}$$

Where:
O = observed frequency,
E = expected frequency.

Degrees of freedom = (number of rows – 1) × (number of columns – 1)

STEP 4 Reject or accept your null hypothesis
How you do this will depend on whether you are using critical values for your statistic, or *P* values generated by a statistical software package on a computer.

Using critical-value tables

First you need to look up the critical value of chi-square, given the critical significance level you have chosen in step 2 and the degrees of freedom you have calculated in step 3, in a chi-square critical-values table (Appendix II, Table AII.1). Then you need to compare your value for chi-square with this critical value. Finally, you should decide to reject or accept your null hypothesis according to the rule given in Box 6.7.

> **Box 6.7** Decision using critical values for a two-way classification chi-square test
>
> If $X^2 \geq \chi^2_{critical} \rightarrow$ reject $H_0 \rightarrow$ significant result.
> If $X^2 < \chi^2_{critical} \rightarrow$ accept $H_0 \rightarrow$ non-significant result.

Using *P* values on computer output

Find *P* on the computer output and make your decision according to the rule given in Box 6.8.

> **Box 6.8** Decision using *P* values for a two-way classification chi-square test
>
> If $P \leq \alpha \rightarrow$ reject $H_0 \rightarrow$ significant result.
> If $P > \alpha \rightarrow$ accept $H_0 \rightarrow$ non-significant result.

6.4.3 **Worked example: by hand**

We are going to run through the four steps of a two-way chi-square test using the data on elephant groupings described in Section 6.2.2.

STEP 1 State the null hypothesis (H_0)

H_0 = There is no difference between the observed frequency distribution by group type and season, and that expected based on no association between group type and season. There is no association between group type and season.

STEP 2 Choose a critical significance level (α)
We will use 5% (0.05).

STEP 3 Calculate the test statistic

The formula is given in Box 6.6 (Section 6.4.2). The easiest way to apply the data to this formula is to use a table like the one shown in Table 6.8. The observed values come from Table 6.3b. The expected values come from multiplying row totals by column

Table 6.7 Contingency table (4 × 2) for Mikumi National Park elephant data (with expected values in brackets).

		First set of categories: group type				
		Solitary bull	Bull group	Family group	Family with bull	Total
Second set of categories: season	Dry	43 (60.05)	4 (9.34)	196 (173.93)	7 (6.67)	250
	Wet	92 (74.95)	17 (11.66)	195 (217.07)	8 (8.33)	312
	Total	135	21	391	15	562

totals and dividing by the grand total. These are shown in Table 6.7, which is a repeat of Table 6.3b with expected values in brackets. For example, the expected value for solitary bulls in the dry season is 250 (the row total), multiplied by 135 (the column total), divided by 562 (the grand total) which gives 60.05.

Our chi-square statistic is the sum of the final column in Table 6.8, which is 19.3. Since there are two rows and four columns of numbers in our frequency contingency table (Table 6.3b or Table 6.7), the degrees of freedom is 1 × 3 which equals 3.

STEP 4 Reject or accept your null hypothesis using critical value tables

The critical value of chi-square at a 5% critical significance level with three degrees of freedom is 7.8 (from Appendix II Table AII.1). Since the critical value 7.8 is smaller than the value of 19.30 that we calculated, we must reject the null hypothesis according to Box 6.7. In short:

$$X^2\ (19.30) > \chi^2_{critical}\ (7.8) \rightarrow \text{reject } H_0 \rightarrow \text{significant result.}$$

Rejecting the null hypothesis leads us to reject that there is no difference between the observed frequency distribution and the expected frequency distribution based on no association. In other words, there was an association between the type of group and season for the frequency of sightings of elephant groups in the study area in

Table 6.8 Table for calculating the chi-square statistic for a two-way classification chi-square test. For Cell ID the number before the underscore indicates the column in the first set of categories and the number after the underscore is the row in the second set of categories: for example, cell 1_2 contains the data for family groups in the wet season.

Cell ID	Observed frequency (*O*)	Expected frequency (*E*)	$O-E$	$(O-E)^2$	$(O-E)^2/E$
1_1	43	60.05	−17.05	290.82	4.84
1_2	4	9.34	−5.34	28.53	3.05
1_3	196	173.93	22.07	486.98	2.80
1_4	7	6.67	0.33	0.11	0.02
2_1	92	74.95	17.05	290.82	3.88
2_2	17	11.66	5.34	28.53	2.45
2_3	195	217.07	−22.07	486.98	2.24
2_4	8	8.33	−0.33	0.11	0.01
Total	**562.00**	**562.00**	**0.00**	**1612.88**	**19.30**

Mikumi National Park in 1998/1999 (two-way classification chi-square: $X^2_3 = 19.30$, $N = 562$, $P < 0.05$).

6.4.4 **Worked example: using SPSS**

online resource centre

We are going to run through the four steps of a two-way chi-square test using the data on elephants introduced in Section 6.2.2, but this time using the SPSS computing package.

Figure 6.6 shows how the example data need to be entered into SPSS. In newer versions of SPSS (version 10 onwards) the **Data Editor** window has a **Variable View** and a **Data View** tab; Figure 6.6a shows the former and Figure 6.6b the latter. Appendix I has notes on how to enter data into SPSS.

(a)

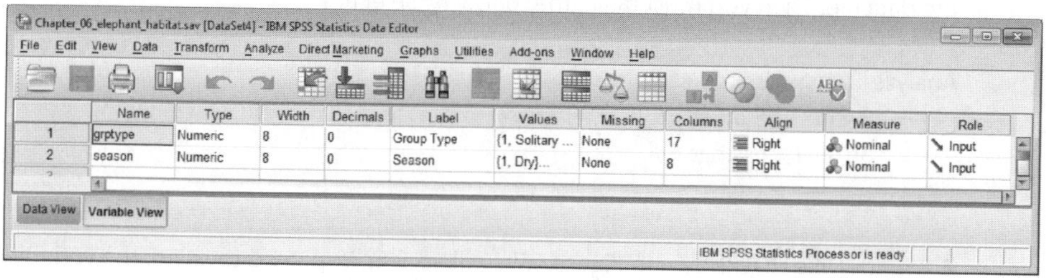

(b)

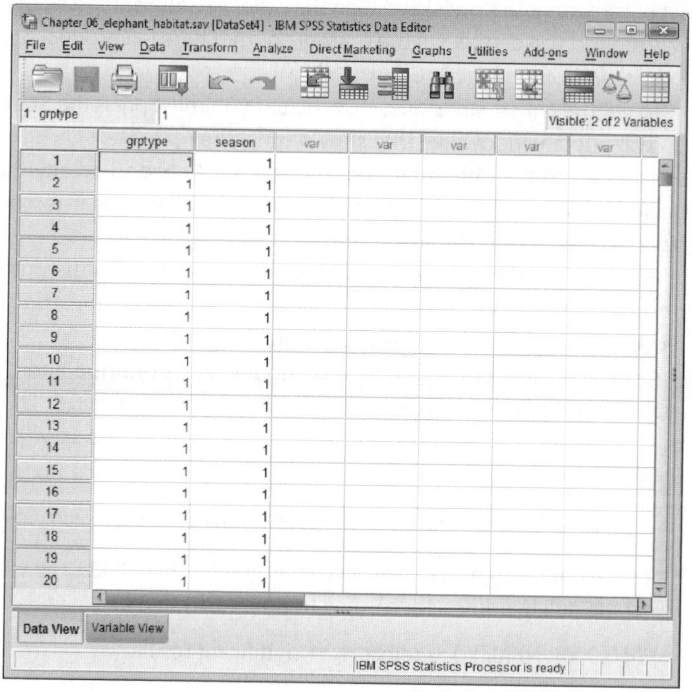

Figure 6.6 Example data in SPSS, Mikumi's elephants. (a) Variable View. (b) Data View.

Notice how Figure 6.6b resembles Table 6.3a. The data are in the form of raw measures of the items sampled. SPSS calculates the frequencies as part of the procedure. If your data are already in the form of a frequency distribution, as in Table 6.3b, your best bet is to use the 'by-hand' route (Section 6.4.3).

STEP 1 State the null hypothesis (H_0)

H_0: There is no difference between the observed frequency distribution by group type and season, and that expected based on no association between group type and season. There is no association between group type and season.

STEP 2 Choose a critical significance level (α)

We will use 5% (0.05).

STEP 3 Calculate the test statistic

To get SPSS to conduct a two-way chi-square test on your data you must first open the data file. Then you must make the following selection:

Analyze
→ Descriptive Statistics
→ Legacy Dialogs
→ Crosstabs...

A window like that shown in Figure 6.7a will appear. You need to select the variable from the list on the left which contains the data you want to analyse. First send one set of categories to **Row(s)** and then the other to **Column(s)**. It does not matter which way around you do this but the example output here is a product of putting *Season* in **Row(s)** and *Group Type* in **Column(s)**.

To get the chi-square statistic calculated you must press the **Statistics** button and up will pop a window like that shown in Figure 6.7b. Next click on the small white square next to the **Chi-square** option. You don't need to do anything about the expected values as they are calculated automatically based on no association, once you have selected the chi-square option.

It is also useful to get expected frequencies displayed in the output as well as the observed ones. To do this you need to press the **Cells** button and, under **Counts** select **Expected** as well as **Observed** (Figure 6.7c).

Once you have done all this, you should click **Continue** and then **OK** and in a matter of moments you will get output looking like Figure 6.8. The key elements of this output have been annotated and can be summarized as: $X^2_3 = 19.30$, N = 562, $P < 0.001$.

STEP 4 Reject or accept your null hypothesis using *P* values on computer output

Since the critical significance level 0.05 is greater than 0.000, the value of *P*, we must reject the null hypothesis according to Box 6.8. In short:

$P(< 0.000) < \alpha\ (0.05) \rightarrow$ reject $H_0 \rightarrow$ significant result.

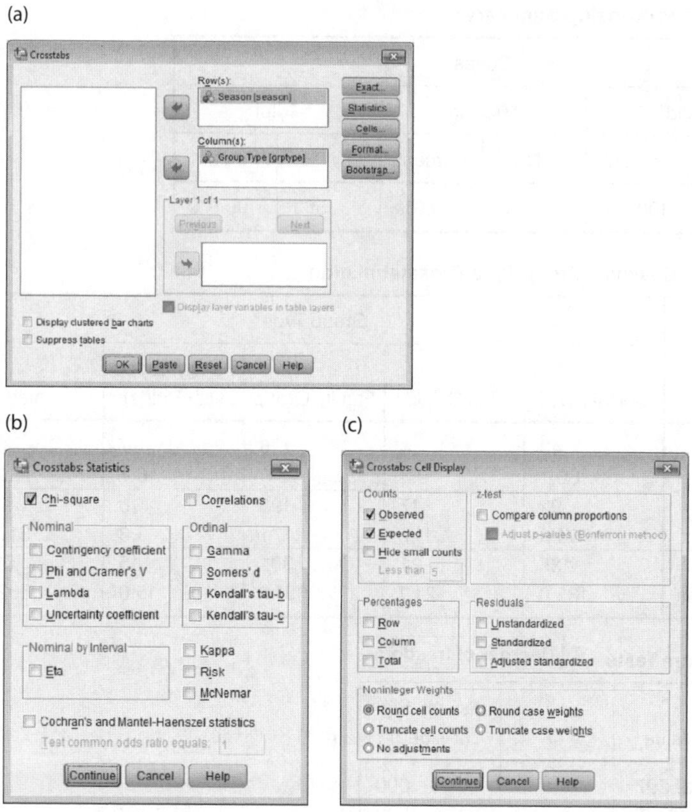

Figure 6.7 Conducting a two-way classification chi-square test using SPSS. (a) Crosstabs dialogue window. (b) Crosstabs Statistics dialogue window. (c) Crosstabs Cell Display dialogue window.

Rejecting the null hypothesis leads us to reject that there is no difference between the observed frequency distribution and the expected frequency distribution based on no association. In other words, there was an association between the type of group and season for the frequency of sightings of elephant groups in the study area in Mikumi National Park in 1998/1999 (two-way classification chi-square: $X^2_3 = 19.30$, $N = 562$, $P = 0.000$.

6.4.5 **Literature link: treatment alliance**

Gavin *et al.* (1999) J. Ped. Psy. 24: 355–365.

online resource centre

The ability of a patient and their doctor to create a positive working relationship is known as treatment alliance. Gavin *et al.* (1999) studied factors promoting treatment alliance for cases of adolescents with severe, long-term asthma. They began their study with 60 subjects who had been admitted to their hospital. One year after the patients were discharged they attempted to collect follow-up data, but only managed to get information on half the subjects. They called the 30 people on whom they obtained follow-up data, participants and the 30 people on whom they did not, non-participants.

Case Processing Summary

	Cases					
	Valid		Missing		Total	
	N	Percent	N	Percent	N	Percent
Season * Group Type	562	100.0%	0	0.0%	562	100.0%

Season * Group Type Crosstabulation

			Group Type				
			Solitary Bull	Bull Group	Family Group	Family Group with Bull(s)	Total
Season	Dry	Count	43	4	196	7	250
		Expected Count	60.1	9.3	173.9	6.7	250.0
	Wet	Count	92	17	195	8	312
		Expected Count	74.9	11.7	217.1	8.3	312.0
	Total	Count	135	21	391	15	562
		Expected Count	135.0	21.0	391.0	15.0	562.0

Total sample size (N)

Statistic (X^2) **Chi-Square Tests** Degrees of freedom

	Value	df	Asymp. Sig. (2-sided)
Pearson Chi-Square	19.297[a]	3	.000
Likelihood Ratio	20.075	3	.000
Linear-by-Linear Association	14.519	1	.000
N of Valid Cases	562		

P

[a]0 cells (0.0%) have expected count less than 5. The minimum expected count is 6.67.

This footnote is useful for determining whether you have a problem with small expected values.

Figure 6.8 SPSS output for a two-way classification chi-square test with key information annotated.

As a preliminary part of the study they compared the participants and non-participants according to demographic, child psychological, family, and medical factors. These included the following variables, measured at a nominal level, on which they performed a two-way chi-square test to see if there was an association between these factors, and whether the subjects participated in the follow-up study or not:

- Gender of the adolescent (male or female).
- Race of the adolescent (Caucasian or non-Caucasian).

Variable	Participant % ($n = 30$)	Non-participant % ($n = 30$)	Statistic[a]
Gender (% male)	46.7	53.3	0.27
Race (% Caucasian)	83.3	66.7	2.22
Biological parents still married (%)	56.7	53.3	0.07
Mother alone (%)	26.7	26.7	0.00

[a]Pearson's chi-square test; all *P* values are non-significant.
Source: modified version of Table 1 from Gavin *et al.* (1999), by permission from Oxford University Press.

Table 6.9 Percentage of participants and non-participants according to four factors listed under variable.

- Whether the adolescent's parents were still married (yes or no).
- Whether the adolescent was from a single-parent family (yes or no).

They reported their findings in their Methods section rather than their Results section, as this was a preliminary part of their study describing their subjects. This is what they wrote:

> As can be seen from Table 1, there were no significant differences between follow-up participants and non-participants on demographic, child, family, or medical variables previously obtained during hospitalization at our institution.

Take a look at Table 6.9, which is a modified version of their Table 1. Notice that it states explicitly the name of the test (see the footnotes), sample sizes (see the column headings), and the value of the statistic (final column). The other two key pieces of information, the degrees of freedom and the *P* value, can be deduced.

Let me explain how degrees of freedom and *P* value can be deduced using sex against participation. Gender has two categories (male and female) and participation has two categories (participating and non-participating), therefore we are talking about a 2×2 contingency table and hence one degree of freedom (Box 6.6). They don't actually say what critical significance level they use, but it is probably safe to assume that they used the traditional 5% level. Given this, we can deduce that all the *P* values are more than 0.05, because a footnote at the bottom of the table tells us that 'all *P* values are non-significant'.

They give us enough information in the test to repeat the chi-square analysis ourselves. Again, let us use gender to see how this could be done. We are told that 46.7% of the participants and 53.3% of the non-participants are males. Converting these percentages to frequencies, we get 14 male participants (46.7% of 30) and 16 (53.3% of 30) male non-participants. Since there were 30 participants and 30 non-participants (the sample sizes given in the columns), this means that there must have been 16 female participants and 14 female non-participants. The 2×2 contingency table would have looked like Table 6.10.

In spite of only having one degree of freedom, the authors choose not to use Yates' or Williams' corrections. This is okay. Firstly, the grand total of 60 is reasonably large. Secondly, using one of these corrections could only have made a difference to the

Table 6.10 Contingency table (2×2) derived from Table 1 in Gavin *et al.* (1999).

		First set of categories: participation		
		Participation	Non-participation	Total
Second set of categories: gender	Male	14	16	30
	Female	16	14	30
	Total	30	30	60

findings if the null hypothesis had been rejected. In other words, if they had used a correction, they would still have obtained a non-significant result.

Checklist of key points

- Only use data in the form of frequencies for chi-square tests.
- Chi-square tests are nonparametric.
- There are two main types of chi-square test:
  - One-way classification chi-square tests are appropriate when frequencies are assigned according to a single set of categories producing a simple frequency distribution.
  - Two-way classification chi-square tests are appropriate when frequencies are assigned according to two sets of categories.
- The chi-square test is traditionally the most widely used test of frequencies. There are alternatives (e.g. G tests, Fisher's exact test, Kolmogorov–Smirnov test, binary logistic GLZM), but you can find a chi-square test to adequately cover almost all your needs in this area.
- Chi-square tests should only be used on unrelated data: counts should contribute to only one category, or combination of categories. A McNemar test is a test of frequencies for related data.
- Potential problems with chi-square tests are as follows:
  - Chi-square tests are unreliable if any expected value is 0 or more than one-fifth of the expected values are less than 5. Merging cells and combining counts can be a way of overcoming this problem.
  - Chi-square tests can be unreliable if the degrees of freedom is 1. Yates' or William's correction factors may be useful in this context.
- The four steps for a one-way chi-square are as follows:
  1. State H_0: there is no difference between the observed frequencies and those expected based on...
  2. Choose a critical significance level: typically, $\alpha = 0.05$.
  3. Calculate the statistic: calculate X^2 and the degrees of freedom according to Box 6.2.
  4. Reject or accept H_0: reject H_0 if $X^2 \geq \chi^2_{critical}$ (Box 6.3) or if $P \leq \alpha$ (Box 6.4).
- The four steps for a two-way chi-square are as follows:
  1. State H_0: there is no difference between the observed two-way frequency distribution and that expected based on no association between the two sets of categories.

2. Choose a critical significance level: typically $\alpha = 0.05$.
3. Calculate the statistic: calculate X^2 and the degrees of freedom according to Box 6.6.
4. Reject or accept H_0: reject H_0 if $X^2 \geq \chi^2_{critical}$ (Box 6.7), or if $P \leq \alpha$ (Box 6.8).

- If your data are already in the form of a frequency distribution, one-way or two-way, then you are better off following the by-hand method, rather than using SPSS.

Self-help questions

1. Which of the following describe data in the form of frequencies?
   (a) The number of pregnant women falling into the following categories: non-smoker, light smoker, heavy smoker.
   (b) The percentage of pregnant women falling into the following categories: non-smoker, light smoker, heavy smoker.
   (c) The mean number of cigarettes smoked per day by 20 pregnant women.
   (d) The hair colour (measured as red, brown, black, or blonde) of 100 people from Great Britain.
   (e) The number of leaves on 20 geranium plants grown for 3 weeks in sunlight.
   (f) The number of geranium plants grown in sunlight for three weeks with between 0 and 10 leaves, 11 and 20 leaves, or 21 or more leaves. *Link back: Section 3.3 and Section 6.1.1.*
2. For each of the following features say whether it applies to a one-way chi-square, a two-way chi-square, both, or neither:
   (a) Parametric test. *Link back: Section 6.1.4.*
   (b) Nonparametric test. *Link back: Section 6.1.4.*
   (c) Numbers plugged into the formula for the statistic are frequencies. *Link back: Section 6.1.1.*
   (d) Numbers plugged into the formula for the statistic are percentages. *Link back: Section 6.1.1.*
   (e) Numbers plugged into the formula for the statistic are means and standard deviations. *Link back: Section 6.1.1.*
   (f) Compares an observed frequency distribution to one based on a specified theoretical distribution. *Link back: Section 6.1.*
   (g) Compares an observed frequency distribution to one based on no association between two sets of categories. *Link back: Section 6.1.2.*
   (h) Should not be used if one or more expected values are 0. *Link back: Section 6.1.3.*
   (i) Should be used with caution if more than one-fifth of the expected values are less than 5. *Link back: Section 6.1.3.*
   (j) Should consider using a correction factor if the null hypothesis is accepted with only one degree of freedom, especially if the grand total is less than 20. *Link back: Section 6.1.3.*
   (k) Should not be used when an item contributes to more than one cell in the frequency table. *Link back: Sections 6.3.1 and 6.4.1.*
   (l) Used on unrelated data. *Link back: Sections 6.3.1 and 6.4.1.*
   (m) Used for related data. *Link back: Sections 6.1.4, 6.3.1, and 6.4.1.*
   (n) Uses a statistic called the Pearson chi-square. *Link back: Section 6.1.5.*
   (o) Uses a statistic called the likelihood chi-square. *Link back: Section 6.1.5.*
   (p) Reject the null hypothesis if the X^2 value calculated is less than the $\chi_{critical}{}^2$ value looked up in a table. *Link back: Sections 6.3.2 and 6.4.2.*
   (q) Reject the null hypothesis if *P* on the SPSS printout is less than 0.05. *Link back: Sections 6.3.2 and 6.4.2.*
3. A researcher introduced life-sized plastic animals into a squirrel monkey enclosure at a wildlife park and recorded whether the monkeys vocalized or not during the following five minutes. He did 20 trials each for four different plastic animals (a snake, a raptor, a rabbit, and a duck). He used SPSS to perform a two-way chi-square test to investigate whether *Link back Section 6.3.4; also see Appendix III*

Chi-Square Tests

	Value	df	Asymp. Sig. (2-sided)
Pearson Chi-Square	29.899[a]	3	.000
Likelihood Ratio	32.632	3	.000
Linear-by-Linear Association	22.982	1	.000
N of Valid Cases	80		

[a]0 cells (.0%) have expected count less than 5. The minimum expected count is 9.00.

Figure 6.9 SPSS output referred to in self-help question 3.

there was an association between vocalization (categorized as occurred or not occurred) and the type of plastic animal introduced. His output is shown in Figure 6.9.

In his results he wrote, 'There was an association between the occurrence of vocalization and the type of plastic animal introduced into the enclosure (___).' The bracket, which has been left blank here, contained the key statistical information to support his statement. Write down what you would have put in the brackets.

7 Tests of difference: two unrelated samples

Chapter Aims

The subject of this chapter is tests of differences on two samples of unrelated data. The parametric *t*-test and the nonparametric Mann–Whitney U test are presented. The chapter starts by comparing these tests and dealing with general points relating to them. The four steps of null hypothesis significance testing introduced in Chapter 5 are applied to both the *t*-test and the Mann–Whitney U test. Firstly, this is done in a general context and then using example data, worked through by hand and using SPSS. An example of the use of each test from the literature is also presented.

7.1 Introduction to the *t*- and Mann–Whitney U tests

This chapter is all about testing to see if a difference between two samples of unrelated data (Section 2.3.1) is due to chance alone, or to chance plus something of potential biological interest. There are parametric and nonparametric ways of doing this. We will be looking at the parametric *t*-test and the nonparametric Mann–Whitney U test. In this introduction I am going to touch on a few general points about these tests:

- The type of data on which *t*- and Mann–Whitney U tests work.
- Similarities and differences between a *t*-test and a Mann–Whitney U test.
- The parametric criteria as applied to a *t*-test.
- Alternatives to and extensions of the *t*-test and the Mann–Whitney U test.

7.1.1 **Variables and levels of measurement needed**

Neither the *t*-test nor the Mann–Whitney U test work if the data in the samples are measured at a nominal level. When you look at your data you need to see numbers representing scale or ordinal levels of measurement. The minimum level of measurement needed for the dependent variable is ordinal for a Mann-Whitney U test and scale for a *t*-test. The independent variable can be measured at any level provided that it only distinguishes two categories. The two categories of the independent variable identify the two samples.

Look at the data set in Table 7.1: what do you see? There are numbers representing the length of brine shrimps belonging to one of two categories: medium salinity and high salinity. Brine shrimp length is the dependent variable and salinity is the independent variable. The numbers you see are on the scale level of measurement: not nominal, not frequencies. Two samples are distinguished by the salinity of the water in which the brine shrimps have been maintained: medium or high. Salinity is measured at an ordinal level. This data set has all the features described in the previous paragraph.

Table 7.1 Brine shrimp body length. Brine shrimp length measured in millimetres after four weeks at either medium salinity ($n = 21$) or high salinity ($n = 16$).

Brine shrimp length (mm)	
Medium salinity	**High salinity**
5.5	6.0
6.0	7.0
5.0	7.5
7.0	6.0
5.5	7.5
6.0	8.0
7.0	11.0
8.0	9.0
6.0	8.0
8.0	11.0
6.0	8.0
7.0	8.0
6.0	7.0
7.0	7.0
6.0	7.0
8.0	9.0
6.0	
7.0	
7.5	
6.0	
7.5	

7.1.2 Comparison of *t*- and Mann-Whitney U tests

Both the *t*-test and the Mann-Whitney U test are used to look for difference between two samples. For example, these tests could both be used to assess whether the length of brine shrimps in the two salinity conditions is different for data like those in Table 7.1. The two samples in Table 7.1 are (i) the length of brine shrimps from a medium-salinity condition and (ii) the length of brine shrimps from a high-salinity condition.

A *t*-test and a Mann-Whitney U test are used when data are unrelated. Look at Table 7.1 and ask yourself if you moved the 5.5 datum at the top of the medium salinity column to the bottom, would you then need to move the 6.0 at the top of the high salinity column down as well? The answer is no. It does not matter what numbers are on the same rows.

The *t*-test is a parametric test and the Mann-Whitney U test is a nonparametric test. This means that for a *t*-test to work properly, parametric criteria must apply (Sections 5.3 and 7.1.3). Since it is a parametric test, the *t*-test will only work when the data in your samples are measured at scale level. The Mann-Whitney U test will work with either ordinal or scale data. The similarities of and differences between the *t*-test and the Mann-Whitney U test are summarized in Table 7.2.

7.1.3 The *t*-test and the parametric criteria

A *t*-test is a parametric test and therefore you must be satisfied that the data meet the parametric criteria (Section 5.3). In the case of the *t*-test this means that:

1. Each sample must be approximately normally distributed.
2. The variances of the two samples must be similar.

Table 7.2 Comparison of the *t*-test and Mann-Whitney U test.

	***t*-Test**	**Mann-Whitney U test**
Similarities	Tests of difference of central tendency Two samples compared Data in samples unrelated	
Differences	Parametric scale data only	Nonparametric scale or ordinal data

To check point 1 in the list above you could draw histograms for each sample and see if they looked reasonably normal. To check point 2 you could calculate and compare the standard deviations of the two samples. Alternatively, you can just assume that these criteria apply, if you have no reason to suspect that they do not (Section 5.3.2).

7.1.4 Alternatives to *t*- and Mann-Whitney U tests

We will consider how to test a difference between two samples when the data are related in the next chapter, Chapter 8. The equivalent of the *t*-test for related data is the paired *t*-test and the equivalent of the Mann–Whitney test for related data is the Wilcoxon signed-rank test. If you have more than two samples you want to explore differences between, you will need the analysis-of-variance techniques dealt with in Chapter 9, or GLZM, which is the subject of Chapter 12.

7.2 Example data: dem bones

In this section I am going to introduce the data set that I will be using in the worked examples in Sections 7.3 and 7.4 for a *t*-test and a Mann–Whitney U test, respectively. I have chosen a data set that meets the parametric criteria so that we can use it for both tests. Using the same data set for both tests emphasizes that the basic data structure required for these analyses is the same. However, you need to remember that if parametric assumptions are not met then you can pursue the nonparametric alternative. If you have the option of either a parametric or nonparametric test, you should do the parametric test, because it is more powerful (Section 5.2.1); there is no need to do both.

The example data for this chapter are on human bones and came to my attention through the work of Liz Smith. Liz conducted research at the Medical Research Council (MRC) Human Nutrition Research Laboratory in Cambridge. She was part of a team of scientists interested in the role diet plays in chronic disease. In particular, Liz looked at how the things we eat might interact with our hormones to make us more prone to the bone disease osteoporosis.

Osteoporosis is characterized by a loss of bone mass, which leaves sufferers at increased risk of breaking arms, legs, and other bones. This can be immensely painful and prevents people from leading happy, independent lives, not to mention that it costs the National Health Service billions of pounds annually. A clue that hormones might be involved in this disease comes from the observation that there is a strong sex bias among osteoporosis sufferers. Women are much more susceptible than men, with one in three women over the age of 50 affected. One of the first things researchers did to try to find out what was going on was to investigate the bone density of healthy patients. Some of the data they collected are presented in Table 7.3.

There are two samples in Table 7.3: a sample of bone density for 20 females and a sample of bone density for 20 males. All subjects were over 50 years old. The numbers in these samples are grams per square centimetre (g/cm^2) and therefore at the scale level of measurement. Bone density is thought to be dependent on sex and so bone density is the dependent variable. Sex is the independent variable, measured as male or female, which is nominal.

Table 7.3 Example data: dem bones. Bone-density measurement in grams per square centimetre of the neck of the femur for healthy men ($n = 20$) and women ($n = 20$) over 50 years of age.

Bone-density measurement (g/cm^2)	
Females	**Males**
0.972	0.905
0.732	1.016
0.874	0.873
0.943	0.74
1.024	0.861
0.755	0.817
0.779	0.897
1.007	0.962
0.816	0.851
0.755	0.821
0.871	0.763
0.721	0.876
0.727	0.944
0.796	0.993
0.612	0.774
0.775	0.785
0.849	0.892
0.773	1.076
0.649	0.888
0.865	0.865

Source of data: courtesy of the MRC Human Nutrition Research Laboratory.

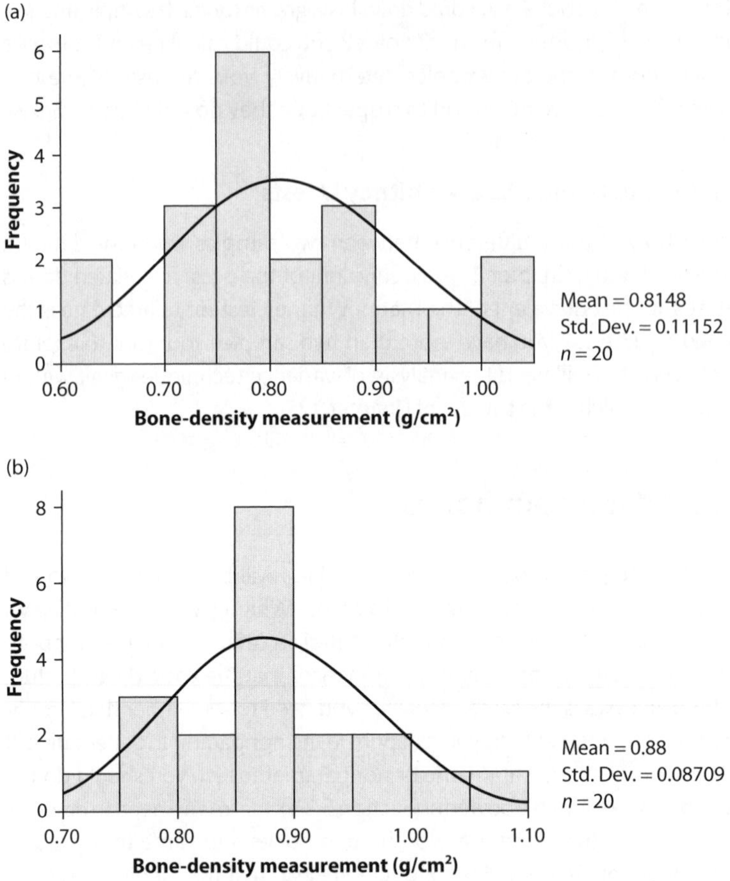

Figure 7.1 Histograms of example data: dem bones. Normal distributions are indicated by the curved lines. (a) Bone density for females ($n = 20$). (b) Bone density for males ($n = 20$).

As you can see from Figure 7.1, each sample is approximately normally distributed and the standard deviations, 0.04 and 0.11, are within a factor of 10 of each other. These data therefore fulfil the parametric criteria for a *t*-test (Sections 5.3.2 and 7.1.3).

Errorplots or boxplots would both be good ways of presenting these data. Errorplots with confidence intervals are presented here in Figure 7.2.

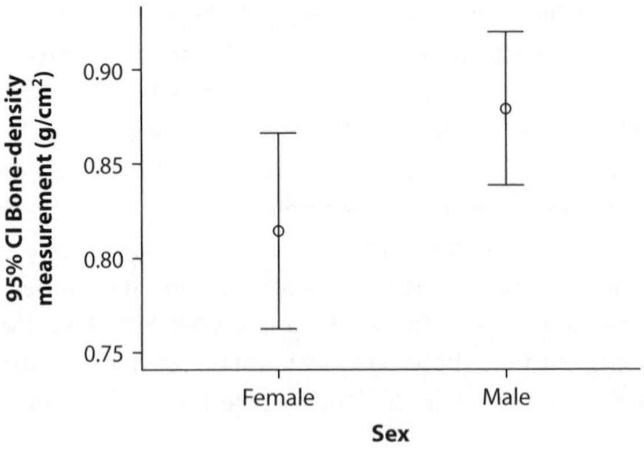

Figure 7.2 Bone density for females ($n = 20$) and males ($n = 20$): means with 95% confidence intervals. (Refer back to Section 4.2 for guidance on interpretation.)

7.3 *t*-Test

The topic of this section is the *t*-test, which is often called the Student's *t*-test or the *t*-test for independent samples. The name Student's *t*-test comes from the chap who first described this test, who published under the name of Student, although this was not his real name. We are going to cover the following aspects of the *t*-test:

- When to use and when not to use a *t*-test.
- The four steps as applied to a *t*-test.
- Worked example of a *t*-test by hand.
- Worked example of a *t*-test using SPSS.
- An example of the use of a *t*-test in the literature.

The data used in the worked examples come from work at the MRC Human Nutrition Research Laboratory, introduced in Section 7.2.

7.3.1 **When to use**

A *t*-test is a parametric test, so you must be satisfied that your data fulfil the parametric criteria (Sections 5.3 and 7.1.3) before using this test. It is a parametric test for assessing whether the difference between two unrelated samples can be accounted for by sample error alone. Issues concerning when and when not to use this test were covered in the introduction to this chapter (Sections 7.1.1 and 7.1.2). They are summarized in Box 7.1.

Box 7.1 When to use a *t*-Test

Use this test when:

- You are looking for a *difference* between *two* samples.
- Data in the samples are measured at the *scale* level.
- Data are *unrelated*.

Do not use this test when:

- You want to compare frequency distributions (Chapter 6).
- Data do not fulfil the parametric criteria (Sections 5.3.2 and 7.1.3).

7.3.2 **Four steps of a *t*-test**

Here are the four steps of a null hypothesis significance test outlined in Chapter 5, specifically applied to a *t*-test.

STEP 1 State the null hypothesis (H_0)

The null hypothesis for a *t*-test takes the general form:

H_0: There is no difference between the populations from which two unrelated samples come.

STEP 2 Choose a critical significance level (α)
Typically this is 5% (0.05).

STEP 3 Calculate the test statistic
For a *t*-test the statistic is *t*. The formula for *t* in a *t*-test is given in Box 7.2. The degrees of freedom equal the sum of the sample sizes minus 2.

Box 7.2 Formula for *t* in a *t*-test

$$t = \sqrt{\frac{[(\bar{G}-\bar{y}_1)n_1 + (\bar{G}-\bar{y}_2)n_2](n_1+n_2-2)}{\left[\sum(y-\bar{y}_1)^2 + \sum(y-\bar{y}_2)^2\right]}}$$

Where: $\bar{G}$ = grand mean, $\bar{y}_1$ = mean of first sample, $\bar{y}_2$ = mean of second sample, n_1 = size of first sample, n_2 = size of second sample.

Degrees of freedom $= n_1 + n_2 - 2$

STEP 4 Reject or accept your null hypothesis
How you do this will depend on whether you are using critical values for your statistic or *P* values generated by a statistical software package on a computer.

Using critical-value tables

First you need to look up the critical value of *t* given the critical significance level you have chosen in step 2, and the degrees of freedom calculated in step 3 (Appendix II, Table AII.2). Then you need to compare the value of *t* that you have calculated in step 3 with this critical value. Finally, you should decide to reject or accept your null hypothesis according to the rule in Box 7.3.

Box 7.3 Decision using critical values for *t*-test

If $t \geq t_{critical} \rightarrow$ reject $H_0 \rightarrow$ significant result.
If $t < t_{critical} \rightarrow$ accept $H_0 \rightarrow$ non-significant result.

Using *P* values on computer output

Find *P* on the computer output and make your decision according to the rule in Box 7.4.

Box 7.4 Decision using *P* value for *t*-test

If $P \leq \alpha \rightarrow$ reject $H_0 \rightarrow$ significant result.
If $P > \alpha \rightarrow$ accept $H_0 \rightarrow$ non-significant result.

7.3.3 **Worked example: by hand**

We are going to run through the four steps of a *t*-test using the data introduced in Section 7.2.

STEP 1 State the null hypothesis (H_0)

H_0: There is no difference in bone density between males and females over 50 years old.

STEP 2 Choose a critical significance level (α)
We will use 5% (0.05).

STEP 3 Calculate the test statistic

The formula is given in Box 7.2. The easiest way to apply the data to this formula is to use a table like the one shown in Table 7.4. The data in Table 7.4 are the same as in Table 7.3, except that they have been rounded up to two decimal places to make life easier. All the numbers you need are in Table 7.4 except for the grand mean ($\bar{G}$). An easy way of getting this is to add the sum of y_1 and y_2 together and then divide this number by the total of n_1 plus n_2. For our example data this works out as follows:

$$\bar{G} = \left(\sum y_1 + \sum y_2\right) / (n_1 + n_2) = (16.30 + 17.60)/(20 + 20)$$
$$= 33.90/40 = 0.85$$

Table 7.4 Table for calculating *t* in a *t*-test. Letters D and E correspond to the equation given in the text.

	Bone density of females (g/cm²)	$\bar{y}_1$	$y_1 - \bar{y}_1$	$(y_1 - \bar{y}_1)^2$	Bone density of males (g/cm²)	$\bar{y}_2$	$y_2 - \bar{y}_2$	$(y_2 - \bar{y}_2)^2$
	0.97	0.81	−0.16	0.02	0.91	0.88	−0.03	0.00
	0.73	0.81	0.08	0.01	1.02	0.88	−0.14	0.02
	0.87	0.81	−0.06	0.00	0.87	0.88	0.01	0.00
	0.94	0.81	−0.13	0.02	0.74	0.88	0.14	0.02
	1.02	0.81	−0.21	0.04	0.86	0.88	0.02	0.00
	0.76	0.81	0.06	0.00	0.82	0.88	0.06	0.00
	0.78	0.81	0.04	0.00	0.90	0.88	−0.02	0.00
	1.01	0.81	−0.19	0.04	0.96	0.88	−0.08	0.01
	0.82	0.81	0.00	0.00	0.85	0.88	0.03	0.00
	0.76	0.81	0.06	0.00	0.82	0.88	0.06	0.00
	0.87	0.81	−0.06	0.00	0.76	0.88	0.12	0.01
	0.72	0.81	0.09	0.01	0.88	0.88	0.00	0.00
	0.73	0.81	0.09	0.01	0.94	0.88	−0.06	0.00
	0.80	0.81	0.02	0.00	0.99	0.88	−0.11	0.01
	0.61	0.81	0.20	0.04	0.77	0.88	0.11	0.01
	0.78	0.81	0.04	0.00	0.79	0.88	0.09	0.01
	0.85	0.81	−0.03	0.00	0.89	0.88	−0.01	0.00
	0.77	0.81	0.04	0.00	1.08	0.88	−0.20	0.04
	0.65	0.81	0.17	0.03	0.89	0.88	−0.01	0.00
	0.87	0.81	−0.05	0.00	0.87	0.88	0.01	0.00
Mean	0.81				0.88			
Sample size	20				20			
Sum	16.30			0.24	17.60			0.14
				D				E

The best way to tackle the equation for calculating t for a t-test is to break it down into the components labelled A–E as shown below:

$$t = \sqrt{\frac{[\overbrace{(\bar{G}-\bar{y}_1)^2 n_1}^{A} + \overbrace{(\bar{G}-\bar{y}_2)^2 n_2}^{B}]\overbrace{(n_1+n_2-2)}^{C}}{[\underbrace{\sum(y-\bar{y}_1)^2}_{D} + \underbrace{\sum(y-\bar{y}_2)^2}_{E}]}}$$

You can get D and E directly from Table 7.4. To get A, B, and C you will need the grand mean ($\bar{G}$), as calculated above, plus the means of each sample ($\bar{y}_1$ and $\bar{y}_2$), and the sample size of each sample (n_1 and n_2), which you will find in Table 7.4. For our example data it works out as follows:

$$A = (0.85 - 0.81)^2 \times 20 = 0.03^2 \times 20 = 0.02$$

$$B = (0.85 - 0.88)^2 \times 20 = 0.03^2 \times 20 = 0.02$$

$$C = 20 + 20 - 2 = 38$$

$$D = 0.24$$

$$E = 0.14$$

Plugging these numbers into the equation for t for a t-test (Box 7.2) gives

$$t = \sqrt{\frac{(0.02 + 0.02) \times 38}{0.24 + 0.14}} = \sqrt{\frac{1.62}{0.38}} = \sqrt{4.26}$$

Since the square root of 4.26 can be either +2.06 or −2.06, the answer to this equation could be + or −2.06. That is $t = \pm 2.06$ (to two decimal places).

We will just be using what's called the absolute number for t in step 4. That is, we will ignore the sign and just look at the number, 2.06. We will also need the degrees of freedom associated with this value. Applying the formula in Box 7.2 to our example data, this works out as

$$df = n_1 + n_2 - 2 = 20 + 20 - 2 = 38$$

So the important information can be summarized as: $t_{38} = 2.06$

STEP 4 Reject or accept your null hypothesis using critical-value tables

The critical value of t at a 5% critical significance level with 38 degrees of freedom is 2.02 (Appendix II, Table AII.2). Since the critical value 2.02 is smaller than 2.06 that we calculated, we must reject the null hypothesis according to Box 7.3. In short:

$$t\,(2.06) > t_{critical}\,(2.02) \rightarrow \text{reject } H_0 \rightarrow \text{significant result.}$$

We conclude that there is a difference between the bone density of males and females over 50 years old (t test: $t_{38} = 2.06$, $P < 0.05$).

(a)

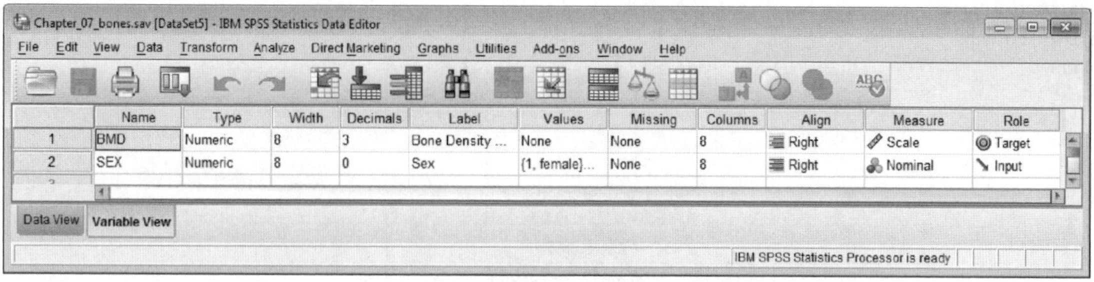

(b)

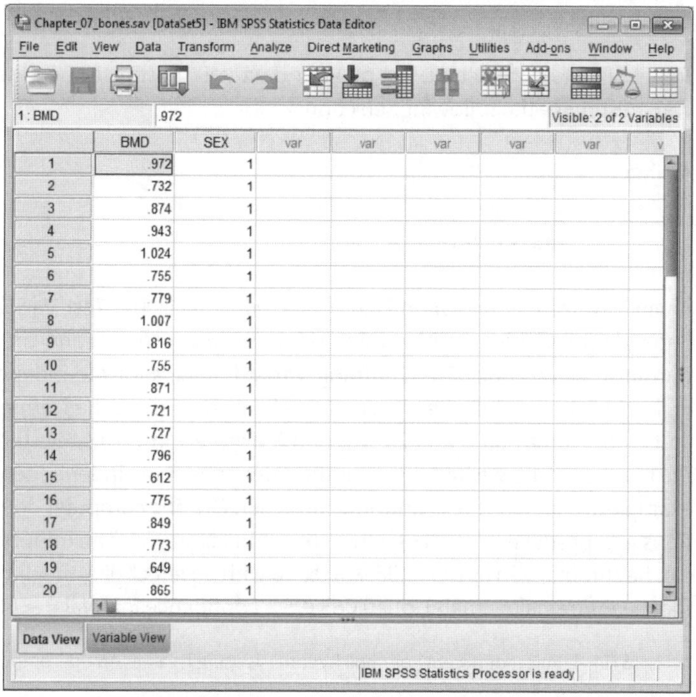

Figure 7.3 Example data in SPSS: dem bones. (a) Variable View. (b) Data View.

7.3.4 **Worked example: using SPSS**

online resource centre

We are going to repeat doing a *t* test on the bone data (Section 7.2), but this time we are going to use SPSS. To do this the data must be entered into SPSS, as shown in Figure 7.3. In newer versions of SPSS (version 10 onwards) the **Data Editor** window has a **Variable View** tab and a **Data View** tab. Figure 7.3a shows the former and Figure 7.3b the latter. Appendix I has notes on how to enter data into SPSS.

STEP 1 State the null hypothesis (H_0)

H_0: There is no difference in bone density between males and females over 50 years old.

STEP 2 Choose a critical significance level (α)
We will use 5% (0.05).

(a) (b)

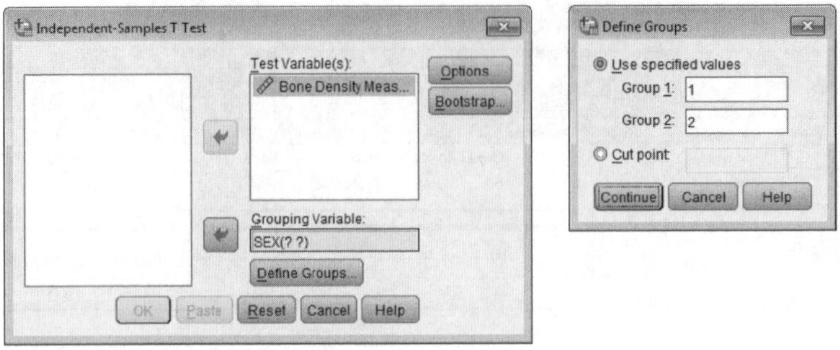

Figure 7.4 Conducting a *t*-test using SPSS. (a) Main dialogue window. (b) Define Groups dialogue window.

STEP 3 Calculate the test statistic

To get SPSS to conduct a *t*-test on your data you must first open the data file. Then you must make the following selections:

Analyze
→ Compare Means
→ Independent-Samples T Test ...

A window like that shown in Figure 7.4a will appear. The **Test Variable(s)** box is for dependent variables: highlight *Bone Density Measurement* and press the arrow to transfer it to this box. The **Grouping Variable** box is for the independent variable: highlight *Sex* and press the arrow to transfer it. Then press the **Define Groups** button and type 1 for group 1 and 2 for group 2 (Figure 7.4b). This tells the computer that the number codes we have used for the categories of the independent variable in this example are 1 and 2. Press **Continue** and then **OK**. The computer does all the calculations and produces output like that shown in Figure 7.5. The important information can be summarized as $t_{38} = 2.06$, $P = 0.046$ with an effect size, the difference between the means (female – male), of –0.065 (95% confidence interval = –0.129 to –0.001).

STEP 4 Reject or accept your null hypothesis using *P* values on the computer output

Since the critical significance level 0.05 is larger than the value 0.046 of *P*, we must reject the null hypothesis according to Box 7.4. In short.

$$P\,(0.046) < \alpha\,(0.05) \rightarrow \text{reject } H_0 \rightarrow \text{significant result.}$$

We conclude that there is a difference between the bone density of males and females over 50 years old (*t*-test: $t_{38} = 2.06$, $P = 0.046$; difference = 0.065, 95% C.I. = –0.129 to –0.001).

7.3.5 **Literature link: silicon and sorghum**

online resource centre

Hattori *et al.* (2003) Plant Cell Physiol. 44: 743–749.

Sorghum is one of the most important cereal crops in the world. The biggest producer is the United States, but millions and millions of tonnes of it are grown globally every

***t*-Test**

Group Statistics

	Sex	N	Mean	Std. Deviation	Std. Error Mean
Bone Density Measurement (g/square cm)	female	20	.81475	.111522	.024937
	male	20	.87995	.087092	.019474

Independent Samples Test

		Levene's Test for Equality of Variances		*t*-test for Equality of Means					95% Confidence Interval of the Difference	
		F	Sig.	*t*	df	Sig. (2-tailed)	Mean Difference	Std. Error Difference	Lower	Upper
Bone Density Measurement (g/square cm)	Equal variances assumed	1.527	.224	–2.061	38	.046	–.065200	.031640	–.129252	–.001148
	Equal variances not assumed			–2.061	35.892	.047	–.065200	.031640	–.129376	–.001024

Annotations: Effect size (Mean Difference); Effect size 95% CI (Lower/Upper, equal variances assumed); Statistic (*t*); Degrees of freedom; *P*

Figure 7.5 SPSS output for a *t*-test with key information annotated.

year, especially in Africa and Asia. Silicon is thought to improve sorghum's resistance to disease and drought by hardening the walls of the root cells. Hattori *et al.* (2003) decided to investigate. As part of their research they assessed the effect of silicon on the growth of roots in sorghum seedlings. They took 100 sorghum seedlings and they let 55 grow in nutrient solution free of silicon (silicon-minus treatment) and 45 in a similar solution, but with silicon added (silicon-plus treatment). After five days they measured the length of the roots of all the plants.

In their Materials and methods section, the authors tell us that 'data were analysed statistically by using a *t*-test to evaluate the effects of silicon'. They do not tell us whether they checked for parametric criteria. In their Results section they tell us that 'sorghum seedlings grown in the presence of silicon (silicon-plus treatment) had significantly longer roots ($P < 0.01$) than those grown in a silicon-free nutrient solution (silicon-minus treatment)'. In support of this statement, they refer readers to a table which is reproduced here in modified form as Table 7.5. The phrase 'significant at 1% level by *t*-test' in Table 7.5 is the equivalent of them saying that they used a 0.01 critical significance level. These authors do not give us any information about the *t*-test other than the *P* value. However, we can work out the degrees of freedom from the sample sizes: $45 + 55 - 2 = 98$.

Table 7.5 Root lengths of sorghum grown in nutrient solution only and nutrient solution with added silicon after five days.

Treatment	Root length (mm)
Silicon-minus	166.6 ± 3.4
Silicon-plus	180.5 ± 3.8
	**

Data are means ± standard error ($n = 55$ for silicon-minus and $n = 45$ for silicon-plus).
**Significant at 1% level by *t*-test.
Source of data: modified version of Table 1 in Hattori *et al.*, 2003.

7.4 Mann-Whitney U test

The topic of this section is the Mann-Whitney U test. Capital letters are used for Mann and Whitney because these are the names of the people who first described this test. This section covers:

- When to use and when not to use a Mann-Whitney U test.
- The four steps as applied to a Mann-Whitney U test.
- Worked examples by hand.
- Worked example using SPSS.
- An example of the use of a Mann-Whitney U test from the literature.

The data used in the worked examples come from Human Nutrition Research, and were introduced in Section 7.2.

7.4.1 When to use

A Mann-Whitney U test is a nonparametric test for assessing whether the difference between two unrelated samples can be accounted for by sample error alone. Issues relating to when and when not to use this test were covered in the introduction to this chapter (Section 7.1). They are summarized in Box 7.5.

Box 7.5 When to use a Mann-Whitney U Test

Use this test when:

- You are looking for a *difference* between *two* samples.
- Data in the samples are measured at the *scale* or *ordinal* levels.
- Data are *unrelated*.

Do not use this test when:

- You want to compare frequency distributions (Chapter 6).

7.4.2 Four steps of a Mann-Whitney U test

Here are the four steps of a null hypothesis significance test outlined in Chapter 5, specifically applied to a Mann-Whitney U test.

STEP 1 State the null hypothesis (H_0)
The null hypothesis for a Mann-Whitney U test takes the general form:

H_0: There is no difference between the populations from which two unrelated samples come.

STEP 2 Choose a critical significance level (α)
Typically this is 5% (0.05).

STEP 3 Calculate the test statistic
For a Mann-Whitney U test the statistic is U. The formula for U is given in Box 7.6. You can use either U_1 or U_2 for U, but you have to know if it's the higher or the lower value and use the appropriate critical value table for your choice in step 4. We will take U as the lower value of either U_1 or U_2 in the worked example. The other values that are particularly important to note are n_1, the size of first sample, and n_2, the size of the second sample.

Box 7.6 Formula for U in a Mann-Whitney U test

$$U_1 = n_1 n_2 + \frac{n_2(n_2+1)}{2} - R_2$$

$$U_2 = n_1 n_2 + \frac{n_1(n_1+1)}{2} - R_1$$

Where: n_1 = size of first sample,
n_2 = size of second sample,
R_1 = sum of the ranks of the first sample,
R_2 = sum of the ranks of the second sample.

You can use either U_1 or U_2 for U, but you have to know if it's the higher or the lower value and use the appropriate critical value table for your choice in step 4.

Check $U_1 + U_2 = n_1 n_2$

STEP 4 Reject or accept your null hypothesis
How you do this will depend on whether you are using critical values for your statistic, or *P* values generated by a statistical software package on a computer.

Using critical-value tables

First you need to look up the critical value of U in Appendix II; Table AII.3a has the values appropriate for use when you are taking the lower of U_1 or U_2 as your U, and Table AII.3b has the values appropriate for use when you are taking the higher of U_1 or U_2 as your U. If you are using a critical-value table from another source, be sure to check whether it is for use with the higher or lower U value.

To find the critical value of U (Appendix II, Table AII.3a or b), you will need the critical significance level you have chosen in step 2 and the sample sizes calculated in step 3. Then you need to compare your value for the value of U you have calculated in step 3 with this critical value. Finally, you should decide to reject or accept your null hypothesis according to the rule in Box 7.7a, if you are using the lower of U_1 or U_2 as your U, or Box 7.7b, if you are using the higher of U_1 or U_2 as your U.

Box 7.7a Decision using critical values for the Mann-Whitney U test (using lower U value of U_1 or U_2 as U)

If $U \leq U_{critical} \rightarrow$ reject $H_0 \rightarrow$ significant result,
If $U > U_{critical} \rightarrow$ accept $H_0 \rightarrow$ non-significant result.

Box 7.7b Decision using critical values for the Mann-Whitney U test (using higher value of U_1 or U_2 as U)

If $U \geq U_{critical} \rightarrow$ reject $H_0 \rightarrow$ significant result.
If $U < U_{critical} \rightarrow$ accept $H_0 \rightarrow$ non-significant result.

Using *P* values on computer output

Find *P* on the computer output and make your decision according to the following rule in Box 7.8.

Box 7.8 Decision using *P* values for the Mann-Whitney U test

If $P \leq \alpha \rightarrow$ reject $H_0 \rightarrow$ significant result.
If $P > \alpha \rightarrow$ accept $H_0 \rightarrow$ non-significant result.

7.4.3 **Worked example: by hand**

We are going to run through the four steps of a Mann-Whitney U test using the data introduced in Section 7.2.

STEP 1 State the null hypothesis (H_0)

H_0: There is no difference in bone density between males and females over 50 years old.

STEP 2 Choose a critical significance level (α)
We will use 5% (0.05).

STEP 3 Calculate the test statistic

The formula is given in Box 7.6. The only vaguely tricky bit is assigning the ranks to get R_1 and R_2. The numbers are ranked according to where they would come if the two samples were combined and listed in ascending order. The smallest number is ranked 1 and so on, regardless of which sample it originated from. Tied ranks are given the median of the rank positions they cover. Study Table 7.6 carefully if you are unsure. For example, there are two values of 0.755, occupying the positions of ranks 7 and 8, so both are ranked as 7.5 in the final column. There are also two values of 0.865, occupying positions 22 and 23, so they are both ranked 22.5 in the final column.

To find R_1 and R_2 it is a good idea to list the ranks of the two samples in different columns, as in Table 7.7. All the numbers you need to calculate U_1 and U_2 using the formulae in Box 7.6 are in Table 7.7. For the example data the working goes as follows:

$$U_1 = (20 \times 20) + \frac{20(20+1)}{2} - 489.5 = 120.5$$

$$U_2 = (20 \times 20) + \frac{20(20+1)}{2} - 330.5 = 279.5$$

Table 7.6 Table illustrating ranking of data for calculating U.

y_1/y_2	Sample	Rank	Rank ignoring ties	Rank adjusted for ties for Mann-Whitney U
0.612	1	1	1	1
0.649	1	2	2	2
0.721	1	3	3	3
0.727	1	4	4	4
0.732	1	5	5	5
0.740	2	6	6	6
0.755	1	7	7	7.5
0.755	1	7	8	7.5
0.763	2	9	9	9
0.773	1	10	10	10
0.774	2	11	11	11
0.775	1	12	12	12
0.779	1	13	13	13
0.785	2	14	14	14
0.796	1	15	15	15
0.816	1	16	16	16
0.817	2	17	17	17
0.821	2	18	18	18
0.849	1	19	19	19
0.851	2	20	20	20
0.861	2	21	21	21
0.865	1	22	22	22.5
0.865	2	22	23	22.5
0.871	1	24	24	24
0.873	2	25	25	25
0.874	1	26	26	26
0.876	2	27	27	27
0.888	2	28	28	28
0.892	2	29	29	29
0.897	2	30	30	30
0.905	1	31	31	31
0.943	1	32	32	32
0.944	2	33	33	33
0.962	2	34	34	34
0.972	1	35	35	35
0.993	2	36	36	36
1.007	1	37	37	37
1.016	1	38	38	38
1.024	1	39	39	39
1.076	2	40	40	40

	y_1	y_2	Rank y_1	Rank y_2
	0.972	0.905	35	31
	0.732	1.016	5	38
	0.874	0.873	26	25
	0.943	0.740	32	6
	1.024	0.861	39	21
	0.755	0.817	7.5	17
	0.779	0.897	13	30
	1.007	0.962	37	34
	0.816	0.851	16	20
	0.755	0.821	7.5	18
	0.871	0.763	24	9
	0.721	0.876	3	27
	0.727	0.944	4	33
	0.796	0.993	15	36
	0.612	0.774	1	11
	0.775	0.785	12	14
	0.849	0.892	19	29
	0.773	1.076	10	40
	0.649	0.888	2	28
	0.865	0.865	22.5	22.5
Sum			**330.5** (R_1)	**489.5** (R_2)
Sample size	**20** (n_1)	**20** (n_2)		

Table 7.7 Table for calculating U in the Mann-Whitney U test.

We can check our calculations by comparing the sum of U_1 and U_2 and seeing if it is the same as the product of n_1 and n_2 (that is, n_1 times n_2). It is a good idea to do this because it's a good way of catching errors in calculation. For our example data this check would be:

$$U_1 + U_2 = 120.5 + 279.5 = 400$$

$$n_1 n_2 = 20 \times 20 = 400$$

In this example we are going to use the higher of these two values, U_2, as our U statistic. For a Mann-Whitney U test we do not calculate degrees of freedom. It is the sizes of the two samples that we need to note. So, the important information is: $U = 279.5$, $n_1 = 20$, $n_2 = 20$.

STEP 4 Reject or accept your null hypothesis using critical-value tables

The critical value of U (when working with the higher value of U_1 or U_2) at a 5% critical significance level with $n_1 = 20$ and $n_2 = 20$ is 273 (Appendix II, Table AII.3b). Since the critical value 273 is less than the value 279.5 that we calculated, we must reject the null hypothesis according to Box 7.7b. In short:

$$U\,(279.5) > U_{critical}\,(273) \rightarrow \text{reject } H_0 \rightarrow \text{significant result.}$$

We conclude that there is a difference between the bone density of males and females over 50 years old (Mann-Whitney U test: $U = 279.5$, $n_1 = 20$, $n_2 = 20$, $P < 0.05$).

7.4.4 **Worked example: using SPSS**

online resource centre

We are going to repeat doing a Mann-Whitney U test on the bone data (Section 7.2), but this time we are going to use SPSS. To do this you must enter the data into SPSS in the same format as for the *t*-test. You should refer back to Section 7.3.4, especially Figure 7.3, to refresh your memory if needed.

STEP 1 State the null hypothesis (H_0)

H_0: There is no difference in bone density between males and females over 50 years old.

STEP 2 Choose a critical significance level (α)
We will use 5% (0.05).

STEP 3 Calculate the test statistic

To get SPSS to conduct a Mann-Whitney U test on your data, you must first open the data file. Then you must make the following selections:

Analyze
→ Nonparametric Tests
→ Independent Samples...

In the window that appears you should see three tabs across the top: **Objective**, **Fields**, and **Settings**. Under **Objective** you should select the **Customize analysis** option, as shown in Figure 7.6a.

Under **Fields**, you can go with the default **Use predefined roles** option. Alternatively, you can choose the **Use custom field assignments** option and highlight *Bone Density Measurement* and press the arrow to transfer it to the **Test Field** box. Also, highlight *Sex* and press the arrow to transfer it to the **Groups** box. It should look like Figure 7.6b.

(a)

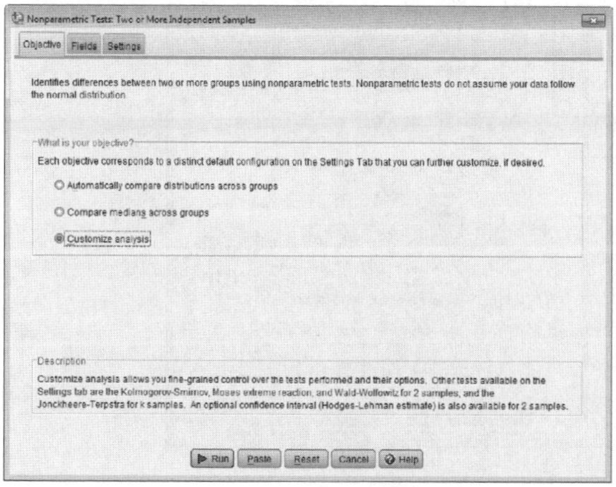

(b)

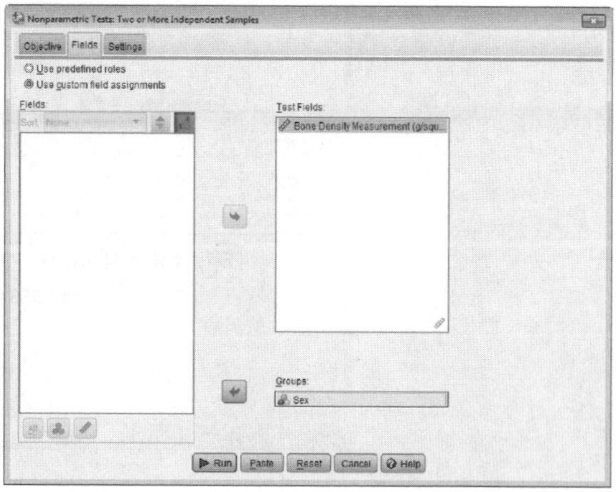

Figure 7.6 Conducting a Mann-Whitney U test using SPSS. (a) Objective tab. (b) Fields tab. (c) Settings tab.

Figure 7.6 (Continued)

(c)

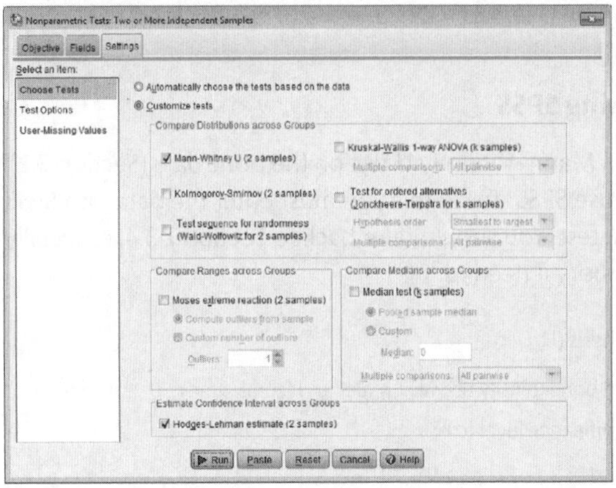

Under **Settings**, select **Customize tests**, then **Mann–Whitney U (2 samples)** in the **Compare distributions across Groups** area and **Hodges-Lehman estimates (2 samples)** in the **Estimate Confidence Interval across Groups** area. You need to do the latter to get an effect size with 95% CI. Look at Figure 7.6c to see how it should now appear.

Once you have gone through all three tabs, press **Run** at the bottom of any one of them. The computer does all the calculations and produces a table labelled **Hypothesis Test Summary**. Double-click on this table to bring up the **Model Viewer** window. At the bottom left of this window is a **View** drop-down menu from which you should select **Confidence Interval Summary View**. What you should now see is Figure 7.7. The important information can be summarized as U = 279.5, n_1 = 20,

Figure 7.7 SPSS output for a Mann–Whitney U test with key information annotated.

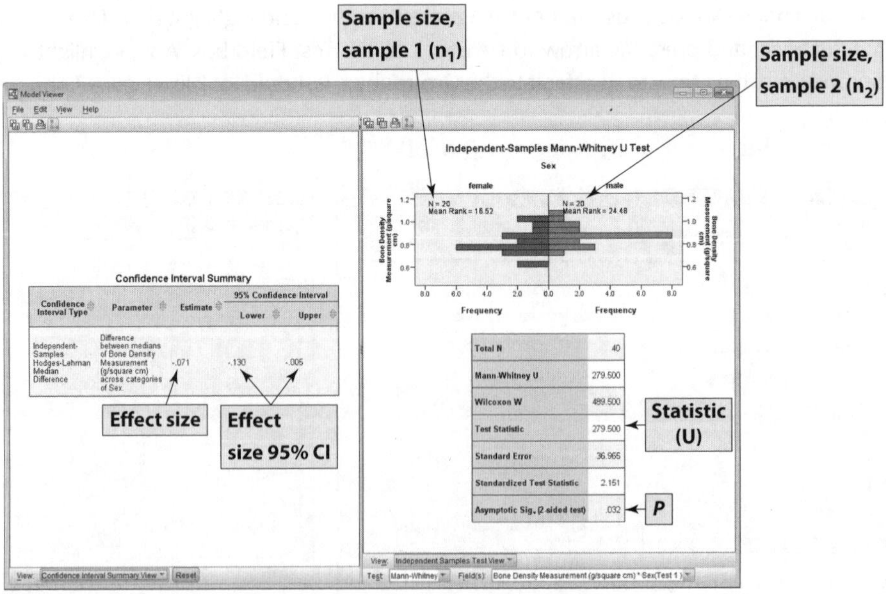

$n_2 = 20$, $P = 0.032$, with an effect size, the difference between the means (female-male) of −0.071 (95% CI 0.130 to −0.005). Note that using this route, the U value that SPSS reports is the higher value of U_1 or U_2, but if you use the legacy dialogue boxes it's the lower value.

STEP 4 Reject or accept your null hypothesis using *P* values on the computer output

Since the critical significance level of 0.05 is larger than the value 0.032 for *P*, we must reject the null hypothesis according to Box 7.8. In short,

$P\,(0.032) < \alpha\,(0.05) \rightarrow$ reject $H_0 \rightarrow$ significant result.

We conclude that there is a difference between the bone density of males and females over 50 years old (Mann–Whitney U test: U = 279.5, $n_1 = 20$, $n_2 = 20$, $P = 0.032$, difference(female-male) = −0.071 (95% CI −0.130 to −0.005)).

7.4.5 **Literature link: Alzheimer's disease**

Mosimann *et al.* (2004) Brain (OUP) 127: 431–438.

online resource centre

The many unpleasant symptoms of Alzheimer's disease include a variety of visual problems. Mosimann *et al.* (2004) investigated some aspects of these by using a clock-reading test. The test involved subjects being asked to read out the time on a clock (the sort with hands, not digital) and then to press a button to see the next clock. If they took longer than 8 seconds the next clock was presented anyway. A relevant region of interest (ROI) on the clock face for each time was defined and, among other things, the length of time it took for people to fix their gaze in this area was recorded. They called this variable the 'time until first fixation inside ROI' and it was measured in milliseconds.

The following excerpts from the statistics section of Mosimann *et al.*'s methods are particular noteworthy: 'All data were tested for normal distribution (Kolmogorov-Smirnov test). Distribution and dispersion measures for parametric data were calculated as mean and SD, and for non-parametric data were calculated as median and range. ... and two-group comparison was made either with parametric (*t*-tests ...) or non-parametric tests (Mann–Whitney tests...). A *P*-value of <0.05 was considered statistically significant, and all reported *P*-values were two-tailed.'

The Kolmogorov–Smirnov test is an alternative to inspecting histograms as a way of checking data for normality. SD is short for standard deviation and they use this as the variability measure with means in association with *t*-tests (which they refer to as *t*-tests for independent samples). For nonparametric data, such as those from the variable 'time to first fixation' considered here, they used medians and ranges in association with Mann–Whitney U tests (which they refer to as Mann–Whitney tests).

Their findings were: 'In Alzheimer's disease patients, the time until the first fixation inside the ROI was longer (Mann–Whitney test: $P < 0.001$) ... compared with controls.' In support of this statement they refer people to a table, reproduced in a modified version here as Table 7.8. In the methods section we are told that the number of subjects taking part in the research is 24 with Alzheimer's and 24 without. If we assume that all the subjects took part in the clock-reading test, then $n_1 = 24$ and $n_2 = 24$. We are not told the value of U.

Table 7.8 Analysis of visual exploration during clock reading. ROI means region of interest.

	Alzheimer's patients		Control group		*P*
	Median	**Range**	**Median**	**Range**	
Time until first fixation inside ROI (ms)	520	340–1727	357	269–569	<0.001

Source of data: modified extract from Table 2 in Mosimann *et al.* (2004).

Checklist of key points

- The *t*-test and the Mann–Whitney U test are used to test for differences between two samples when the data are unrelated.
- You should consider these tests when:
  - the numbers that you see when you look at your data all come from the dependent variable and are at least of an ordinal level of measurement;
  - two samples of the dependent variable are distinguished by two categories of the independent variable.
- The minimum level of measurement needed for the dependent variable for a *t*-test is scale, and for a Mann–Whitney U test it is ordinal.
- The *t*-test is parametric and the Mann–Whitney U test is nonparametric. The parametric criteria for the *t*-test are that each sample is normally distributed and that the variances of the two samples are similar.
- The four steps in tests of difference for two unrelated samples are:
  1. State H_0: there is no difference between the populations.
  2. Choose a critical significance level: typically $\alpha = 0.05$.
  3. Calculate the statistic: For the *t*-test calculate *t* and the degrees of freedom according to Box 7.2. For the Mann–Whitney U test, calculate U and the sample sizes according to Box 7.6.
  4. Reject or accept H_0. For a *t*-test reject H_0 if $t \geq t_{critical}$ (Box 7.3) or if $P \leq \alpha$ (Box 7.4). For a Mann–Whitney U test, reject H_0 if $U \leq U_{critical}$ when using lower U (Box 7.7a), $U \geq U_{critical}$ when using higher U (Box 7.7b), or if $P \leq \alpha$ (Box 7.8).
- The effect size can be estimated for both the *t*-test and the Mann–Whitney U test as the the difference between the samples, and should be reported along with its 95% confidence intervals.

Self-help questions

1. For each of the following features say whether it applies to a *t*-test, Mann–Whitney U test, both, or neither.
   (a) Parametric test. *Link back: Section 7.1.*
   (b) Nonparametric test. *Link back: Section 5.1.*
   (c) Numbers plugged into the formula for the statistic are frequencies. *Link back: Section 7.1.*
   (d) Numbers plugged into the formula for the statistic include the sum of ranks. *Link back: Section 7.4.2.*
   (e) Numbers plugged into the formula for the statistic include means. *Link back: Section 7.3.2.*
   (f) Used for assessing differences between more than two samples. *Link back: Sections 7.3.1 and 7.4.1.*
   (g) Used on related data. *Link back: Sections 7.3.1 and 7.4.1.*

t-Test

Group Statistics

	Music	N	Mean	Std. Deviation	Std. Error Mean
Time spent moving (% day 6am–6pm)	Rock	12	51.9167	12.60922	3.63997
	Classical	13	34.8462	11.80287	3.27353

Independent Samples Test

		Levene's Test for Equality of Variances		t-test for Equality of Means						
									95% Confidence Interval of the Difference	
		F	Sig.	*t*	df	Sig. (2-tailed)	Mean Difference	Std. Error Difference	Lower	Upper
Time spent moving (% day 6am–6pm)	Equal variances assumed	.072	.790	3.497	23	.002	17.07051	4.88197	6.97138	27.16965
	Equal variances not assumed			3.487	22.498	.002	17.07051	4.89544	6.93101	27.21002

Figure 7.8 SPSS output referred to in self-help question 7.2.

(h) Used on unrelated data. *Link back: Sections 7.3.1 and 7.4.1.*
(i) Uses a statistic called *t*. *Link back: Section 7.3.2.*
(j) Uses a statistic called U. *Link back: Section 7.4.2.*
(k) Reject the null hypothesis if the calculated statistic is greater than the critical value of the statistic looked up in a table. *Link back: Sections 7.3.2 and 7.4.2.*
(l) Reject the null hypothesis if the calculated statistic is less than the critical value of the statistic looked up in a table. *Link back: Sections 7.3.1 and 7.4.2.*
(m) Reject the null hypothesis if *P* on the SPSS printout is less than 0.05. *Link back: Sections 7.3.2 and 7.4.2.*

2. A researcher working on the welfare of farmed animals played rock music to some pigs and classical music to others and recorded their behaviour over a 12-hour period. She used SPSS to perform a *t*-test to investigate whether there was a difference between the percentage of time pigs spent moving when played rock music compared with classical music. Her SPSS output appears in Figure 7.8.
(a) What is the key information she would need to report? *Link back: Section 7.4.4.*
(b) Comment on her use of a *t*-test rather than a Mann-Whitney U test. *Link back: Sections 7.1.3 and 7.3.1.*

8 Tests of difference: two related samples

Chapter Aims

The subject of this chapter is tests of differences on two samples of related data. The parametric paired *t*-test and the nonparametric Wilcoxon signed-rank test are presented. The chapter starts by comparing these tests and dealing with general points relating to them. The four steps of null hypothesis significance testing introduced in Chapter 5 are applied to both the paired *t*-test and the Wilcoxon signed-rank test. This is first done in a general context and then using example data worked through by hand and using SPSS. An example of the use of each test in the literature is also presented.

8.1 Introduction to paired *t*- and Wilcoxon signed-rank tests

This chapter parallels the previous chapter, but deals with related data rather than unrelated data. Remember that you will have related data when you collect the data in your two samples that can be paired up in some way, such as coming from the same individual (Section 2.3.1). This chapter is all about assessing to see whether a difference between two samples of related data is due to chance alone, or to chance plus something of potential biological interest. There are parametric and nonparametric ways of doing this. We will be looking at the parametric paired *t*-test and the nonparametric Wilcoxon signed-rank test (otherwise known as the Wilcoxon matched-pairs test). In this introduction I am going to emphasize some general points about these tests:

- The types of data on which paired *t*-tests and Wilcoxon signed-rank tests work.
- Similarities and differences between a paired *t*-test and a Wilcoxon signed-rank test.
- The parametric criteria as applied to a paired *t*-test.
- Alternatives to and extensions of paired *t*-tests and Wilcoxon signed-rank tests.

8.1.1 Variables and levels of measurement needed

In terms of level of measurement (Section 2.3.2), the paired *t*-test and the Wilcoxon signed-rank test are like the *t*-test and the Mann–Whitney U test we considered in the last chapter. The data from the dependent variable, that is the numbers in the samples, must be measured at least at the ordinal level for a Wilcoxon signed-rank test and at the scale level for a paired *t*-test. Any level of measurement will work for the independent variable, provided that only two categories are defined.

A suitable data set is presented in Table 8.1. The numbers in Table 8.1 are measures of the pressure inside people's eyeballs, which is technically known as intraocular pressure (IOP). IOP is the dependent variable and is measured at the scale level. The independent variable is the thickness of the cornea of the eyeball relative to its partner eyeball. Cornea thickness is measured at the centre and is technically known as the central corneal thickness (CCT). There are two ordinal categories of this variable: high CCT and low CCT. These categories distinguish the two samples. For each pair of eyeballs belonging to a single person, the one with the thickest cornea has been assigned to high CCT and the other to low CCT.

Table 8.1 Intraocular pressure (IOP; measured in millimetres of mercury, mmHg) and central cornea thickness (CCT). Each row contains data from the two eyes of a single person. IOP values for the eye in the pair with the highest CCT value are in the right column, and with the lowest CCT value, in the left column.

IOP (mmHg)	
Low CCT	**High CCT**
20.0	14.2
13.9	13.8
18.3	15.8
21.1	33.4
20.1	20.3
24.4	19.9
20.2	14.3
11.6	11.4
28.2	25.1
18.5	24.1

Source of data: courtesy of Pinakin Gunvant and Daniel O'Leary.

8.1.2 Comparison of paired *t*- and Wilcoxon signed-rank tests

Both the paired *t*-test and the Wilcoxon signed-rank test are used to assess for a difference between two related samples. Look again at the IOP example in Table 8.1. The data are related because the people measured had two eyeballs. This is a repeated design, because the related measures in the different samples are from the same person (Section 2.3.1). One way of checking with yourself that you understand that these data are related is to think, if we moved the 20.0 to the bottom of the left-hand column in Table 8.1, would we have to move the 14.2 to the bottom of the right-hand column? The answer is that yes we would, in order to keep data from the same pair of eyes together. In this example we have 10 pairs of related data.

The two samples in the IOP example are the pressure in eyeballs with lower corneal curvature (low CCT) and the pressure in eyeballs with higher corneal curvature (high CCT). A paired *t*-test or a Wilcoxon signed-rank test could be used to assess whether there is a significant difference between the pressure in eyeballs with lower corneal curvature and that in eyeballs with higher corneal curvature.

The paired *t*-test and Wilcoxon signed-rank test differ in that the paired *t*-test is a parametric test and the Wilcoxon signed-rank test is a nonparametric test. Since the paired *t*-test is parametric, the data must fulfil special criteria if the test is to be valid

	Paired *t*-test	**Wilcoxon signed-rank test**
Similarities	Tests of difference Two samples compared Data in samples related	
Differences	Parametric scale data only	Nonparametric scale or ordinal data

Table 8.2 Comparison of the paired *t*-test and Wilcoxon signed-rank test.

(Sections 5.3 and 8.1.3.). As you hopefully remember, data can only possibly fulfil parametric criteria if they are measured at the scale level. On the other hand, the Wilcoxon signed-rank test can work when the dependent variable has been measured at either the ordinal or scale level.

The similarities and differences between the paired *t*- and the Wilcoxon signed-rank tests are summarized in Table 8.2.

8.1.3 The paired *t*-test and the parametric criteria

A paired *t*-test is a parametric test and therefore you must be satisfied that the data meet the parametric criteria before proceeding (Section 5.3). In the case of the paired *t*-test, this means that the differences between the related data points should be normally distributed. Note that this is different from the *t*-test in the previous chapter where the data in the samples needed to be normal.

For our IOP data in Table 8.1 you would firstly find the differences between the pressures in the eyes of each person by subtracting the pressure in their high CCT eyeball from the pressure in their low CCT eyeball. Then, secondly, you would draw a histogram to check to see if these differences were reasonably normally distributed. Alternatively, you could just assume that the differences were normally distributed if you had no reason to think otherwise (Section 5.3.2).

8.1.4 Alternatives to and extensions of the paired *t*- and Wilcoxon signed-rank tests

In Chapter 7 we considered what to do if you need to test a difference between two samples and the data are unrelated. The equivalent of the paired *t*-test for unrelated data is the *t*-test and the equivalent of the Wilcoxon signed-rank test for unrelated data is the Mann–Whitney U test. If you have more than two samples that you want to explore differences between, you will need the analysis-of-variance techniques dealt with in Chapter 9, although related versions are not dealt with in detail in this book. There are versions of the GLZM techniques, introduced in Chapter 12, that can look for differences between two or more related samples.

8.2 Example data: bighorn ewes

In this section I am going to introduce the dataset that I will be using in the worked examples in Sections 8.3 and 8.4 for a paired *t*-test and a Wilcoxon signed-rank test, respectively. I have chosen a data set that meets the parametric criteria so that we can use it for both parametric and nonparametric tests. Using the same data set for both tests emphasizes that the basic data structure required for a paired *t*-test and a Wilcoxon signed-rank test is the same. However, you need to remember that if parametric assumptions are not met, then you must pursue the nonparametric alternative. If you have the option of either parametric or nonparametric testing, you should do the parametric test because it is more powerful (Section 5.3.1): there is no need to do both.

Ewe ID	% Time spent grazing in autumn		Difference, $y_1 - y_2$
	Without lamb, y_1	With lamb, y_2	
10	72.00	55.50	16.5
168	62.35	43.80	18.55
227	55.77	66.80	–11.03
801	59.98	68.00	–8.02
805	51.60	57.88	–6.27
820	61.48	61.90	–0.42
823	52.57	45.40	7.17
837	52.50	56.67	–4.17
842	56.43	73.30	–16.87
853	60.13	77.50	–17.37
864	48.60	63.53	–14.92
883	42.90	54.50	–11.6
899	53.50	55.80	–2.3
945	70.43	91.10	–20.67
953	47.10	64.05	–16.95
967	50.08	71.40	–21.33

Source of data: courtesy of Dr Kathreen Ruckstuhl.

Table 8.3 Example data: grazing ewes. The time spent grazing by bighorn ewes with and without lambs.

The example data for this chapter are on the behaviour of sheep. From 1994 to 1996, Kathreen Ruckstuhl collected data on bighorn sheep in the foothills of the Rocky Mountains in Canada. One aspect of this project focused on the effect of reproduction on the foraging behaviour of bighorn ewes. Energetically, producing milk is the most demanding period of the reproductive process for mammalian females. Kathreen wondered if ewes have to spend more time eating when they are lactating in order to compensate for this demand. To explore this idea she looked to see if there was a difference in the time ewes spent grazing when they did and did not have lambs. Table 8.3 presents data collected during the autumns of the study period.

There are two samples in Table 8.3: a sample of time spent grazing by ewes without lambs and a sample of time spent grazing by ewes with lambs. The data in these samples are the percentage of the day spent in this activity, and it is a scale level of measurement. The method for obtaining these percentages is described in Ruckstuhl and Festa-Bianchet (1998). Time spent grazing, measured as a percentage (scale level), is the dependent variable. Reproductive status, measured as 'without lamb' or 'with lamb' (nominal level), is the independent variable.

The data in Table 8.3 are related through the identity (ID) of the ewe. For example, ewe 10 grazed for 72% of the time in year(s) when she was without a lamb and 55.5% of the time in year(s) when she was with a lamb. Since the related data items are from the same animal, this is a repeated-measures design (Section 2.3.1). Kathreen collected data on many more than 16 ewes, but only had 16 cases where she had data for the same ewe under the two conditions. Errorplots or boxplots can be used to present these data. Boxplots are presented in Figure 8.1.

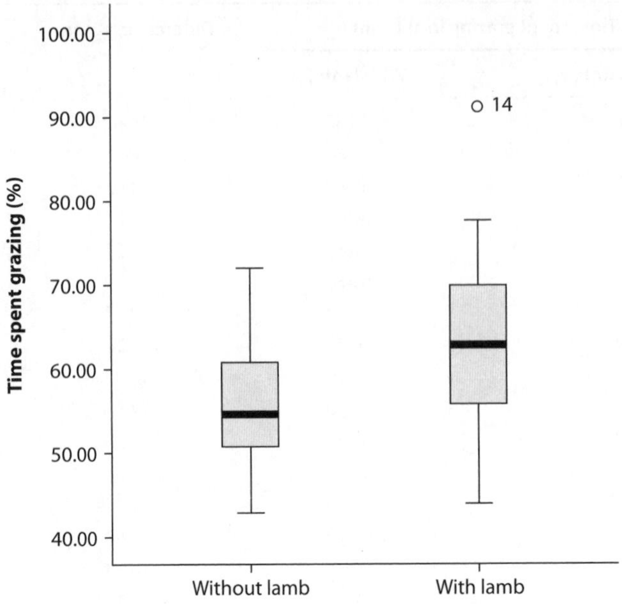

Figure 8.1 Time spent grazing by bighorn ewes with and without lambs ($n = 16$). (Refer back to Section 3.4.2 for guidance on interpretation.)

The differences between the pairs of data are shown in the final column in Table 8.3. Normally you would need to do something called an arcsine transformation on percentage data to make sure that they meet the parametric criteria (Section 5.3.3). However, if you look at Figure 8.2, you can see that these differences are close enough to being normally distributed for us not to have to worry in this case, and we can say that these data met the criteria for parametric analysis (Sections 5.3 and 8.1.3).

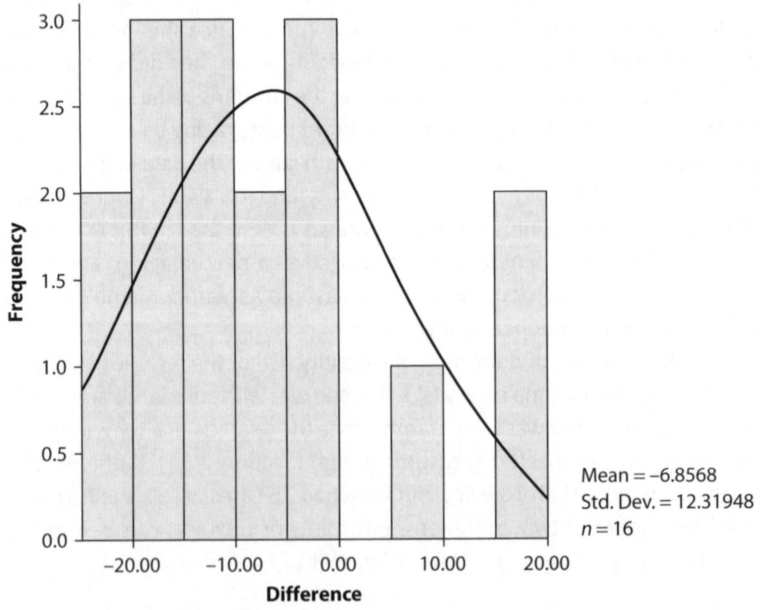

Figure 8.2 Histogram of the differences between the times ewes spent grazing with and without lambs ($n = 16$). The curved line indicates a normal distribution.

Kathreen's study of bighorn sheep gave her insights into one of behavioural ecology's big questions: why do male and female mammals typically hang out in different groups? She suggested that it happens because males and females have to spend different amounts of time in activities like feeding and moving. Kathreen thought that this could make it difficult for them to stay in synchrony and that separate, same-sex groups therefore emerge. Recent research offers a lot of support to Kathreen's research hypothesis (Ruckstuhl and Neuhaus, 2002).

8.3 Paired *t*-test

The topic of this section is the paired *t*-test, sometimes called 'the *t*-test for dependent samples'. The following are covered:

- When to use and when not to use a paired *t*-test.
- The four steps as applied to a paired *t*-test.
- Worked example by hand.
- Worked example using SPSS.
- An example of the use of a paired *t*-test from the literature.

The data used in the worked examples come from Kathreen Ruckstuhl's work and were introduced in Section 8.2.

8.3.1 When to use

A paired *t*-test is a parametric test for assessing whether the difference between two related samples can be accounted for by sample error alone. Issues relating to when and when not to use this test were covered in the introduction to this chapter (Sections 8.1.1 and 8.1.2). They are summarized in Box 8.1.

Box 8.1 When to use a paired *t*-test

Use this test when:

- You are looking for a *difference* between *two* samples.
- Data in the samples are measured at the *scale* level.
- Data are *related*.

Do not use this test when:

- You want to compare frequency distributions (Chapter 6).
- Data do not fulfil the parametric criteria (Sections 5.3 and 8.1.3).

8.3.2 Four steps of a paired *t*-test

A paired *t*-test is a parametric test so you must be satisfied that the data fulfil the parametric criteria (Sections 5.3 and 8.1.3) before proceeding. Once you have done this you can start to go through the four steps of a null hypothesis significance test, as outlined in Chapter 5, specifically applied to a paired *t*-test. These are as follows.

STEP 1 State the null hypothesis (H_0)

The null hypothesis for a paired *t*-test takes the general form:

H_0: There is no difference between the populations from which two related samples come.

STEP 2 Choose a critical significance level (α)

Typically this is 5% (0.05).

STEP 3 Calculate the test statistic

For a paired *t*-test the statistic is *t*. This is the same statistic used for the unrelated-samples *t*-test described in the previous chapter. The formula for *t* in a paired *t*-test is given in Box 8.2. The degrees of freedom equal the number of pairs minus 1. The number of pairs will be the size of either sample.

Box 8.2 Formula for *t* in a paired *t*-test

$$t = \frac{\bar{D} \times \sqrt{n}}{\sqrt{\frac{\sum(D-\bar{D})^2}{n-1}}}$$

Where:
$D = y_1 - y_2$, n = number of pairs

Degrees of freedom $n - 1$

STEP 4 Reject or accept your null hypothesis

How you do this will depend on whether you are using critical values for your statistic or *P* values generated by a statistical software package on a computer.

Using critical value tables

First you need to look up the critical value of *t* given the critical significance level you have chosen in step 2 and the degrees of freedom calculated in step 3 (Appendix II, Table AII.2). Then you need to compare the value of *t* you have calculated in step 3 with this critical value. Finally, you should decide to reject or accept your null hypothesis according to the rule in Box 8.3.

Box 8.3 Decision using critical values for a paired *t*-test

If $t \geq t_{critical} \rightarrow$ reject $H_0 \rightarrow$ significant result.

If $t < t_{critical} \rightarrow$ accept $H_0 \rightarrow$ non-significant result.

Using *P* values on computer output

Find *P* on the computer output and make your decision according to the rule in Box 8.4.

8.3.3 **Worked example: by hand**

We are going to run through the four steps of a paired *t*-test using the data introduced in Section 8.2.

Box 8.4 Decision using *P* value for a paired *t*-test

If $P \leq \alpha \rightarrow$ reject $H_0 \rightarrow$ significant result.

If $P > \alpha \rightarrow$ accept $H_0 \rightarrow$ non-significant result.

STEP 1 State the null hypothesis (H_0)

H_0: There is no difference between the times spent grazing by ewes when they do and do not have lambs.

STEP 2 Choose a critical significance level (α)
We will use 5% (0.05).

STEP 3 Calculate the test statistic

The formula is given in Box 8.2. The easiest way to apply the data to this formula is to use a table like the one shown in Table 8.4. Three key numbers are identified in Table 8.4, as A, B, and C. These plug into the equation for *t* for a paired *t*-test given in Box 8.2, as follows:

A B

$$t = \frac{\bar{D} \times \sqrt{n}}{\sqrt{\frac{\sum (D - \bar{D})^2}{n - 1}}}$$

B C

With our example data this gives you the following:

$$t = \frac{-6.86 \times \sqrt{16}}{\sqrt{\frac{2276.64}{16 - 1}}}$$

This boils down to

$$t = \frac{27.43}{\sqrt{151.78}}$$

Since the square root of 151.78 can be either +12.32 or −12.32, the answer to 27.43 divided by the square root of 151.78 could be + or −2.23. That is:

$$t = \pm 2.23$$

We will just be using what is called the absolute number for *t* in step 4. That is, we will ignore the sign and just look at the number, 2.23. We will need the degrees of freedom associated with this value. Since there are 16 pairs of numbers, there are

Ewe ID	% Time spent grazing (in autumn)					
	Without lamb, y_1	With lamb, y_2	$D\ (y_1 - y_2)$	$\bar{D}$	$D - \bar{D}$	$(D - \bar{D})^2$
10	72.00	55.50	16.50	−6.86	23.36	545.54
168	62.35	43.80	18.55	−6.86	25.41	645.51
227	55.77	66.80	−11.03	−6.86	−4.17	17.41
801	59.98	68.00	−8.02	−6.86	−1.16	1.35
805	51.60	57.88	−6.28	−6.86	0.58	0.33
820	61.48	61.90	−0.42	−6.86	6.44	41.43
823	52.57	45.40	7.17	−6.86	14.03	196.75
837	52.50	56.67	−4.17	−6.86	2.69	7.22
842	56.43	73.30	−16.87	−6.86	−10.01	100.26
853	60.13	77.50	−17.37	−6.86	−10.51	110.53
864	48.60	63.53	−14.93	−6.86	−8.07	65.18
883	42.90	54.50	−11.60	−6.86	−4.74	22.50
899	53.50	55.80	−2.30	−6.86	4.56	20.77
945	70.43	91.10	−20.67	−6.86	−13.81	190.80
953	47.10	64.05	−16.95	−6.86	−10.09	101.87
967	50.08	71.40	−21.32	−6.86	−14.46	209.18
		Mean	−6.86 (A)			142.29
		Sample size (n)	16 (B)			16
		Sum (Σ)	−109.71			2276.64 (C)

Table 8.4 Table for calculating t in a paired t-test. See text for how key numbers A–C relate to calculating t.

15 degrees of freedom. Applying the formula in Box 8.2 to our example data this can be worked out as:

$$\text{df} = n_1 - 1 = 16 - 1 = 15$$

So the important information can be summarized as: $t_{15} = 2.23$.
In this latter case the degrees of freedom are shown as a subscript.

STEP 4 Reject or accept your null hypothesis using critical-value tables

The critical value of t at the 5% critical significance level with 15 degrees of freedom is 2.13 (Appendix II, Table AII.2). Since the critical value 2.13 is smaller than 2.23 which we calculated, we must reject the null hypothesis according to Box 8.3. In short:

$$t(2.23) > t_{\text{critical}}(2.13) \rightarrow \text{reject } H_0 \rightarrow \text{significant result.}$$

We can conclude that there is a difference between the time ewes spent grazing when they did not have a lamb, compared with the time they spent grazing when they did have a lamb (paired t-test: $t_{15} = 2.23$, $P < 0.05$).

(a)

(b)

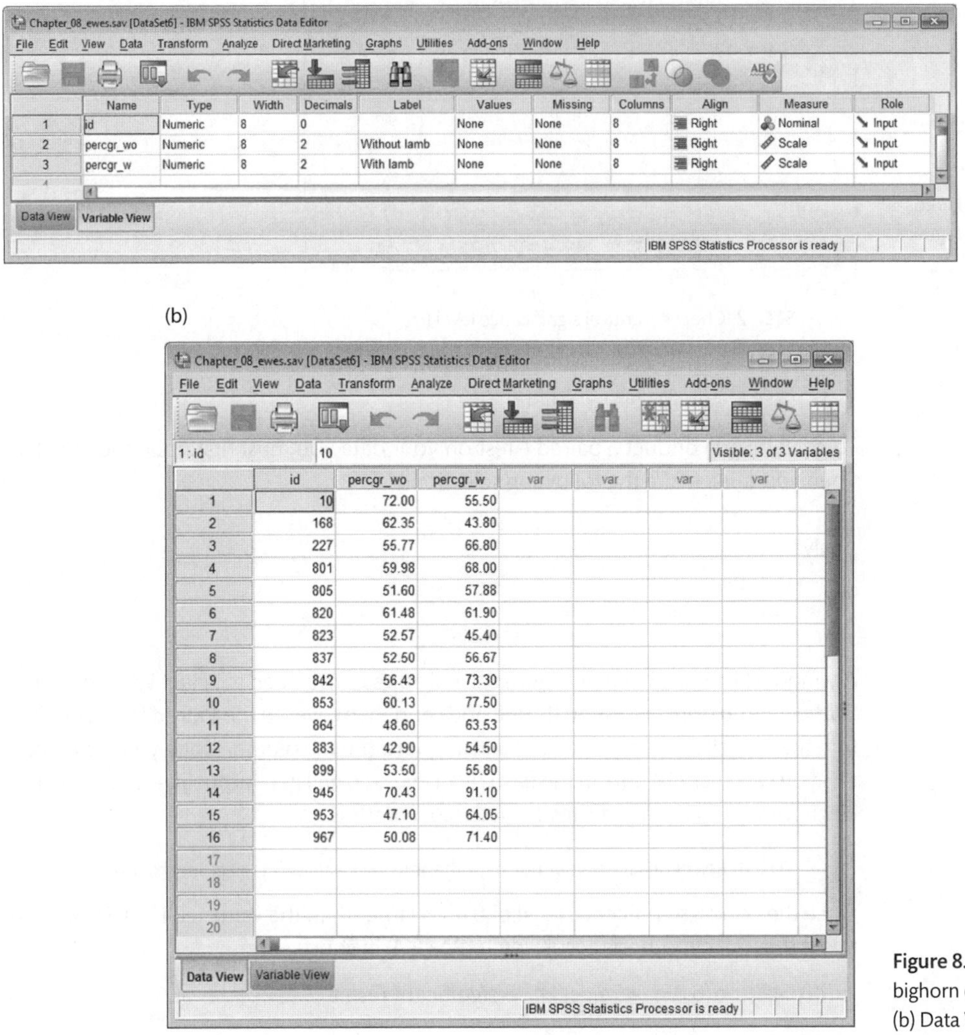

Figure 8.3 Example data in SPSS: bighorn ewes. (a) Variable View. (b) Data View.

8.3.4 **Worked example: using SPSS**

online resource centre

We are going to repeat doing a paired *t*-test on the bighorn ewe data (Section 8.2), but this time we are going to use SPSS. To do this the data must be entered into SPSS, as shown in Figure 8.3. In newer versions of SPSS (version 10 onwards), the **Data Editor** window has a **Variable View** tab and a **Data View** tab. Figure 8.3a shows the former and Figure 8.3b the latter. Appendix I has notes on how to enter data into SPSS.

STEP 1 State the null hypothesis (H_0)

H_0: There is no difference between the time spent grazing by bighorn ewes when they do and do not have lambs.

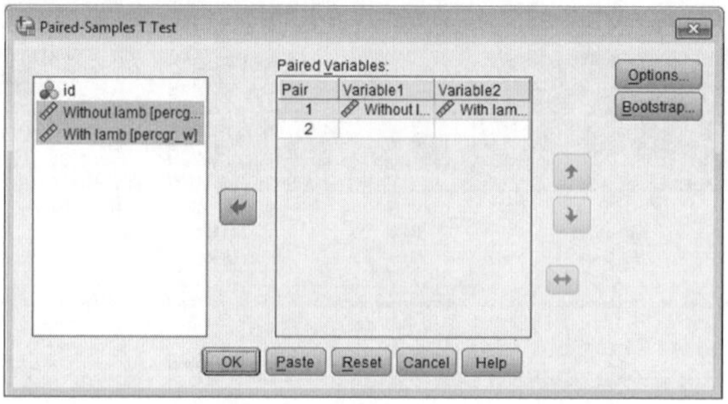

Figure 8.4 Conducting a paired *t*-test using SPSS. Main dialogue window.

STEP 2 Choose a critical significance level (α)
We will use 5% (0.05).

STEP 3 Calculate the test statistic

To get SPSS to conduct a paired *t*-test on your data you must first open the data file. Then you must make the following selections:

Analyze
→ Compare Means
→ Paired-Samples T Test...

A window like that shown in Figure 8.4 will appear. You need to select *With lamb* and *Without lamb* from the list on the left and send them over to the **Paired Variables** box together, by clicking the arrow. You just ignore the id column. Once you have done this you can click **OK** and, in the flash of an eye, you will get output like Figure 8.5. The key information is: Paired *t*-test: $t_{15} = 2.226$, $P = 0.042$.

STEP 4 Reject or accept your null hypothesis using *P* values on the computer output.

Since the critical significance level of 0.05 is larger than the value 0.042 of *P*, we must reject the null hypothesis according to Box 8.4. In short:

$P(0.042) < \alpha(0.05) \rightarrow$ reject $H_0 \rightarrow$ significant result.

We can conclude that there is a difference between the time ewes spent grazing when they did not have a lamb, compared with the time they spent grazing when they did have a lamb (paired *t*-test: $t_{15} = 2.226$, $P = 0.042$).

8.3.5 **Literature link: slug slime**

online resource centre

Jordaens *et al.* (2003) J. Moll. Stud. 69: 285–288 (OUP).

Slugs produce mucus for many reasons, one of which is to communicate with other slugs of their own and different species. Jordaens *et al.* (2003) conducted some preliminary research on the effects of mucus on the behaviour of the terrestrial land slug *Deroceras laeve*.

For one of their experiments, Jordaens *et al.* lined a plastic box with damp paper and poplar leaves, one half containing the mucus of another slug of the same species

T-Test

Paired Samples Statistics

		Mean	N	Std. Deviation	Std. Error Mean
Pair 1	Without lamb	56.0880	16	7.99046	1.99761
	With lamb	62.9448	16	11.95437	2.98859

Paired Samples Correlations

		N	Correlation	Sig.
Pair 1	Without lamb & With lamb	16	.288	.280

Paired Samples Test

		Paired Differences							
					95% Confidence Interval of the Difference				
		Mean	Std. Deviation	Std. Error Mean	Lower	Upper	*t*	df	Sig. (2-tailed)
Pair 1	Without lamb – With lamb	−6.85677	12.31948	3.07987	−13.42136	−.29219	−2.226	15	.042

Effect size

Effect size 95% CIs

Statistic (*t*)

Degrees of freedom

P

Figure 8.5 SPSS output for a paired *t*-test with key information annotated.

and the other half containing mucus of another slug of a different species. The test slug was then introduced to this box and its behaviour recorded for one hour in complete darkness using the night-time-viewing option on a digital video camera. At the end of the hour they recorded the position of the slug, calculated the time it had spent in each of the two halves, and the percentage of time spent in each half where it had been active. They explain that 'Activity was expressed as (total time an individual was crawling in area/total time spent in area) × 100% and arcsin square root transformed to meet the parametric assumptions of normality' (Section 5.3.3).

Jordaens *et al.* say that they found 'a tendency for activity to be higher on the area with conspecific mucus (74.37 ± 27.94% versus 86.76 ± 24.20%; paired *t*-test: $t = (-1.93$; df = 27; $P = 0.06$; Figs 2C and 3C)'. Notice that they use the word 'tendency'. They do not actually state what critical significance level they used, but their wording here suggests that they had 0.05 in mind. Instead of using the 0.05 as an absolute cut-off, they have evaluated their *P* value as just slightly over the critical level and therefore conclude a tendency rather than non-significance. It is not uncommon for people to describe a value in the range 0.05–0.10 as indicating a tendency, or being suggestive, rather than outright non-significant. Look back at Table 5.1 for a reminder on this.

The '74.37 ± 27.94%' reported is the mean time spent active in the area with the mucus of the other species, plus or minus the standard deviation. You can tell it is the

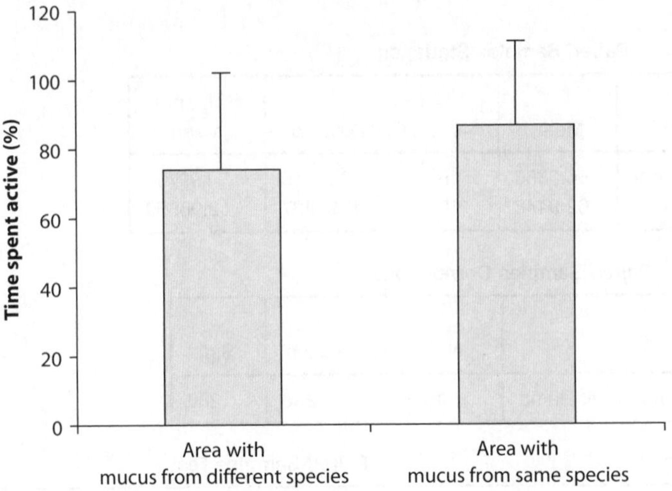

Figure 8.6 Time that slugs spent active on substrates with mucus of a different slug from a different species and from the same species: means (standard deviations) are shown. Based on data from Jordaens *et al.* (2003).

standard deviation by referring to their Figure 3C, which appears here, modified, in Figure 8.6. An alternative would have been for them to use boxplots.

It is worth noting that they have used standard deviation as a measure of variability. Standard error or confidence intervals (Section 4.2) would perhaps have been better alternatives because samples are being compared (Section 4.2.3).

8.4 Wilcoxon signed-rank test

The topic of this section is the Wilcoxon signed-rank test. We use a capital letter for Wilcoxon because it is the name of a person, the person who is famous for first describing this test. This test is also known as the Wilcoxon matched-pairs test. This section covers:

- When to use and when not to use a Wilcoxon signed-rank test.
- The four steps as applied to a Wilcoxon signed-rank test.
- Worked examples by hand.
- Worked examples using SPSS.
- An example of the use of a Wilcoxon signed-rank test in the literature.

The data used in the worked examples come from Kathreen Ruckstuhl's work and were introduced in Section 8.2.

8.4.1 **When to use**

You can use a Wilcoxon signed-rank test when your data are suited to a paired *t*-test, except that they do not fulfil, or cannot be assumed to fulfil, the criteria for parametric testing (Section 5.3). It is a nonparametric test for assessing whether the difference between two related samples can be accounted for by sample error alone. Issues relating to when and when not to use this test were covered in the introduction to this chapter (Sections 8.1.1 and 8.1.2). They are summarized in Box 8.5.

Box 8.5 When to use a Wilcoxon signed-rank test

Use this test when:

- You are looking for a *difference* between *two* samples.
- Data in the samples are measured at the *scale* or *ordinal* levels.
- Data are *related*.

Do not use this test when:

- You want to compare frequency distributions (Chapter 6).

8.4.2 **Four steps of a Wilcoxon signed-rank test**

Below are the four steps of a null hypothesis significance test outlined in Chapter 5, specifically applied to a Wilcoxon signed-rank test.

STEP 1 State the null hypothesis (H_0)
The null hypothesis for a Wilcoxon signed-rank test takes the general form:

H_0: There is no difference between the populations from which two related samples come.

STEP 2 Choose a critical significance level (α)
Typically this is 5% (0.05).

STEP 3 Calculate the test statistic
The statistic used for the Wilcoxon signed-rank test is T. The formula for this statistic is shown in Box 8.6. You can use either T^- or T^+ for T but you have to know if it's the higher or the lower value and use the appropriate critical value table for your choice in step 4. The way we are going to do it in SPSS uses the higher value of T^- or T^+ as T so this is what we will use in the worked example. It doesn't matter if you use $(y_1 - y_2)$ rather than $(y_2 - y_1)$, but if we do it this way, we can compare the outcomes by hand and using SPSS. In addition to T^- and T^+, SPSS also calculates a statistic called Z (or standardized test statistic) which is derived from T. You don't really need to know any more about Z, except that sometimes people report Z rather than T if they have done the calculation using SPSS. The other value that is particularly important to note is *N*. This is the number of pairs minus the number of pairs where y_2 and y_1 are the same, i.e. where $y_2 - y_1 = 0$.

Box 8.6 Formula for T in a Wilcoxon signed-rank test

$T^- = \Sigma \text{ Rank } |D|$ for negative values of *D*;
$T^+ = \Sigma \text{ Rank } |D|$ for positive values of *D*;
You can use either T^- or T^+ for T, but you have to know if it's the higher or the lower value and use the appropriate critical value table for your choice in Step 4..

Where: $D = (y_2 - y_1)$
The two straight lines either side of the *D*s mean this refers to the absolute value, i.e. its value regardless of its sign.
Therefore, Rank $|D|$ is the rank of *D* irrespective of the sign of *D*.

n = number of pairs, and
N = the number of values of *D* that are not zero.

STEP 4 Reject or accept your null hypothesis

How you do this will depend on whether you are doing the calculations by hand or using a statistical computing package.

Using critical-value tables

First you need to look up the critical value of T in Appendix II, Table AII.4b has the values appropriate for use when you are taking the higher of T^- or T^+ as your T. (Table AII.4a has the values appropriate for use when you are taking the lower of T^- or T^+ as your T). If you are using a critical-value table from another source you should check whether it's for use with the higher or lower value, but it will almost always be for use with the lower value.

To find the critical value of T (Appendix II Table AII.4a or b) you will need the critical significance level you have chosen in step 2 and the value of *N* you calculated in step 3. Then you need to compare the value of T you have calculated in step 3 with this critical value. Finally you should decide to reject or accept your null hypothesis according to the rule in Box 8.7a, if you are using the lower of T^- or T^+ as your T, or Box 8.7b, if you are using the higher of T^- or T^+ as your T.

Box 8.7a Decision using critical values for a Wilcoxon signed-rank test using lower of T^- or T^+ as T

If $T \leq T_{critical} \rightarrow$ reject $H_0 \rightarrow$ significant result.

If $T > T_{critical} \rightarrow$ accept $H_0 \rightarrow$ non-significant result.

Box 8.7b Decision using critical values for a Wilcoxon signed-rank test using higher of T^- or T^+ as T

If $T \geq T_{critical} \rightarrow$ reject $H_0 \rightarrow$ significant result.

If $T < T_{critical} \rightarrow$ accept $H_0 \rightarrow$ non-significant result.

Using *P* values on computer output

Find *P* on the computer output and make your decision according to the rule in Box 8.8.

Box 8.8 Decision using *P* value for a Wilcoxon signed-rank test

If $P \leq \alpha \rightarrow$ reject $H_0 \rightarrow$ significant result.

If $P > \alpha \rightarrow$ accept $H_0 \rightarrow$ non-significant result.

8.4.3 Worked example: by hand

We are going to run through the four steps of a Wilcoxon signed-rank test using the data introduced in Section 8.2.

Table 8.5 Table for calculating T in a Wilcoxon signed-rank test.

Ewe ID	% Time spent grazing (in autumn)						
	Without lamb, y_1	With lamb, y_2	$D\ (y_2 - y_1)$	$\lvert D \rvert$	Rank $\lvert D \rvert$	Ranks of negative *D* values	Ranks of positive *D* values
10	72.00	55.50	−16.50	16.50	10	10	0
168	62.35	43.80	−18.55	18.55	14	14	0
227	55.77	66.80	11.03	11.03	7	0	7
801	59.98	68.00	8.02	8.02	6	0	6
805	51.60	57.88	6.28	6.28	4	0	4
820	61.48	61.90	0.42	0.42	1	0	1
823	52.57	45.40	−7.17	7.17	5	5	0
837	52.50	56.67	4.17	4.17	3	0	3
842	56.43	73.30	16.87	16.87	11	0	11
853	60.13	77.50	17.37	17.37	13	0	13
864	48.60	63.53	14.93	14.93	9	0	9
883	42.90	54.50	11.60	11.60	8	0	8
899	53.50	55.80	2.30	2.30	2	0	2
945	70.43	91.10	20.67	20.67	15	0	15
953	47.10	64.05	16.95	16.95	12	0	12
967	50.08	71.40	21.32	21.32	16	0	16
					Sum (Σ)	**29**	**107**
						T^-	T^+

N = number of non-zero *D* values = 16

STEP 1 State the null hypothesis (H_0)

H_0: There is no difference between the times spent grazing by bighorn ewes when they do and do not have lambs.

STEP 2 Choose a critical significance level (α)
We will use 5% (0.05).

STEP 3 Calculate the test statistic

The formula is given in Box 8.6. The easiest way to apply the data to this formula is to use a table like the one shown in Table 8.5. Notice that calculations are based on ranks. The smallest difference is given rank 1, the next-smallest difference rank 2, and so on. There are no tied values in this example: that is, no values that are the same. If there were you would take the median rank of the tied values. In the Mann-Whitney U test in the previous chapter, ranks were of raw data, while for the Wilcoxon signed-rank test it is the differences in the paired raw data that are ranked. Otherwise the principles of ranking are the same in the two procedures. In this example we will use the higher of T^- or T^+ as T which is T^+, at 107, rather than T^-, at 29.

For a Wilcoxon signed-rank test it is the number of non-zero differences, N, rather than the degrees of freedom that is the other bit of information we will use in step 4. So, the important information for this example is: $T = 107$, $n = 16$, $N = 16$.

STEP 4 Reject or accept your null hypothesis using critical value tables

The critical value of T (when working with the higher of T^- or T^+) at the 5% critical significance level when $N = 16$ is 107 (Appendix II, Table AII.4b). Since the critical value, 107, is equal to the T value that we calculated, 107, we must reject the null hypothesis according to Box 8.7b. In short:

$$T(107) = T_{critical}(107) \rightarrow \text{reject } H_0 \rightarrow \text{significant result.}$$

We can conclude that there is a difference between the time ewes spent grazing when they did not have a lamb compared to the time they spent grazing when they did have a lamb (Wilcoxon signed-rank test: $T = 107$, $n = 16$, $N = 16$, $P < 0.05$).

8.4.4 **Worked example: using SPSS**

online resource centre

We are going to repeat doing a Wilcoxon signed-rank test on the bighorn ewe data (Section 8.2), but this time we are going to use SPSS. To do this the data must be entered into SPSS in the same format as for a paired *t*-test. You need to refer back to Figure 8.3 if you are not sure about this.

STEP 1 State the null hypothesis (H_0)

H_0: There is no difference between the time spent grazing by bighorn ewes when they do and do not have lambs.

STEP 2 Choose a critical significance level (α)
We will use 5% (0.05).

STEP 3 Calculate the test statistic

To get SPSS to conduct a Wilcoxon signed-rank test on your data you must first open the data file. Then you must make the following selections:

Analyze
→ Nonparametric Tests
→ Related Samples...

In the window that appears you should see three tabs across the top: **Objective**, **Fields**, and **Settings**. Under **Objective** you should select the **Customize analysis** option, as shown in Figure 8.7a.

Under **Fields**, you need to select *Without lamb* and *With lamb* from the list on the left and send them over to the **Test Fields** box together. What you should now be looking at appears in Figure 8.7b.

Under **Settings**, select Customize tests, then **Wilcoxon matched-pairs signed-rank (2 samples)** in the **Compare Median Difference to Hypothesized** area and

(a) (b) (c)

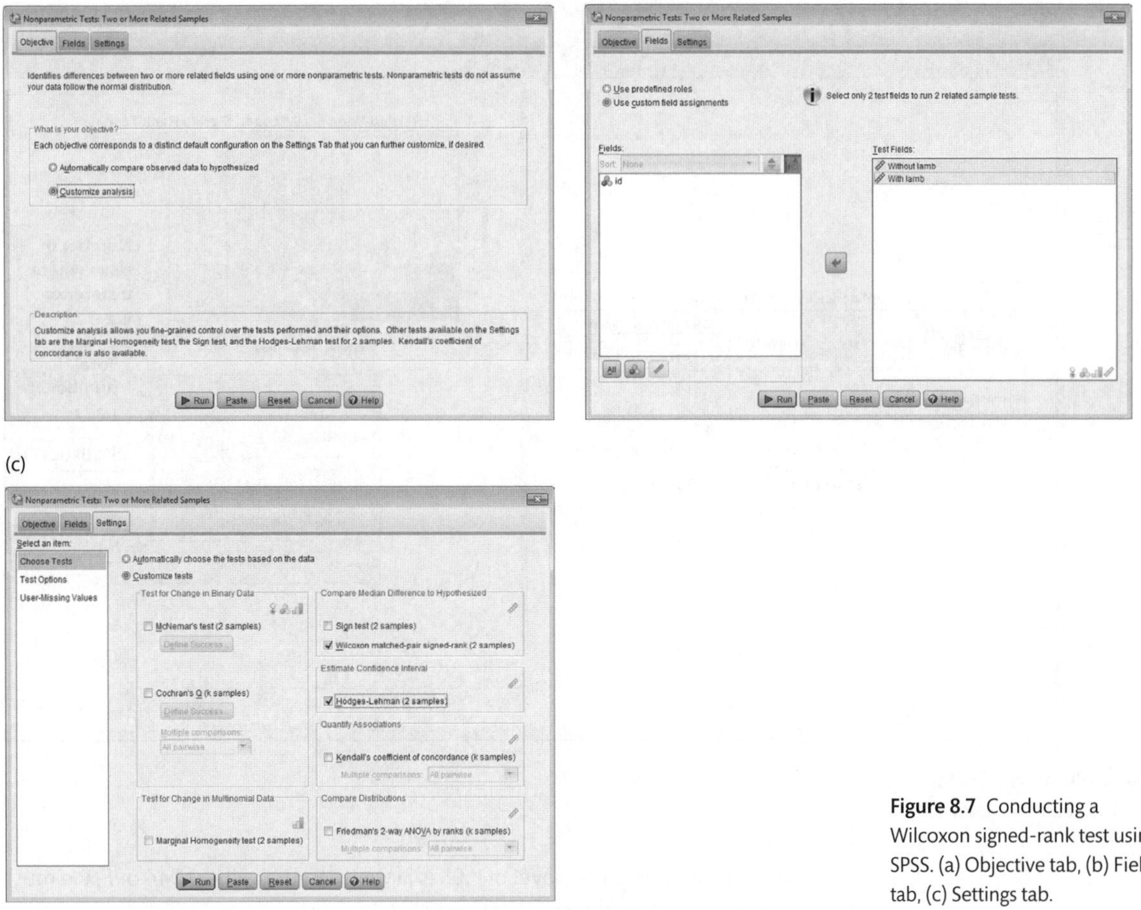

Figure 8.7 Conducting a Wilcoxon signed-rank test using SPSS. (a) Objective tab, (b) Fields tab, (c) Settings tab.

Hodges-Lehman (2 samples) in the **Estimate Confidence Interval** area. You need to do the latter to get an effect size with 95% confidence interval. Look at Figure 8.7c to see how it should now appear.

Once you have gone through all three tabs, press **Run** at the bottom of any one of them. The computer does all the calculations and produces a table labelled **Hypothesis Test Summary**. Double click on this table to bring up the **Model Viewer** window. At the bottom left of this window is a **View** drop-down menu from which you should select **Confidence Interval Summary View**. What you should now see is Figure 8.7d. The important information for this example is: $T = 107$, $n = 16$, $N = 16 - 0$, $P = 0.044$, with an effect size, the difference = 8.629 (95% CI 0.225 to 14.196). Where N is the total number of pairs minus the number of pairs for which the difference is 0. In this case $16 - 0 = 16$.

STEP 4 Reject or accept your null hypothesis using P values from the computer output

(d)

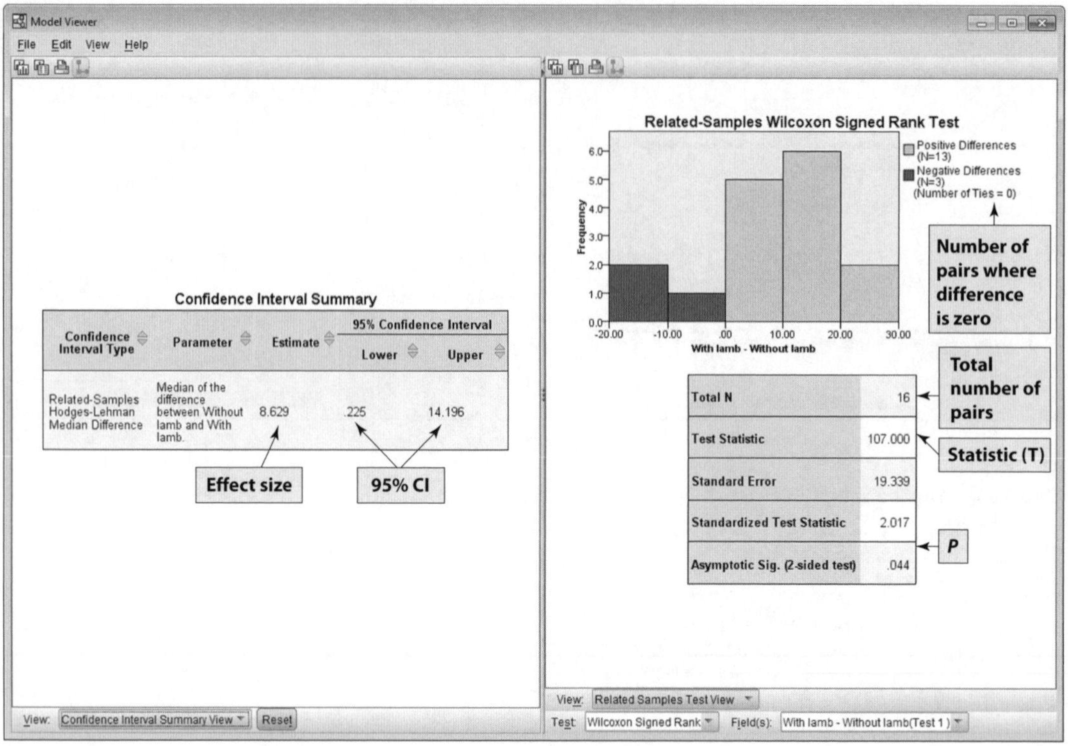

Figure 8.7(d) SPSS output for a Wilcoxon signed-rank test with key information annotated.

Since the critical significance level of 0.05 is larger than the value 0.044 of *P*, we must reject the null hypothesis according to Box 8.8. In short:

$P(0.044) < \alpha(0.05) \rightarrow$ reject $H_0 \rightarrow$ significant result.

We can conclude that there is a difference between the time ewes spent grazing when they did not have a lamb, compared with the time they spent grazing when they did have a lamb (Wilcoxon signed-rank test: $T = 107$, $n = 16$, $N = 16$, $P = 0.044$, difference = 8.629 (95% CI 0.225 to 14.196)).

8.4.5 **Literature link: head injuries**

online resource centre

Heitger *et al.* (2004) Brain 127: 575–590 (OUP).

The technical name for a bump on the head is a closed head injury (CHI). In a study to investigate the brain damage caused by mild closed head injuries, Heitger *et al.* (2004) looked at eye and arm movements. They took 30 patients with mild closed head injuries and matched each one with someone who did not have a closed head injury, but who was of the same age and sex and had had the same number of years

Table 8.6 Speed of arm movement in people with (CHI) and without (control) closed head injuries.

Measure	CHI ($n = 30$)		Control ($n = 30$)		*P* level
	Mean	SD	Mean	SD	
Basic motor function: movement peak velocity (mm/s)	821	246	949	225	0.033

Source of data: modified extract from Table 3 in Heitger *et al.* (2004).

of formal education. The 30 people without closed head injuries were the controls. Heitger *et al.* (2004) tested all 60 subjects using a variety of tests developed to monitor the progress of, for example, people recovering from strokes. One of these tests measured the fastest speed, or peak velocity, at which the patient could move their arm (Table 8.6).

In the statistical analysis section of their methods, Heitger *et al.* (2004) state: 'Most measures displayed considerable non-normality and skewed distributions. Hence, a nonparametric Wilcoxon matched-pairs statistic was used for comparing the CHI group with controls. Differences between the groups were considered significant at a two-tailed *P* value of ≤ 0.05.'

This last sentence could be rephrased as: a critical significance level of 0.05 was used. In other words, it corresponds to step 2 of the null hypothesis significance testing procedure. The Wilcoxon matched-pairs test is another name for the Wilcoxon signed-rank test.

In their results Heitger *et al.* (2004) state that 'the CHI group had reduced arm movement peak velocity'. They refer readers to Table 3, with results of a number of different tests. This table is reproduced here in a modified form in Table 8.6 and includes the values of sample size, *n* (30) and *P* (0.033). We are not told if there are any pairs for which the difference is zero, so we must assume that there are none. In other words, that $n = N = 30$. Furthermore the authors chose not to report either the T or the Z statistic. A significant difference between the peak velocity of closed-head-injury patients and the control group was concluded because the *P* value of 0.033 was less than the critical significance level of 0.05.

The statistical result only tells us that there is a difference between the two samples. However, we can tell the direction of this difference from the descriptive statistics in Table 8.6. The mean for the control group was 949 mm/s compared with 821 mm/s in the closed-head-injury group. It is worth noting that these authors reported the means and their standard deviations, but as they used a non-parametric test they should have considered reporting medians and interquartile ranges instead.

Checklist of key points

- The paired *t*-test and the Wilcoxon signed-rank test are used to test for differences between two samples when the data are related.

- You should consider these tests when:
  - the numbers that you see when you look at your data all come from the dependent variable and are at least of an ordinal level of measurement;
  - two samples of the dependent variable are distinguished by two categories of the independent variable.
- The minimum level of measurement needed for the dependent variable for a paired *t*-test is scale and for a Wilcoxon signed-rank test it is ordinal.
- The paired *t*-test is parametric and the Wilcoxon signed-rank test is nonparametric. The parametric criteria for the paired *t*-test are that the differences between the related data points are normally distributed.
- The four steps for tests of difference of two related samples are as follows.
  1. State H_0: H_0 = there is no difference between the populations.
  2. Choose a critical significance level: typically $\alpha = 0.05$.
  3. Calculate your statistic. For a paired *t*-test, calculate *t* and the degrees of freedom according to Box 8.2. For a Wilcoxon signed-rank test, calculate T and the number of non-zero pairs according to Box 8.6.
  4. Reject or accept H_0. For a paired *t*-test, reject H_0 if $t \geq t_{critical}$ (Box 8.3) or if $P \leq \alpha$ (Box 8.4). For a Wilcoxon signed-rank test, reject H_0 if $T \leq T_{critical}$ using lower T (Box 8.7a), or if $T \geq T_{critical}$ using higher T (Box 8.7b) or if $P \leq \alpha$ (Box 8.8).

Self-help questions

1. For each of the following features say whether it applies to a paired *t*-test, Wilcoxon signed-rank test, both, or neither.
   (a) Parametric test. *Link back: Section 8.1*
   (b) Nonparametric test. *Link back: Section 8.1*
   (c) Numbers in the formula for the statistic are frequencies. *Link back: Sections 8.3.1 and 8.4.1*
   (d) Numbers plugged into the formula for the statistic are based on the ranks of the differences between related data in two samples. *Link back: Section 8.4.2*
   (e) Used if the differences between more than two samples need to be assessed. *Link back: Section 8.1*
   (f) Used on unrelated data. *Link back: Sections 8.1.1 and 8.1.2*
   (g) Used on related data. *Link back: Sections 8.1.1 and 8.1.2*
   (h) Uses a statistic called *t*. *Link back: Section 8.3.2*
   (i) Uses a statistic called T or Z. *Link back: Section 8.4.2*
   (j) Reject the null hypothesis if the calculated statistic is greater than the critical value of the statistic looked up in a table. *Link back: Sections 8.3.2 and 8.4.2*
   (k) Reject the null hypothesis if the calculated statistic is less than or equal to the critical value of the statistic looked up in a table. *Link back: Section 8.4.2*
   (l) Reject the null hypothesis if *P* on the SPSS printout is less than 0.05. *Link back: Sections 8.3.2 and 8.4.2*
2. A researcher working in the tropics recorded an index of baboon-food availability in ten 1-hectare plots in the dry season and again in the wet season. She used SPSS to perform a Wilcoxon signed-rank test to investigate whether there was a difference between her fruiting index in these two seasons. The output she got is shown in Figure 8.8. What is the key information she would need to report? *Link back: Section 8.4.4*

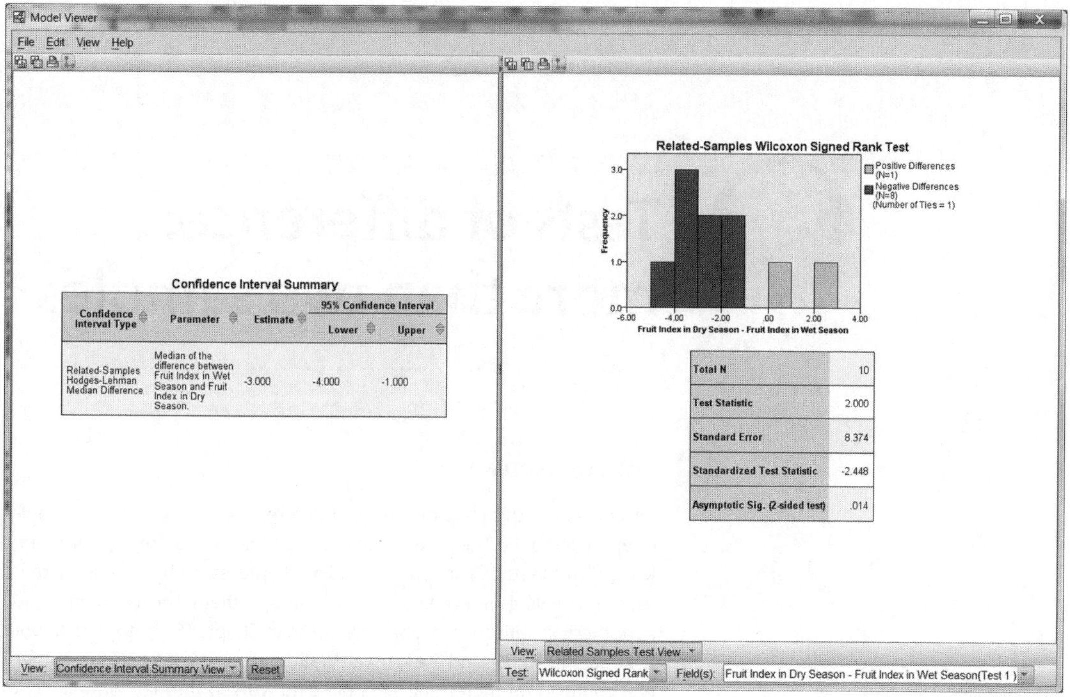

Figure 8.8 SPSS output referred to in self-help question 2.

9 Tests of difference: more than two samples

Chapter Aims

The subject of this chapter is tests of differences on two or more samples of unrelated data. The parametric one-way Anova and the nonparametric Kruskal–Wallis Anova are presented. The chapter starts by comparing these tests and dealing with general points relating to them. The four steps of null hypothesis significance testing introduced in Chapter 5 are applied to both the one-way Anova and the Kruskal–Wallis Anova. This is first done in a general context and then using example data worked through with SPSS. An example of the use of each test in the literature is also presented.

9.1 Introduction to one-way and Kruskal–Wallis Anova tests

Chapters 7 and 8 were about testing for differences between two samples of data collected using at least an ordinal level of measurement. Chapter 7 considered the situation when these samples were unrelated and Chapter 8 when they were related. The same issue is being considered in this chapter, as in Chapters 7 and 8: is the difference between samples due to chance alone or to chance plus something of potential biological interest? This chapter will expand your skills by showing you what to do when you have more than two samples.

In short, this chapter is all about testing to see whether a difference between two or more samples is due to chance alone or chance plus something of biological interest. Some tests work on unrelated data and others work on related data. We will focus here mainly on unrelated data, and only briefly mention the tests for related data. There are parametric and nonparametric tests for doing this. We will be looking at the parametric one-way Anova and the nonparametric Kruskal–Wallis Anova. By the way, Anova stands for analysis of variance.

In this introduction I am going to touch on a few general points about these tests:

- The type of data on which one-way Anova and Kruskal-Wallis Anova work.
- Similarities and differences between a one-way Anova and the Kruskal-Wallis Anova.
- The parametric criteria as applied to a one-way Anova.
- Alternatives to and extensions of one-way Anova and the Kruskal-Wallis Anova.
- Special terminology used for Anova: the language of Anova.
- The issue of multiple comparisons.

9.1.1 **Variables and levels of measurement needed**

The structure of the data on which you would use both a one-way Anova and a Kruskal-Wallis Anova are as for the *t*-test and the Mann-Whitney U test considered in Chapter 7, except that there can be more than two samples involved. Neither test works if the data in the samples are measured at a nominal level, but Kruskal-Wallis can cope with ordinal data. Both work with scale data. The independent variable can technically be measured at any level provided it can be used to distinguish a small number of categories. If there are only two categories for the independent variable, you should consider the tests in Chapter 7. It would not be wrong to do Anovas on these data, but two-sample tests are mathematically simpler and have been used traditionally in preference to Anovas under such circumstances.

Look at the data set in Table 9.1. It has all the features just mentioned. The numbers represent the percentage of healthy aphids in a petri dish responding to 'juice' from an aphid at various stages of infection by a fungal disease. The juice contains a chemical called a pheromone, which normally acts as an alarm to others and is obtained by squeezing the aphid gently. The percentage of healthy aphids responding is the dependent variable and is at the scale level of measurement. There are four samples of this dependent variable distinguished by the stage of infection of the aphid squeezed: pre-infection stage (uninfected), early stage of infection, late stage of infection, and post-death stage. Stage of infection is the independent variable; it is measured at an ordinal level but has only four categories. The squeezing procedure was repeated five times for each condition with different aphids giving a size of five for each sample.

Stage of infection	**Response of healthy aphids (%)**			
	Pre-infection	**Early**	**Late**	**Post-death**
	77.80	14.30	0.00	37.50
	100.00	0.00	14.30	0.00
	60.00	0.00	2.00	6.00
	87.50	0.00	0.00	0.00
	75.00	0.00	0.00	0.00

Source of data: courtesy of Helen Roy, Jason Baverstock, Keith Chamberlain, and Judith Pell.

Table 9.1 Response of healthy aphids to an aphid infected with a fungal disease.

Table 9.2 Comparison of the one-way Anova and the Kruskal-Wallis Anova.

	One-way Anova	Kruskal-Wallis Anova
Similarities	Tests of difference Two or more samples compared Data in samples unrelated	
Differences	Parametric Scale data only	Nonparametric Scale or ordinal data

9.1.2 Comparison of one-way and Kruskal-Wallis Anovas

The one-way Anova and the Kruskal-Wallis Anova are both used to look for differences between samples. For example, these tests could potentially be used to assess whether the percentage of healthy aphids responding to the juice of aphids at different stages of fungal infection is significantly different; the data set is presented in Table 9.1. There are four samples, one for each category of the independent variable, infection stage. The data in the four samples shown are unrelated and we have no reason to link the numbers in the different columns (Section 2.3.1).

The one-way Anova is a parametric test and the Kruskal-Wallis is a nonparametric test. This means that for a one-way Anova to work properly, the parametric criteria must be met (Sections 5.3 and 9.1.3). Since it is parametric, the one-way Anova test will only give reliable results when the data in your samples are measured at the scale level. The Kruskal-Wallis test is reliable with either ordinal or scale data.

The similarities and differences between the one-way Anova and the Kruskal-Wallis Anova are summarized in Table 9.2.

9.1.3 One-way Anova and the parametric criteria

A one-way Anova is a parametric test and therefore you must be satisfied that the data meet the parametric criteria (Section 5.3). In the case of one-way Anova this means that:

1. Each sample must be approximately normally distributed.
2. The variances of the different samples must be similar.

You can draw histograms and compare standard deviations to assess each of these respectively. Alternatively, you can just assume that these criteria are met if you have no reason to suspect that they are not (Section 5.4.4).

9.1.4 Alternatives to and extensions of one-way and Kruskal-Wallis Anovas

One-way and Kruskal-Wallis Anovas are for use with unrelated data. The names of the equivalent tests for related data are repeated-measures one-way Anova and Friedman Anova for the parametric and nonparametric situations respectively. As I said earlier, we will not be covering these in detail in this book. However, you should be able to apply the principles you have learned to understand these techniques; see Selected further reading: Siegel and Castellan (1988) or Sokal and Rohlf (2012).

Anova is short for **analysis of variance**. If people use the term Anova without stating if they are talking about the parametric or nonparametric alternatives, you can assume that they are referring to parametric Anova. In other words, used without qualification, Anova means parametric Anova. One-way Anova and repeated-measures Anova are part of a huge family of parametric procedures which include the following:

Anova is short for **analysis of variance**.

- Multiway Anova or multifactorial Anova for situations involving more than one independent variable. If two independent variables are involved this would be called two-way Anova, and so on.
- Multivariate Anova (Manova) for situations involving more than one dependent variable.
- Analysis of covariance (Ancova) and multivariate analysis of covariance (Mancova; Section 10.1.6) for the situation in which you need to combine Anova with the technique presented in the next chapter, regression.

Multiway Anova and analysis of covariance are essentially covered by the General Linear Model, as explained in Chapter 12. In fact the General Linear Model can also be used as an alternative to one-way Anova. Multivariate Anova is not covered in detail in this book. When you think you need one of these you can look it up in the index of another text (see Selected further reading). You can apply the basics you learn here to understanding whether the test is appropriate and to carrying it out correctly.

9.1.5 **The language of Anova**

There are some alternative terms used when talking about Anovas, especially parametric ones. These might be confusing if you don't realize that they are referring to the same concepts that you have come across in other tests, so I am going to review this alternative terminology here.

When Anovas are being performed, dependent variables are sometimes called **data variables**. Independent variables typically are called **factors**, **grouping variables**, **explanatory variables**, or even **predictor variables**. The term one-way indicates that a single independent variable, or factor, is involved. As we saw in Section 9.1.4, there are Anova techniques for the situation where you have more than one factor, and these use the term multiway.

Dependent variables are often called **data variables** in the context of an Anova. Independent variables are generally called **factors** when conducting Anova. Alternatively they are referred to as **grouping**, **explanatory**, or **predictor variables**.

The categories of independent variable that are used to distinguish the samples are typically referred to as **treatments**. In the example data in Table 9.1, the factor is stage of infection and the treatments are pre-infection, early, late, and post-death. We say that the four samples come from four different treatments.

The categories of the independent variable used to distinguish different samples are generally called **treatments** in Anova.

One final thing, do you remember what **variance** is? If not you can reread Section 3.2.2. To get variances you divide **sum of squares** by degrees of freedom (Section 3.2.2). In Anova you often see the term **mean square** used instead of variance.

Sum of squares is short for sum of squared deviations (from the mean). **Mean square** is short for mean squared deviation (from the mean), alternatively known as **variance** (Section 3.2.2).

9.1.6 Multiple comparisons

If you get a significant difference when you use an Anova to test for differences between three or more samples, all this tells you is that at least two of the samples differ by more than chance alone. You are probably thinking: well that's easy, I'll just do a two-sample test, for example a *t*-test, between all the possible pairwise combinations, and this will tell me where my differences lie. Unfortunately this is a major no no as far as statisticians are concerned and so you should never solve the problem this way. It's to do with the philosophy statisticians have about what they call multiple comparisons. This is all I'll say here, but you can consult Sokal and Rohlf (2012) in Selected further reading if you are curious to find out more.

This problem of knowing where the difference lies when you have a significant result is a general issue when you have more than two samples. There are three alternative solutions:

1. Apply a Bonferroni correction.
2. Carry out post hoc multiple comparison using, for example, a Bonferroni, least significant difference, or Tukey's honestly significant difference test.
3. Produce errorplots (parametric situation) or boxplots (nonparametric situations), and look for the samples where the confidence intervals or interquartile ranges overlap least.

Again, that's all I am going to say here, but you can follow these options up in various other texts (see the Selected further reading section).

9.2 Example data: nitrogen levels in reeds

In this section I am going to introduce the data set that I will be using in the worked examples in Sections 9.3 and 9.4 for a one-way Anova and a Kruskal–Wallis Anova. I have chosen a data set that can be assumed to meet the appropriate parametric criteria, so that we can use it both for parametric and nonparametric tests. Using the same data set for both tests emphasizes that the basic data structure required for a one-way Anova and a Kruskal–Wallis Anova is the same. However, you need to remember that if parametric assumptions are not met, then you must pursue the nonparametric alternative. Furthermore, if you have the option of either parametric or nonparametric tests, you should do the parametric test because it is more powerful (Section 5.4.1); there is no need to do both.

The naturally occurring micro-organisms in reedbeds feed upon the harmful materials in contaminated water and make it clean again. This feature can be exploited in the treatment of polluted water, including sewage. Reedbeds are also important commercially in the thatching industry and in terms of biodiversity, offering unique habitat to a wide range of vertebrates and invertebrates. The United Kingdom is the home of the most westerly patches of this habitat in Europe. The decline of reedbeds in the United Kingdom is therefore particularly alarming to thatchers, waste-water managers, and conservationists. Research into factors promoting the successful reconstruction and management of reedbeds is of great interest to all of these groups.

Against this background, Deborah Clements decided to take a look at the role of wainscot moths in the growth and persistence of healthy reedbeds (Clements, 2000). The caterpillars of these moths feed on reeds, preventing them from flowering and making them unusable for thatching. On the other hand, caterpillars and moths are a good source of food for birds. Understanding the population biology of the wainscot moth is therefore an important consideration in the management of reedbeds. As part of her research, Deborah explored the sensitivity of the caterpillars to variation in the levels of different nutrients in reeds. As a first stage to this she compared nutrient levels in reeds from different areas within her study sites. The data we are going to use here are on the nitrogen levels in reeds measured at three different sites, identified as Site 1, Site 2, and Site 3, within a reedbed near Fowlmere in Cambridgeshire in 1996 (Table 9.3). Site 3 was an area close to the reedbed's water source, while Site 1 and Site 2 were further away. The water source was a natural spring, high in nitrogen from the agricultural runoff of surrounding farmland.

Table 9.3 Example data: nitrogen levels in reeds. Nitrogen content of reeds (as a percentage of dry weight) from three different areas of a Fowlmere reedbed in 1996.

Nitrogen content (% dry weights)		
Site 1	**Site 2**	**Site 3**
2.92	3.06	3.41
2.88	2.60	3.23
3.25	2.55	3.93
2.64	2.42	3.74
3.28	2.35	3.18

Source of data: courtesy of Deborah Clements.

There are three samples in Table 9.3: samples of nitrogen levels from each of five plants from three different sites. The numbers in these samples are the nitrogen concentration for each reed plant expressed as a percentage of its dry weight; this is a scale-level measurement. Nitrogen level is the dependent variable and location is the independent variable.

With sizes as small as five for each sample, it is not reasonable to draw histograms to assess whether or not the data are normally distributed. Given that we have no reason to think that the data are not normally distributed, we can just assume that they are (Section 5.3.2), and this is what we are going to do for this example. We are not going to transform the data, even though this would make it more likely that they would then comply with parametric criteria. With percentage data, the usual procedure is the arcsine transformation (Section 5.3.3). Using these small samples of untransformed percentage data is not the most cautious approach and you should only consider conducting a nonparametric test under such circumstances. When she analysed her nitrogen-level data, Deborah worked with a larger data set and used arcsine transformation. Nevertheless, it will suffice for our purposes to use this subset of the data in its untransformed form. The standard deviations of the samples, 0.27, 0.28, and 0.33, are within the range for us to accept that the parametric criteria are met in terms of homogeneity of variances. Errorplots or boxplots would both be good ways of presenting these data. Errorplots with confidence intervals are presented here in Figure 9.1.

9.3 One-way Anova test

The topic of Section 9.3 is the one-way Anova. It covers:

- When to use and when not to use a one-way Anova.
- The four steps as applied to a one-way Anova.

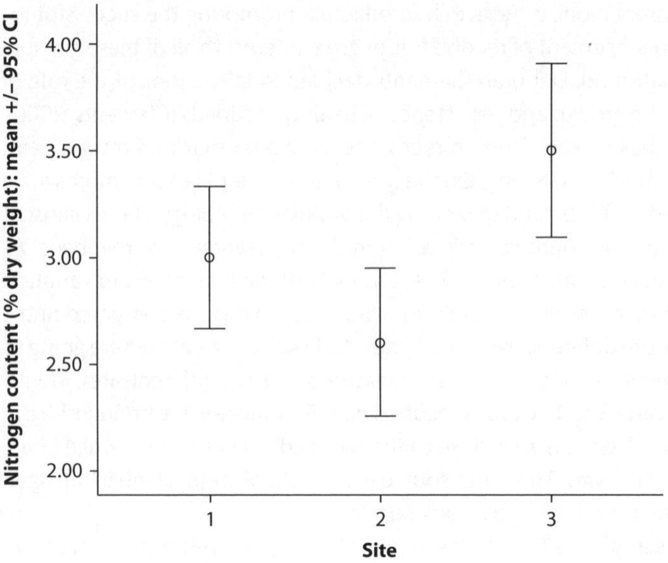

Figure 9.1 Nitrogen content of reeds from three different sites within the Fowlmere reedbed ($n_1 = 5$, $n_2 = 5$, $n_3 = 5$): means with 95% confidence intervals. (Refer back to Section 4.2.3 for guidance on interpretation.)

- A worked example using SPSS.
- An example of the use of a one-way Anova from the literature.

The data used in the worked examples come from the work of Deborah Clements, introduced in Section 9.2.

9.3.1 **When to use**

A one-way Anova is a parametric test, so you must check that your data fulfil the parametric criteria (Section 9.1.3) before using these tests on them. It is a parametric test for assessing whether the difference between two or more unrelated samples can be accounted for by sample error alone. Issues relating to when and when not to use this test were covered in the introduction to this chapter (Section 9.1). They are summarized in Box 9.1.

Box 9.1 When to use a one-way Anova

Use this test when

- You are looking for a *difference* between *two or more* samples.
- Data in the samples are measured at the *scale* level.
- Data are *unrelated*.

Do not use this test when

- You want to compare frequency distributions (Chapter 6).
- Data do not fulfil the parametric criteria (Sections 5.3 and 9.1.3).

9.3.2 Four steps of a one-way Anova

Here are the four steps of a null hypothesis significance test outlined in Chapter 5, specifically applied to a one-way Anova.

STEP 1 State the null hypothesis (H_0)
The null hypothesis for a one-way Anova test takes the general form:

H_0: There is no difference between the populations from which the samples come.

STEP 2 Choose a critical significance level (α)
Typically this is 5% (0.05).

STEP 3 Calculate the test statistic
For an Anova the statistic is F. The F statistic is a ratio of the variance between the treatments and the variance within the treatments. The ratio is expressed by dividing the between-treatment variance by the within-treatment, or error, variance. This is summarized in general terms in Box 9.2a and the formulae are given in Box 9.2b: remember that variance is otherwise known as mean square.

Box 9.2 (a) Calculating the F statistic for an Anova: words; (b) Calculating the F statistic for an Anova: formulae

(a)

Source (of variance)	Sum of squares	Degrees of freedom	Mean square (or variance)	F
Between treatments	$SS_{between}$	$DF_{between}$	$\frac{SS_{between}}{DF_{between}}$	$\frac{MS_{between}}{MS_{within}}$
Within treatments (Error)	SS_{within}	DF_{within}	$\frac{SS_{within}}{DF_{within}}$	
Total	SS_{total}	DF_{total}		

(b)

Source (of variance)	Sum of squares	Degrees of freedom	Mean square (or variance)	F
Between treatments	$\sum n(\bar{y}-\bar{G})^2$	$k-1$	$\frac{\text{Column 2}}{\text{Column 3}}$	$\frac{\text{Row 2}}{\text{Row 3}}$
Within treatments	$\sum (y-\bar{y})^2$	$N-k$	$\frac{\text{Column 2}}{\text{Column 3}}$	
Total	$\sum (y-\bar{G})^2$	$N-1$		

Where SS = sum of squares, DF = degrees of freedom, MS = mean square, k = number of treatments, N = the total sample size (the sum of the sample sizes for each treatment), $\bar{G}$ = the grand mean (the mean of the combined data from all treatments), $\bar{y}$ = a treatment mean, n = a sample size, $\sum n(\bar{y}-\bar{G})^2$ = the sum of the squared differences between each treatment mean and the grand mean multiplied by the sample size of that treatment, $\sum (y-\bar{y})^2$ = the sum of the squared differences between each y and its treatment mean, and $\sum (y-\bar{G})^2$ = the sum of the squared differences between each y and the grand mean.

The variance between the treatments reflects the variability between data in different samples. This will be a result of background random variation, plus any effect of the treatment. The variance within the treatments reflects background random variation only. If the treatments do not have any effect the between- and within-treatment variances will be similar and F will be close to 1. We would expect them to be similar and not the same, even if the samples are from the same population as proposed by the null hypothesis, because of sample error. The Anova test is a way of objectively deciding how big F has to be before we are happy to reject the null hypothesis.

There are two different degrees of freedom to report in association with an F statistic: one is associated with the between-treatments variance and the other with the within-treatments variance. The between-treatments degrees of freedom is one less than the number of treatments. The within-treatments degrees of freedom is the total sample size minus the number of treatments.

The *t*-test is actually a special case of the one-way Anova that you can use as an alternative when the between-treatments degrees of freedom are equal to one. When you have one degree of between-treatment freedom, *t* equals the square root of F. The *t*-test is traditionally used instead of an Anova when there are only two samples because of its greater ease of calculation. In these days of computers, however, the *t*-test offers no real advantage.

STEP 4 Reject or accept your null hypothesis

How you do this will depend on whether you are doing the calculations by hand or using a statistical computing package.

Using critical-value tables

First you need to look up in the critical value of F given the critical significance level you have chosen in step 2 and the degrees of freedom calculated in step 3 (Appendix II, Table II 5). Then you need to compare your value for the value of F calculated in step 3 with this critical value. Finally, you should decide to reject or accept your null hypothesis according to the rule in Box 9.3.

Box 9.3 Decision using critical values for a one-way Anova

If $F \geq F_{critical} \rightarrow$ reject $H_0 \rightarrow$ significant result.
If $F < F_{critical} \rightarrow$ accept $H_0 \rightarrow$ non-significant result.

Using *P* values on computer output

Find *P* on the computer output and make your decision according to the rule in Box 9.4.

Box 9.4 Decision using *P* values for a one-way Anova

If $P \leq \alpha \rightarrow$ reject $H_0 \rightarrow$ significant result.
If $P > \alpha \rightarrow$ accept $H_0 \rightarrow$ non-significant result.

(a)

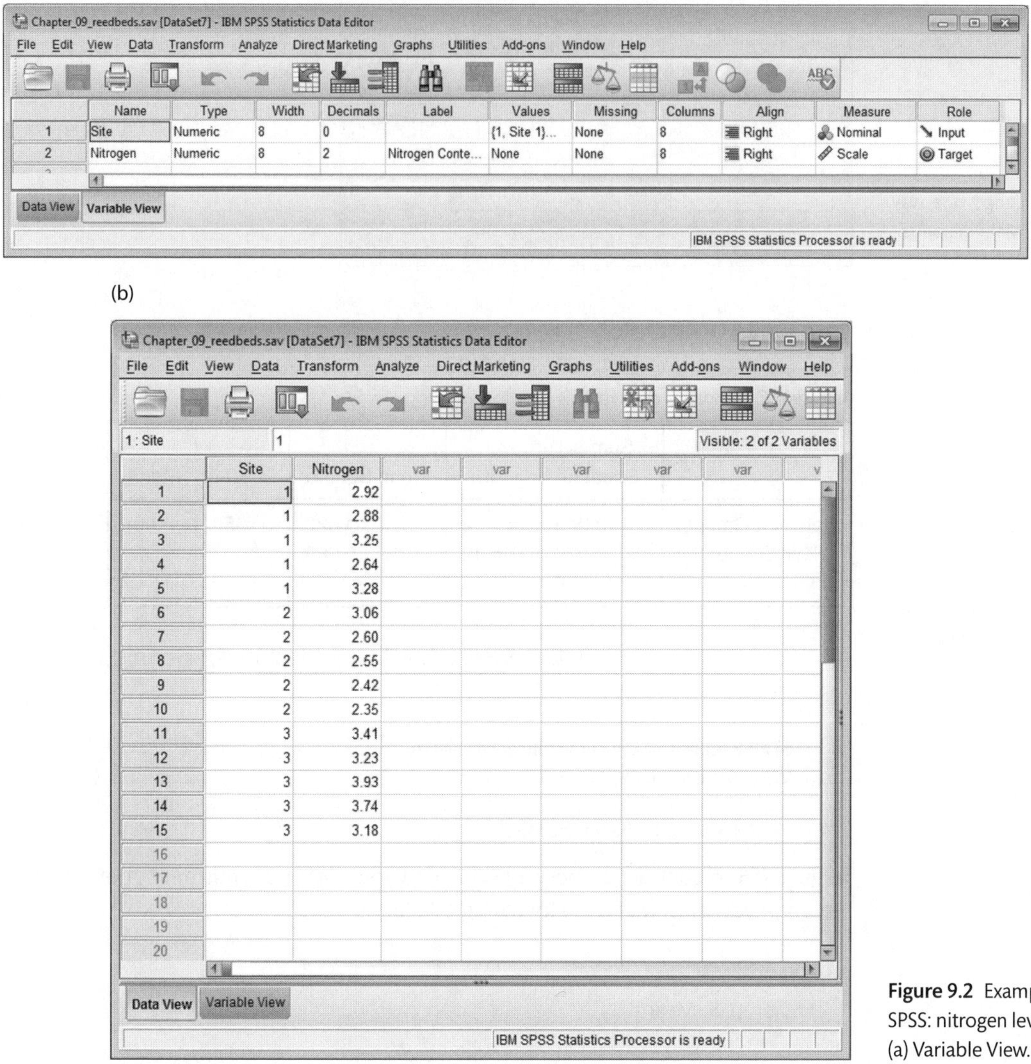

Figure 9.2 Example data in SPSS: nitrogen levels in reeds. (a) Variable View. (b) Data View.

9.3.3 **Worked example: using SPSS**

online resource centre

We are going to do a one-way Anova on the nitrogen levels in reedbed data (Section 9.2) using SPSS. To do this the data must be entered into SPSS as shown in Figure 9.2. In newer versions of SPSS the **Data Editor** window has a **Variable View** tab and a **Data View** tab. Figure 9.2a shows the former and Figure 9.2b the latter. Appendix I has notes on how to enter data into SPSS.

STEP 1 State the null hypothesis (H_0)

H_0: There is no difference between the nitrogen levels in reeds from different areas of the reedbed.

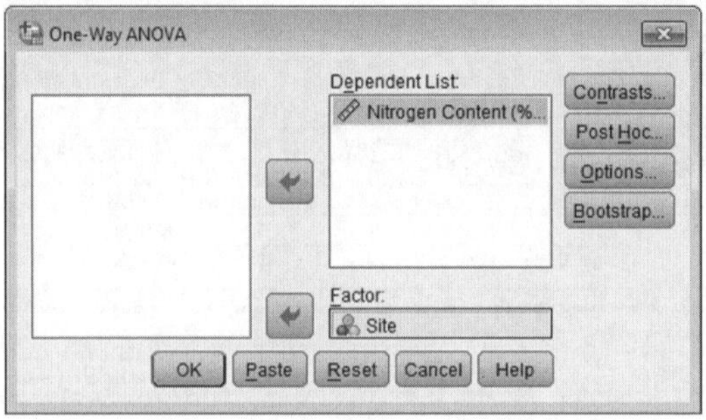

Figure 9.3 Conducting a one-way Anova using SPSS. Main dialogue window.

One-way

ANOVA

Nitrogen Content (% dry weight)

	Sum of Squares	df	Mean Square	F	Sig.
Between Groups	2.043	2	1.022	11.950	.001
Within Groups	1.026	12	.085		
Total	3.069	14			

Degrees of freedom **Statistic (F)** ***P***

Figure 9.4 SPSS output for a one-way Anova with key information annotated.

STEP 2 Choose a critical significance level (α)
We will use 5% (0.05).

STEP 3 Calculate the test statistic

To get SPSS to conduct a one-way Anova on your data you must first open the data file. Then you must make the following selections:

Analyze
→ Compare Means
→ One-Way ANOVA...

This will bring up the **One-Way Anova** dialogue window (Figure 9.3). Select *Nitrogen Content* and press the arrow to transfer it to the **Dependent List** box. Select *Site* and send it to the **Factor** box. Press **OK** and an output like that shown in Figure 9.4 will appear. The important information can be summarized as: $F_{2,12} = 11.950$, $P = 0.001$.

STEP 4 Reject or accept your null hypothesis using *P* values on the computer output.

Since the critical significance level of 0.05 is larger than the value 0.001 of *P*, we must reject the null hypothesis according to Box 9.4. In short:

$P\ (0.001) < \alpha\ (0.05) \rightarrow$ reject $H_0 \rightarrow$ significant result.

We can conclude that nitrogen levels in the reeds differ between the three areas surveyed (one-way Anova: $F_{2,12} = 11.950$, $P = 0.001$).

9.3.4 **Literature link: running rats**

Hutchinson *et al.* (2004) Brain 127: 1403–1414 (OUP).

online resource centre

If a vertebrate injures its spinal cord it often becomes extra sensitive, experiencing extreme pain in response to even mild stimuli. Hutchinson *et al.* (2004) decided to investigate the effect of exercise on the sensitivity of rats with moderate spinal cord injury (SCI). They were prompted to do this by previous studies suggesting that physical activity improves sensory function in healthy animals. Hutchinson *et al.*'s (2004) work was funded by the Christopher Reeve Paralysis Association because of its relevance to treating spinal cord injuries in humans.

Hutchinson and her team put three groups of SCI rats on exercise regimes of 20–25 min per day, 5 days per week, for 7 weeks. Each group of rats did a different type of activity: tread milling (TM), swimming (SW), or standing (ST). A fourth group of SCI rats were given no exercise (No-Ex).

After 7 weeks the scientists made various measurements including the levels of NT-3 in the soleus leg muscle (SOL). NT-3 is a type of neurotrophin. Neurotrophins are biochemicals that are vital to the functioning and maintenance of vertebrate nervous systems. Hutchinson *et al.* (2004) measured NT-3 expression rather than NT-3 directly: that is, they measured the levels of NT-3 indirectly by measuring the amount of NT-3 messenger RNA in the cells.

In the methods section, under Statistical analysis, they tell us that: 'Neurotrophin levels ... were analysed using one-way Anovas with Tukey's post hoc tests.' Tukey's post hoc tests are a way of finding out where differences lie when you have overall significance involving more that two samples. In other words, they are a way of dealing with multiple comparisons (Section 9.1.6).

In their results section, Hutchinson *et al.* (2004) report that, 'All exercise paradigms produced significantly greater NT-3 expression in SOL relative to the sedentary condition (SCI No-Ex: 88.33 $\pm$ 5.41 versus SCI + TM: 133.43 $\pm$ 10.66, $P < 0.01$; SCI + SW: 124.20 $\pm$ 4.84, $P < 0.05$; SCI + ST: 124.33 $\pm$ 8.56, $P < 0.05$ ...).' The *P* values given are from Tukey's tests for the comparison of each exercise regime with the no-exercise control group. We can tell this from the wording in the brackets (SCI No-Ex versus...). The details of the overall Anova (value of F, degrees of freedom, and *P* value) are not given, but we can tell from the results of the Tukey's tests that the one-way Anova must have been significant. The authors refer to a bar graph with error bars in support of their results (their Figure 5b). In this graph the bars represent the mean NT-3 expression in the soleus muscle for each of the three exercise treatments (TM, SW, and ST) and the no exercise (No-Ex) control with the error bars representing standard error. We know this because in the Statistical analysis section of their methods they tell us, 'All data are shown as mean $\pm$ SEM'.

9.4 **Kruskal-Wallis test**

The topic of this section is the Kruskal-Wallis Anova. This section covers:

- When to use and when not to use a Kruskal-Wallis Anova.

- The four steps as applied to a Kruskal-Wallis Anova.
- Worked examples using SPSS.
- An example of the use of a Kruskal-Wallis Anova from the literature.

The data used in the worked examples come from Deborah Clements' work and were introduced in Section 9.2.

9.4.1 **When to use**

A Kruskal-Wallis Anova is a nonparametric test for assessing whether the difference between two or more unrelated samples can be accounted for by chance alone. Issues relating to when and when not to use this test were covered in the introduction to this chapter (Section 9.1). They are summarized in Box 9.5.

Box 9.5 When to use a Kruskal-Wallis Anova

Use this test when:

- You are looking for a *difference* between *two or more* samples.
- Data in the samples are measured at *ordinal* or *scale* level.
- Data are *unrelated.*

Do not use this test when:

- You want to compare frequency distributions (Chapter 6).

9.4.2 **Four steps of a Kruskal-Wallis test**

Here are the four steps of a null hypothesis significance test outlined in Chapter 5, applied specifically to a Kruskal-Wallis Anova.

STEP 1 State the null hypothesis (H_0)
The null hypothesis for a Kruskal-Wallis Anova test takes the general form:

H_0: There is no difference between the populations from which the samples come.

STEP 2 Choose a critical significance level (α)
Typically this is 5% (0.05).

STEP 3 Calculate the test statistic
For a Kruskal-Wallis Anova the statistic is *H*, alternatively known as KW. The formula is shown in Box 9.6. The calculation of *H* involves using ranks. The principles of ranking in a Kruskal-Wallis Anova are the same as those used in a Mann-Whitney U test (Chapter 7). We do not calculate degrees of freedom with *H*, but need to note the size of each of the samples we are dealing with.

Box 9.6 Formula for H (otherwise known as KW) in a Kruskal-Wallis Test

$$H = \frac{12}{N(N+1)} \sum n(\bar{R} - \bar{G})^2$$

Where: N = total sample size = $n_1 + n_2 + n_3 + \cdots$, n = sample size for samples 1, 2, 3, and so on, $\bar{R}$ = mean of ranks for samples 1, 2, 3, and so on, $\bar{G}$ = mean of all ranks.

Note that for sample sizes of at least $n = 5$ for each of three or more samples, H approximates the chi-square distribution with $k - 1$ degrees of freedom, where k = number of samples.

When the calculation involves larger amounts of data, for example sample sizes of five or more for three or more samples, the H statistic approximates the chi-square (χ^2) statistic, with the number of samples minus one degrees of freedom ($k - 1$). We met chi-square in Chapter 6.

STEP 4 Reject or accept your null hypothesis
How you do this will depend on whether you are doing the calculations by hand or using a statistical computing package.

Using critical-value tables

First you need to look up the critical value of H given the critical significance level you have chosen in step 2 and the sample sizes you have noted in step 3 (Appendix II, Table AII.6). Then you need to compare your value for the value of H calculated in step 3 with this critical value. Finally, you should decide whether to reject or accept your null hypothesis according to the rule in Box 9.7a.

Box 9.7a Decision using H (or KW) critical values for Kruskal-Wallis Anova

If $H \geq H_{critical} \rightarrow$ reject $H_0 \rightarrow$ significant result.
If $H \leq H_{critical} \rightarrow$ accept $H_0 \rightarrow$ non-significant result.

Alternatively, if you are working with larger amounts of data, you can look up the critical value of chi-square given the critical significance level you have chosen in step 2 and the degrees of freedom you calculated in step 3 (Appendix II, Table AII.1). Then you need to compare your value of chi-square calculated in step 3 with this critical value of chi-square. Finally, you should decide whether to reject or accept your null hypothesis according to the rule in Box 9.7b.

Box 9.7b Decision using X^2 critical values for Kruskal-Wallis Anova

If $X^2 \geq X^2_{critical} \rightarrow$ reject $H_0 \rightarrow$ significant result.
If $X^2 < X^2_{critical} \rightarrow$ accept $H_0 \rightarrow$ non-significant result.

Using *P* values on computer output

Find *P* on the computer output and make your decision according to the rule in Box 9.8.

Box 9.8 Decision using *P* values for Kruskal-Wallis Anova

If $P \leq \alpha \rightarrow$ reject $H_0 \rightarrow$ significant result.
If $P > \alpha \rightarrow$ accept $H_0 \rightarrow$ non-significant result.

online resource centre

9.4.3 **Worked example: using SPSS**

We are going to do a Kruskal-Wallis test on the reedbed nitrogen-level data (Section 9.2) using SPSS. To do this the data must be entered into SPSS as shown in Figure 9.2, as explained for the worked example on one-way Anova (Section 9.3.3).

STEP 1 State the null hypothesis (H_0)

H_0: There is no difference between the nitrogen levels in reeds from different areas of the reedbed.

STEP 2 Choose a critical significance level (α)
We will use 5% (0.05).

STEP 3 Calculate the test statistic

To get SPSS to conduct a one-way Anova on your data you must first open the data file. Then you must make the following selections:

Analyze
→Nonparametric Tests
→Legacy Dialogs
→K Independent Samples...

This will bring up the dialogue window shown in Figure 9.5a. Select *Nitrogen Content* and click on the arrow to transfer it to the **Test Variable List** box. Select *Site* and send it to the **Grouping Variable** box. Before you can proceed you have to press the **Define Range** button, which brings up the window shown in Figure 9.5b. Type the minimum and maximum values for the number codes used for the **Grouping Variable**: in this case *Site* has a minimum of 1 and a maximum of 3. Press **Continue** to go back to the main dialogue window and then click **OK**. The output produced will look like that shown in Figure 9.6. The important information can be summarized as: $X^2_2 = 9.260$, $n_1 = 5, n_2 = 5, n_3 = 5, P = 0.010$

STEP 4 Reject or accept your null hypothesis using *P* values on the computer output

Since the critical significance level of 0.05 is larger than the value 0.010 of *P*, we must reject the null hypothesis according to Box 9.8. In short:

$P(0.010) < \alpha(0.05) \rightarrow$ reject $H_0 \rightarrow$ significant result.

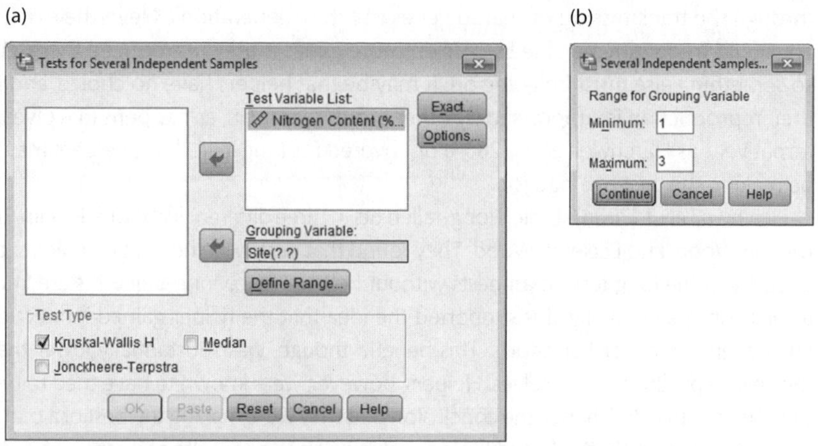

Figure 9.5 Conducting a Kruskal-Wallis Anova using SPSS. (a) Main dialogue window. (b) Define Range dialogue window.

We can conclude that the nitrogen levels in the reeds differed between the three areas surveyed (Kruskal-Wallis: $X^2_2 = 9.260$, $n_1 = 5$, $n_2 = 5$, $n_3 = 5$, $P = 0.010$).

9.4.4 **Literature link: cooperating long-tailed tits**

Hatchwell *et al.* (2004) Behav. Ecol. 15: 1-10 (OUP).

online resource centre

In animals that breed cooperatively, some individuals do not reproduce but assist others to do so instead. These helpers usually assist close relatives and so they benefit

Kruskal-Wallis Test

Ranks

	Site	N	Mean Rank
Nitrogen Content (% dry weight)	Site 1	5	8.20
	Site 2	5	3.60
	Site 3	5	12.20
	Total	15	

Sample size, sample 1 (n_1)
Sample size, sample 2 (n_2)
Sample size, sample 3 (n_3)

Test Statistics[a,b]

	Nitrogen Content (% dry weight)
Chi-Square	9.260
df	2
Asymp. Sig.	.010

Statistic (X^2)
Degrees of freedom
P

[a]Kruskal Wallis Test
[b]Grouping Variable: Site

Figure 9.6 SPSS output for a Kruskal-Wallis Anova with key information annotated.

through the transmission of shared genes to future generations. Nevertheless, helpers would frequently be able to pass on more genes if they reproduced themselves, so something else must be going on. It may be that helpers have no choice, and that their reproduction is suppressed by other group members, as happens in wolves. Alternatively, as Hatchwell *et al.* (2004) discovered for long-tailed tits, helpers may just be making the best of a bad job.

Hatchwell *et al.* (2004) studied long-tailed tits in three different locations: Rivelin River, Melton Wood, and Ecclesall Wood. They found that nests with helpers had higher productivity in the long term than nests without helpers. Since long-tailed tits are known to help kin preferentially, this supported the idea that the helper gained indirect genetic benefit for his or her efforts. This benefit, though, was substantially lower than if they had reproduced themselves. Helpers, however, were known to have tried to breed themselves but failed; hence the conclusion that they were making the best of a bad job.

Hatchwell *et al.* (2004) based their conclusions on the results of a range of statistical analyses, including a Kruskal–Wallis Anova to compare the number of helpers per nest at the different sites. They report that 'Melton Wood had significantly more helpers per nest than the other two sites (Rivelin Valley, 1.66 ± 0.78 helpers, $n = 44$; Melton Wood, 3.90 ± 1.66, $n = 10$; Ecclesall Wood, 1.54 ± 0.78, $n = 13$; Kruskal–Wallis test, $H = 16.8$, $df = 2$, $P < 0.001$,...).' Notice how they have given us information on both the sample sizes—44, 10, and 13—and the degrees of freedom—2. Earlier in the paper, in the methods section, they tell us that 'Means are reported ± 1 SD, unless otherwise stated.' They use SD as an abbreviation for standard deviation. So we know that, for example, 1.66 is the mean number of helpers per nest at Rivelin Valley and 0.78 is the standard deviation around this mean. The authors do not explain why they used a Kruskal–Wallis test instead of a one-way Anova. We must assume that they were not satisfied that their data fulfilled the parametric criteria.

9.5 Model I and model II Anova

If the independent variable is determined by the investigator (**fixed**) then **model I Anova** is appropriate. If it is not (that is, it is **random**), then a **model II Anova** is appropriate. Model I situations are the most common.

You may come across mention of **model I** and **model II** Anovas. In brief, model I requires that the independent variable is **fixed** and the dependent variable is random. If both the dependent and independent variables are **random** then model II applies.

When we say a variable is fixed we mean that its variability is determined by the investigator and that the treatments differ only in ways that the investigator allows. This is the case when the investigator creates the treatments himself, as for the stages of infection in the aphid example in Table 9.1. It is also the case when the investigator selects specific treatments from naturally occurring variation, as for the reedbed example (Section 9.2). Random variables, such as the proportion of aphids responding and the nitrogen levels in reeds, are not determined by the investigator.

The types of Anova we have been dealing with here fall into the model I category. For most of the situations you are likely to find yourself in, model I Anova will serve you fine, but if you want to explore these ideas further there is a full discussion of model I and II Anovas in Sokal and Rohlf (2012; pp. 200–204).

Checklist of key points

- The one-way Anova and the Kruskal–Wallis Anova are used to test for differences between two or more samples when the data are unrelated.
- You should consider these tests when:
  - the numbers that you see when you look at your data all come from the dependent variable and are at least of an ordinal level of measurement;
  - two samples of the dependent variable are distinguished by two or more categories of the independent variable.
- The minimum level of measurement needed for the dependent variable for a one-way Anova is scale, and for a Kruskal–Wallis Anova it is ordinal.
- The one-way Anova is parametric and the Kruskal–Wallis is nonparametric. The parametric criteria for the one-way Anova are that each sample is normally distributed and that the variances of the samples are similar.
- The four steps in tests of difference for two unrelated samples are as follows.
  1. State H_0: there is no difference between the populations.
  2. Choose a critical significance level: typically $\alpha = 0.05$.
  3. Calculate your statistic. For a one-way Anova, calculate F and degrees of freedom according to Box 9.2. For a Kruskal–Wallis Anova, calculate *H* and the sample sizes or degrees of freedom according to Box 9.6.
  4. Reject or accept H_0. For a one-way Anova, reject H_0 if $F \geq F_{critical}$ (Box 9.3), or if $P \leq \alpha$ (Box 9.4). For a Kruskal–Wallis Anova, reject H_0 if $H \geq H_{critical}$ (Box 9.7a), or if $X^2 \geq X^2_{critical}$ (Box 9.7b), or if $P \leq \alpha$ (Box 9.8).

Self-help questions

1. For each of the following features say whether it applies to a one-way Anova, Kruskal–Wallis Anova, both, or neither.
   (a) Parametric test. *Link back: Section 9.1*
   (b) Nonparametric test. *Link back: Section 9.1*
   (c) Numbers plugged into the formula for the statistic are frequencies. *Link back: Section 9.1*
   (d) Numbers plugged into the formula for the statistic are based on ranks. *Link back: Section 9.4.2*
   (e) The statistic is a ratio of between- and within-sample variances. *Link back: Section 9.3.2*
   (f) The statistic has two different degrees of freedom associated with it. *Link back: Section 9.3.2*
   (g) Used for assessing the differences between more than two samples. *Link back: Section 9.1*
   (h) Used on related data. *Link back: Sections 9.1.1 and 9.1.2*
   (i) Used on unrelated data. *Link back: Sections 9.1.1 and 9.1.2*
   (j) Uses a statistic called F. *Link back: Section 9.3.2*
   (k) Uses a statistic called *H*. *Link back: Section 9.4.2*
   (l) Reject the null hypothesis if the calculated statistic is greater than the critical value of the statistic looked up in a table. *Link back: Sections 9.3.2 and 9.4.2*
   (m) Reject the null hypothesis if the calculated statistic is less than the critical value of the statistic looked up in a table. *Link back: Sections 9.3.2 and 9.4.2*
   (n) Reject the null hypothesis if *P* on the SPSS printout is less than 0.05. *Link back: Section 9.3.1*

2. A student conducted a study on the behaviour of caged budgeriars kept singly, in pairs, and large groups. She used SPSS to perform a one-way Anova to investigate whether there was a difference between the rate of pecking of individuals kept under the three different social conditions. Her SPSS output appears in Figure 9.7.
   (a) What is the key information she would need to report?
   (b) Comment on her use of a one-way Anova rather than a Kruskal–Wallis Anova.

Link back: Section 9.3.3
Link back: Sections 9.1.2 and 9.3.1

One-way

ANOVA

Rate of Pecking (no. of pecks per hour)

	Sum of Squares	df	Mean Square	F	Sig.
Between Groups	1213.039	2	606.519	13.405	.000
Within Groups	1221.661	27	45.247		
Total	2434.700	29			

Figure 9.7 SPSS output referred to in self-help question 2.

10 Tests of relationship: regression

Chapter Aims

The subject of this chapter is tests of relationship for a dependent/independent variable pair. Only parametric linear regression is considered. The chapter starts by dealing with general points relating to regression. The additional information that you get with a regression procedure is explained. The four steps of null hypothesis significance testing introduced in Chapter 5 are applied to linear regression. This is first done in a general context and then using example data worked through using SPSS. An example of the use of regression in the literature is also presented.

10.1 Introduction to bivariate linear regression

This chapter considers regression, which is a technique for assessing the relationship between two variables. Nonparametric versions of regression are poorly developed and rarely used, so we are just going to be dealing with parametric techniques in this book. You are very unlikely to be in a situation where nonparametric regression will be of use to you, but if you are feeling curious you can look it up in Sokal and Rohlf (2012).

The parametric technique that we will be exploring in this chapter is bivariate linear regression. The word bivariate tells you that we will be considering regression involving two variables, one from an independent variable and the other from a dependent variable. The word linear tells you that we will be looking at situations where we have a relationship we think might be described by a straight sloping line like the one shown in Figure 10.1. In short, we are going to look at the relationship between two variables to see if they can be described by a straight line.

Bivariate regression analysis involves a null hypothesis significance test which assesses whether the line is sloped, as compared to horizontal (Figure 10.1). A horizontal line means that there is no relationship, whereas a sloping line means that when the independent variable changes so does the dependent variable. We need to use

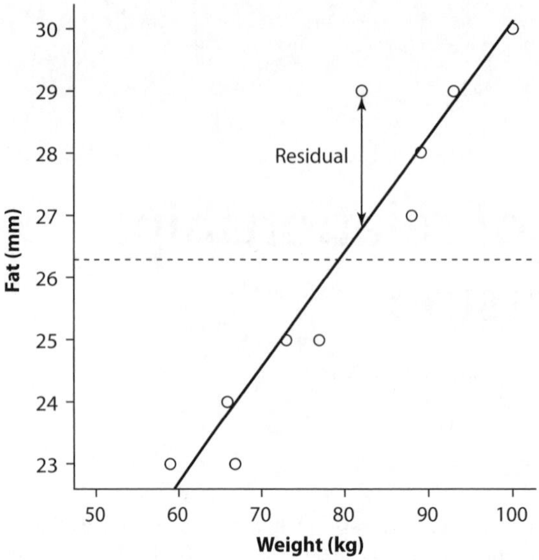

Figure 10.1 Fat plotted against weight in humans ($n = 10$). The sloping line is the line of best fit calculated by the ordinary least squares method. The horizontal line indicates the mean value for fat. The vertical distance between a point and the fit line is called a residual.

a null hypothesis significance test to help us to decide if any apparent slope is due to chance alone, or to chance and something of potential biological interest.

In this introduction I am going to emphasize some general points about bivariate linear regression:

- The type of data on which bivariate linear regression tests work.
- The idea of a linear model.
- The things you can find out by performing a regression analysis.
- The parametric criteria in relation to regression.
- Similarities between and the combining of regression and Anova.
- Alternatives to and extensions of one-way bivariate linear regression.

Table 10.1 The weight and body fat of ten humans.

Weight (kg)	Fat (mm)
89	28
88	27
66	24
59	23
93	29
73	25
82	29
77	25
100	30
67	23

10.1.1 **Variables and levels of measurement needed**

When you look at a data set that is appropriate for bivariate linear regression there will be two samples, one from each variable, numbers in these samples will be scale level, and sample sizes will be the same. Take a look at the data in Table 10.1. These data have these characteristics. Now look back at Table 8.1 which are data suitable for paired *t*-testing. The data should look similar to the regression data in Table 10.1. The key difference is that the numbers in Table 10.1 are not all from the dependent variable. Numbers in one sample will be from the dependent variable, while numbers in the other will be from the independent variable. Samples are not distinguished by categories of the independent variable but by the names of the variables themselves. When we have performed a regression analysis we can say that we have regressed the dependent variable against the independent variable.

In Table 10.1, the independent variable is body weight, measured in kilograms. The dependent variable is body fat in millimetres, estimated from skinfolds. Weight in kilograms and fat in millimetres are both measurements at the continuous scale level. If we performed a regression analysis on these data we would regress body fat against body weight, which would include a regression test on the slope of the line.

10.1.2 **Linear model: scary—not!**

The fancy way of saying that two variables have a relationship that can be summarized by a straight sloping line is to say that they conform to a **linear model**. The linear bit is easy enough to understand: linear as in line. However, the use of the word model makes it sound like something highly complicated and elusive is going on, but a model is simply an equation that describes, or attempts to describe, patterns in data.

The equation of a straight line, $y = bx + c$, is a **linear model**.

The model for a linear relationship between two variables is the equation for a straight line. You may have seen it written something like:

$$y = bx + c$$

where y and x are variables and b and c are called **coefficients** and are constants. The constant b is the slope of the line. The constant c is the value of y when x equals zero and is called the intercept. Written in full, the equation would look like this:

The letters b and c in the equation for a line are constants and are called **coefficients**

Dependent variable = slope × independent variable + intercept

For the example data in Table 10.1, the linear model of the data would be:

Fat = slope × weight + intercept

If a slope is 0, the line is horizontal and the dependent variable disappears from the equation. Since the dependent variable no longer equals anything to do with the supposed independent variable, the two are not related in a linear way.

But how do you find out what values for the slope and the intercept of the line best describe the data? There are two main methods: **ordinary least squares (OLS)** and **maximum likelihood (ML)**. The ordinary least squares method is a lot less complicated to calculate than the maximum likelihood method, although the latter does have advantages under some circumstances. Since both methods will give you the same answer for bivariate linear situations, we shall use ordinary least squares and talk no more about maximum likelihood. If you are curious to learn more, try one of the more advanced texts listed in the Selected further reading section.

Ordinary least squares (OLS) and **maximum likelihood (ML)** are two methods that can be used to find the linear model that best fits your data. For example, ordinary least squares works by minimizing the vertical distance between data points and the line. This distance is known as the **residual** or **error**.

Ordinary least squares works by minimizing the vertical distance between each data point and the line. This distance represents what is known as the **residual** or **error.** Take a look at Figure 10.1 to see this illustrated. The formulae for working out the slope and the intercept based on this method are quite straightforward. I am not going to present them in this book, but if you want to see what they look like, you can find details in another text (see Selected further reading). They involve calculating a **sum of squares;** you will be familiar with this if you have gone through Section 3.2.2.

Sum of squares is the sum of the squared deviations from the mean, $\Sigma(y-\bar{y})^2$ (see Section 3.2.2 on variance).

For the data in Table 10.1, the slope works out at 0.186 and the intercept as 11.571, so the model looks like this:

$$\text{Fat} = 0.186 \times \text{weight} + 11.571$$

We could get this apparent pattern by chance alone. So, is this model really any different from a horizontal line with zero slope? This is what we use a regression test to decide. Since the test is based on a linear model, it is a good idea to check to see if a linear relationship is at least a plausible idea. You can make a visual assessment of this by constructing a scatterplot like that shown in Figure 10.1. If the relationship is not linear, it may be possible to make it so by transforming the data (Section 5.3.4).

10.1.3 **The three regression questions**

Bivariate linear regression analysis involves answering three questions, as follows:

1. What is the linear model that best describes changes in the dependent variable based on the independent variable? To do this you need to find the values of the slope, b, and the intercept, c, in the equation for a line (coefficients). This can be done using formulae based on the ordinary least squares method of finding the line of best fit (Section 10.1.2).
2. Does this model describe a significant amount of the variation in the dependent variable? That is, does the dependent variable have a linear relationship with the independent variable? To answer this you need to conduct a regression test. This is a null hypothesis significance test involving the F statistic.
3. How much variation does the model explain? To do this you calculate something called a coefficient of determination (R^2).

10.1.4 **Added extras: how much is explained, and prediction**

The answer to question 2, introduced in the previous section, uses a null hypothesis significance test called the regression test, involving the F statistic. A regression test runs through the same four steps as all the other tests we have covered. By answering questions 1 and 3 you get a couple of added extras: the ability to make predictions using the linear model and a number that tells you how much variation is being explained, called the coefficient of determination.

The coefficient of determination (R^2) for regression is a measure of how much the dependent variable varies with the independent variable. For example, a coefficient of determination of 0.25 means that 25% of the variation of the dependent variable is accounted for by the independent variable.

We have to answer question 1 to answer questions 2 and 3, but the creation of a model also means that we can predict any value of the dependent variable from the independent value. For example, I have already told you that the line equation for the regression of fat on weight in Table 10.1 is

$$\text{Fat} = 0.186 \times \text{weight} + 11.571$$

According to this model, if a person weighs 70 kg we would predict that their fat would be 24.591 because

$$(0.186 \times 70) + 11.571 = 24.591$$

Of course, it makes no sense to use the model for the prediction if it does not explain a significant amount of the variation. That is, if the answer to question 2 is no, then you should not use your model for prediction. Another pitfall to avoid is using the model to predict values of the independent variable falling outside of the range of those in the data set. For example, based on the data in Table 10.1, it would be bad practice to try to predict the body fat of someone weighing 55 kg.

The answer to question 3 is given by a coefficient of determination, R^2. If, for example, $R^2 = 0.40$, this tells us that 40% of the variation in the dependent variable is a consequence of changes in the independent variable. If $R^2 = 0.04$, this would mean only 4% is explained.

Of course, it only makes sense to consider R^2 if the relationship is significant. If the answer to question 2 is no, do not bother with question 3. It is true that the larger R^2, the more likely the relationship is to be significant. However, a high R^2 does not necessarily mean the relationship will be significant and conversely a low R^2 does not necessarily mean it will not be.

Here is a summary of how these added extras relate to the three questions introduced in Section 10.1.3.

1. What is the linear model that best describes changes in the dependent variable based on the independent variable? This question is answered by finding the coefficients b and c in the equation for a line and can be used for prediction if the answer for question 2 is yes.
2. Does this model describe a significant amount of the variation in the dependent variable? This question is answered using a regression test involving F. If it is yes, then you can use your answer to question 1 for prediction and ask question 3.
3. How much variation does the model explain? This question is answered by looking at the coefficient of determination (R^2). It is only meaningful to ask this question if the answer to question 2 is yes.

10.1.5 **Regression and the parametric criteria**

Since regression is a parametric test you must be satisfied that the data meet the parametric criteria (Section 5.3). The parametric criteria for a regression test are that:

1. The residuals for any value of the independent variable, *x*, must be normally distributed;
2. The variance of the dependent variable should be similar for all values of the independent variable, *y*.

Since typically we only have one value of the dependent variable, *y*, for every value of the independent variable, *x*, we cannot actually check our data for these features. The precedent is that, provided you have no reason to think that the data might not conform to these criteria, you can assume that they do (Section 5.3). In practice, this means that the most important thing to double check before proceeding with a regression test is that the dependent variable is measured at scale level. Strictly speaking the dependent variable ought actually to be continuous scale rather than discrete scale, but this is a grey area (Section 3.3.4).

10.1.6 **Alternatives to and extensions of bivariate linear regression and Anova**

The regression we are dealing with in this chapter involves one dependent and one independent variable, where these two variables have a linear relationship. This basic type of regression can be extended in a number of ways including the examples listed below:

- Multiple regression is used for situations involving more than one independent variable (this parallels multiway Anova, mentioned in Section 9.1.4).
- Multivariate regression is used for situations involving more than one dependent (this parallels multivariate Anova, Manova, mentioned in Section 9.1.4).
- Analysis of covariance, or Ancova for short, is used for situations involving both continuous and categorical (typically nominal level) independent variables in a mixture of Anova and regression. Another way of thinking of Ancova is that it is a way of using regression to assess the effects of a potentially confounding scale-level variable in an Anova.
- Curvilinear regression is needed to explore relationships that are not linear.
- Logistic regression can be used when you have a dependent variable which is measured at the nominal or ordinal level.

Multiregression and analysis of covariance are essentially covered by the General Linear Model, in Chapter 12. In fact the General Linear Model can also be used as an alternative to bivariate linear regression. Logistics regression is covered by the Generalized Linear Model, (Chapters 12 and 13). Although only simple bivariate linear regression is covered in this chapter, a good understanding of this technique will help you grasp these other techniques by reading Chapter 12 and books from the Selected further reading section.

Although we are not going to deal with multiple regression in detail in this chapter, we need to discuss it briefly in order to explain why you sometimes see a *t* statistic reported in association with a bivariate regression, rather than an F statistic. Multiple regression is when more than one independent variable is involved. The principles are the same, but a model can be calculated that includes all the independent variables together, as well as for each independent/dependent variable pair. This means that question 2 in Section 10.1.3 will have two parts:

2a. Does the overall model, including all the independent variables, explain a significant amount of the variation in the dependent variable? To do this you would need to conduct a null hypothesis significance test involving the F statistic.

2b. Does the dependent variable have a linear relationship with a particular independent variable? To do this you need to conduct a null hypothesis significance test on the slope of the line calculated for each independent variable using the *t* statistic.

For a bivariate linear regression, testing the overall model using the F statistic achieves the same as testing the slope of the line using the *t* statistic. It is helpful to know this, as people sometimes report the *t* statistic instead of the F statistic when reporting the results of a bivariate linear regression.

10.2 Example data: species richness

In this section I am going to introduce the data set that I will be using in the worked examples in Section 10.3. Ecological theory suggests that the number of species (which is known as species richness) within a wildlife reserve is dependent on the area of that reserve. The exact nature of the relationship between these two variables (reserve area and species richness) is of great interest to conservationists and wildlife managers. They could use it, for example, to predict the effect on species richness of reducing the size of a reserve.

Data sets useful for testing the relationship between reserve area and biodiversity are few and far between. However, the dedicated work of rangers from the Ghanaian Wildlife Division has generated just such a data set. Every month, rangers on patrol in Ghana's six reserves count the number of large mammals they see. Justin Brashares has calculated species richness for the different reserves based on these counts (Brashares *et al.*, 2001). The data for 1999 are given in Table 10.2.

There are two samples in Table 10.2, one from the dependent variable, species richness, and one from the independent variable, reserve area. Species richness is measured as the number of large mammal species and reserve area is measured as square kilometres. Both variables are measured at the scale level. The data in Table 10.2 are presented as a scatterplot in Figure 10.2. It looks reasonable that a straight line could summarize these data and therefore using a linear model for our regression is reasonable. Since we have no reason to think that these data do not conform to the parametric criteria, we can assume that they do (Section 10.1.5).

Reserve	**Area (km^2)**	**Species richness (no. of species)**
Mole	4840	22
Digya	3126	14
Bui	2074	17
Gbele	565	10
Kalakpa	325	12
Shai	58	5

Source of data: courtesy of Justin Brashares.

Table 10.2 Example data: species richness. Reserve size and species richness in Ghana in 1999.

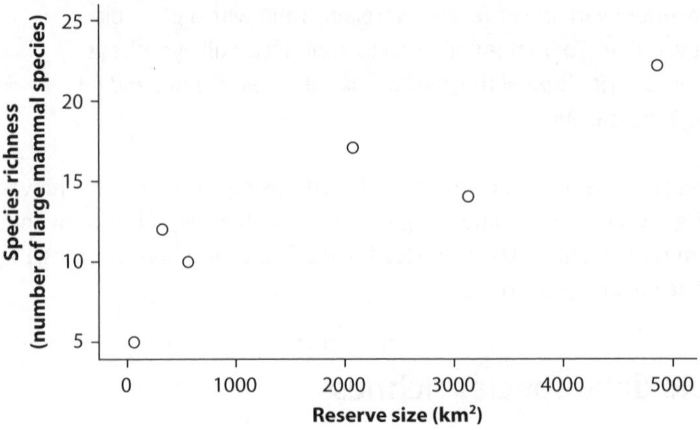

Figure 10.2 Species richness against reserve size ($n = 6$).

10.3 Regression test

This section focuses on the regression test. It covers:

- When to use and when not to use regression.
- The four steps as applied to a regression test.
- A worked example focusing on a regression test using SPSS.
- A worked example focusing on the added extras in a regression analysis using SPSS.
- An example of the use of a regression from the literature.

The data used in the worked examples come from Justin Brashares' work and were introduced in Section 10.2.

10.3.1 When to use

Regression is a test for assessing whether the linear relationship between two variables can be accounted for by sample error alone. Issues relating to when and when not to use this test were covered in the introduction to this chapter (Section 10.1). They are summarized in Box 10.1. A regression test is a parametric test, so you must check that your data fulfil the parametric criteria (Sections 5.3 and 10.1.5) before proceeding. You should also draw a scatterplot to check that a linear model is reasonable.

10.3.2 Four steps of a regression test

Here are the four steps of null hypothesis significance testing outlined in Chapter 5, applied specifically to a regression test.

STEP 1 State the null hypothesis (H_0)

The null hypothesis for a regression test takes the general form:

H_0: The dependent variable is not related to the independent variable in a linear fashion.

Box 10.1 When to use a regression

Use this test when:

- You are looking for a *relationship* between a *dependent variable* and an *independent variable.*
- You believe the relationship might be modelled by a *straight line.*
- Data in the samples are measured at the *scale* level.
- Data are *related.*
- You have a particular interest in *prediction.*

Do not use this test when:

- You want to compare frequency distributions (Chapter 6).
- You cannot identify a dependent and an independent variable (Chapter 11).
- Data do not fulfil the parametric criteria (Sections 5.3 and 10.1.5).

STEP 2 Choose a critical significance level (α)
Typically this is 5% (0.05).

STEP 3 Calculate the test statistic
For a regression test the statistic is F. This is the same statistic used in an Anova (Chapter 9). In the case of regression, the F statistic reflects the variance accounted for by the regression line relative to the remaining variance, otherwise known as an **error** variance or **residual** variance. If the line does not account for any variance, these two values will be the same and F will equal one. This is summarized in general terms in Box 10.2a and the formulae are given in Box 10.2b.

The variance remaining after that caused by the regression line is taken into account is called **residual** or **error**.

F statistics have two different degrees of freedom. In the case of regression, one is associated with the regression variance and the other with the error variance. The regression degrees of freedom is one less than the number of variables which, for the bivariate situation is one (two variables minus one equals one). The error degrees of freedom is the number of related data points minus the number of treatments, which for the bivariate situation will be the number of related data points minus one.

STEP 4 Reject or accept your null hypothesis
How you do this will depend on whether you are doing the calculations by hand with critical-value tables, or using a statistical computing package with *P* values.

Using critical-value tables

First you need to look up the critical value of F given the critical significance level you have chosen in step 2 and the degrees of freedom calculated in step 3 (Appendix II, Table AII.5). Then you need to compare the value of F that you have calculated in step 3 with this critical value. Finally, you should decide whether to reject or accept your null hypothesis according to the rule in Box 10.3.

Using *P* values on computer output

Find *P* on the computer output and make your decision according to the rule in Box 10.4.

Box 10.2 (a) Calculating the F statistic for a regression: words; (b) Calculating the F statistic for a regression: formulae

(a)

Source (of variance)	Sum of squares	Degrees of freedom	Mean square (or variance)	F
Regression	$SS_{regression}$	$DF_{regression}$	$\frac{SS_{regression}}{DF_{regression}}$	$\frac{MS_{regression}}{MS_{error}}$
Residual (Error)	SS_{error}	DF_{error}	$\frac{SS_{error}}{DF_{error}}$	
Total	SS_{total}	DF_{total}		

(b)

Source (of variance)	Sum of squares	Degrees of freedom	Mean square (or variance)	F
Regression	$\Sigma(\hat{y}-\bar{y})^2$	$k - 1$	$\frac{\text{Column 2}}{\text{Column 3}}$	$\frac{\text{Row 2}}{\text{Row 3}}$
Residual (Error)	$\Sigma(y-\hat{y})^2$	$N - k$	$\frac{\text{Column 2}}{\text{Column 3}}$	
Total	$\Sigma(y-\bar{y})^2$	$N - 1$		

Where SS = sum of squares; DF = degrees of freedom; MS = mean square; k = number of variables; N = the sample size (that is, the number of related data points); $\bar{y}$ = the overall mean; $\hat{y}$ = value of y predicted by the linear model; $\Sigma(\hat{y}-\bar{y})^2$ = the sum of the squared differences between each predicted value of y and the overall mean of y; $\Sigma(y-\hat{y})^2$ = the sum of the squared differences between each y and its predicted value; $\Sigma(y-\bar{y})^2$ = the squared sum of the differences between each y and the overall mean of y.

Box 10.3 Decision using critical values for a regression test

If $F \geq F_{critical}$ → reject H_0 → significant result.
If $F < F_{critical}$ → accept H_0 → non-significant result.

Box 10.4 Decision using *P* values for a regression test

If $P \leq \alpha$ → reject H_0 → significant result.
If $P > \alpha$ → accept H_0 → non-significant result.

(a)

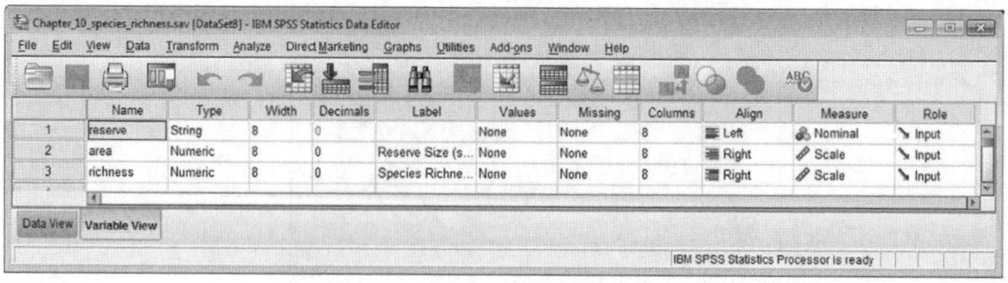

(b)

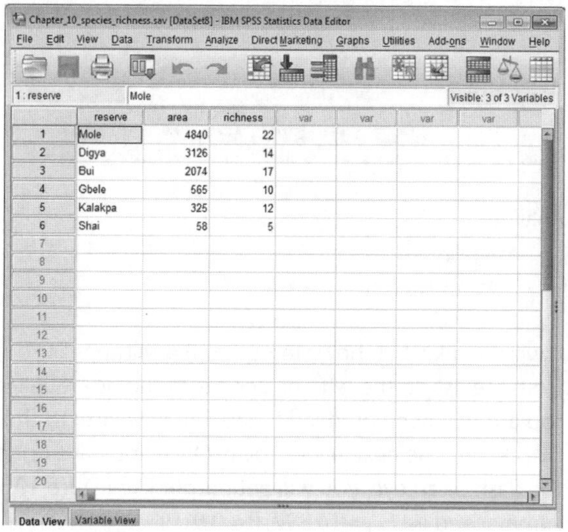

Figure 10.3 Example data in SPSS: species richness. (a) Variable View. (b) Data View.

10.3.3 **Worked example: using SPSS for a regression test**

online resource centre

We are going to conduct a regression test on the species-richness data (Section 10.2) using SPSS. To do this the data must be entered into SPSS as shown in Figure 10.3. In newer versions of SPSS (version 10 onwards), the **Data Editor** window has a **Variable View** tab and a **Data View** tab. Figure 10.3a shows the former and Figure 10.3b the latter. Appendix I has notes on how to enter data into SPSS.

STEP 1 State the null hypothesis (H_0)

H_0: Species richness does not show a linear relationship with reserve size.

STEP 2 Choose a critical significance level (α)
We will use 5% (0.05).

STEP 3 Calculate the test statistic

To get SPSS to conduct a regression test on your data you must first open the data file. Then you must make the following selections:

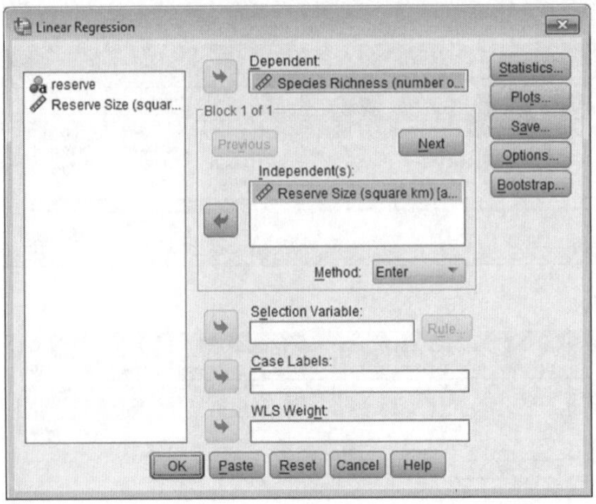

Figure 10.4 Conducting a regression using SPSS. Main dialogue window.

Analyze
→Regression
→Linear...

A window like that shown in Figure 10.4 will appear. You need to select *Species Richness* from the list on the left and send it to the box labelled **Dependent** by pressing the top arrow. Then you need to send *Reserve Size* to the box labelled **Independent(s)** by pressing the next arrow down. Once you have done this you can click **OK** and, in the flash of an eye, you will get output like Figure 10.5. The key information can be summarized as: $F_{1,4} = 14.332$, $P = 0.019$).

STEP 4 Reject or accept your null hypothesis using *P* values on the computer output

Since the critical significance level of 0.05 is larger than the value 0.019 of *P*, we must reject the null hypothesis according to Box 10.4. In short:

$P(0.019) < \alpha(0.05) \rightarrow$ reject $H_0 \rightarrow$ significant result.

We can conclude that species richness is linearly related to reserve size (regression test: $F_{1,4} = 14.332$, $P = 0.019$).

10.3.4 **Worked example: using SPSS to get the added extras**

online resource centre

In Section 10.3.3 we answered the second of our three regression questions (Section 10.1.3): does a linear model describe a significant amount of the variation in the dependent variable? We found the answer in the table labelled Anova on our SPSS output (Figure 10.5). We can get the answer to the other two regression questions from this output using the tables labelled **Coefficients** and **Model Summary** (Figure 10.6).

Question 1: What is the best linear model describing changes in the dependent variable based on the independent variable? To do this you need to find the values of the slope, b, and the intercept, c, in the equation for a line, the general format of which is

$$y = bx + c$$

Regression

Variables Entered/Removed[b]

Model	Variables Entered	Variables Removed	Method
1	Reserve Size (square km)[a]	.	Enter

[a]All requested variables entered
[b]Dependent Variable: Species Richness (number of large mammal species)

Model Summary

Model	R	R Square	Adjusted R Square	Std. Error of the Estimate
1	.884[a]	.782	.727	3.057

[a]Predictors: (Constant), Reserve Size (square km)

ANOVA[b]

Model		Sum of Squares	df	Mean Square	F	Sig.
1	Regression	133.949	1	133.949	14.332	.019[a]
	Residual	37.384	4	9.346		
	Total	171.333	5			

[a]Predictors: (Constant), Reserve Size (square km)
[b]Dependent Variable: Species Richness (number of large mammal species)

Statistic (F) **P**

Degrees of freedom

Coefficients[a]

Model		Unstandardized Coefficients		Standardized Coefficients		
		B	Std. Error	Beta	*t*	Sig.
1	(Constant)	8.310	1.822		4.562	.010
	Reserve Size (square km)	.003	.001	.884	3.786	.019

[a]Dependent Variable: Species Richness (number of large mammal species)

Figure 10.5 SPSS output for a regression with key information annotated.

The values of b and c can be found in the **Coefficients** table on the SPSS output shown in Figure 10.6. For our example data the model, written in full is

$$\text{Species richness} = 0.003 \times \text{reserve area} + 8.310$$

Since our answer to question 2 was yes (Section 10.3.3), we can use this equation to predict values of species richness for any reserve size within the range 58–4840 km^2.

Question 3: How much variation does the model explain? To do this you find the coefficient of determination (R^2). This can be found in the **Model Summary** table as indicated in Figure 10.6.

Regression

Variables Entered/Removed[b]

Model	Variables Entered	Variables Removed	Method
1	Reserve Size (square km)[a]	.	Enter

[a]All requested variables entered
[b]Dependent Variable: Species Richness (number of large mammal species)

QUESTION 3: How much is explained?

Model Summary

Model	R	R Square	Adjusted R Square	Std. Error of the Estimate
1	.884[a]	.782	.727	3.057

[a]Predictors: (Constant), Reserve Size (square km)

Coefficient of determination (R^2)

QUESTION 2: Is the model significant?

ANOVA[b]

Model		Sum of Squares	df	Mean Square	F	Sig.
1	Regression	133.949	1	133.949	14.332	.019[a]
	Residual	37.384	4	9.346		
	Total	171.333	5			

[a]Predictors: (Constant), Reserve Size (square km)
[b]Dependent Variable: Species Richness (number of large mammal species)

Degrees of freedom | Statistic (F) | *P*

QUESTION 1: What is the model ($y = bx + c$)?

Coefficients[a]

Model		Unstandardized Coefficients B	Unstandardized Coefficients Std. Error	Standardized Coefficients Beta	*t*	Sig.
1	(Constant)	8.310	1.822		4.562	.010
	Reserve Size (square km)	.003	.001	.884	3.786	.019

[a]Dependent Variable: Species Richness (number of large mammal species)

Intercept (c) | Slope (b)

Figure 10.6 SPSS output for a regression annotated in relation to the three regression questions.

All the key information from our regression analysis can be summarised as:

Regression: $y = 0.003x + 8.310$; $F_{1,4} = 14.332$, $P = 0.019$; $R^2 = 0.782$

This summary says we did a regression analysis and found that:

- The best model of these data is the equation for the straight line $y = 0.003x + 8.310$.
- This model explains a significant amount of the variation in species richness between reserves of different size ($F_{1,4} = 14.332$, $P = 0.019$).
- Variation in reserve size explains 78.2% of the variation in species richness ($R^2 = 0.782$).

Notice that the answers to our three regression questions appear in reverse order on our SPSS output (Figure 10.6).

10.3.5 **Reporting bivariate linear regression results**

Our regression analysis has generated more information than the other tests we have looked at so far. What you should report will partly depend on the context, but whether the model is significant or not is always going to be relevant. It's also useful for your reader if you always report the coeffcient of determination, as this gives them an indication of the size of the effect at which they are looking. So for our worked example, you could report that species richness is linearly related to reserve size (regression: $F_{1,4} = 14.332$, $P = 0.019$, $R^2 = 0.782$).

In this case you don't really need to report the details of your model, but you could do. In this case you could report that species richness is linearly related to reserve size (Regression: $y = 0.003x + 8.310$; $F_{1,4} = 14.332$, $P = 0.019$; $R^2 = 0.782$). It becomes essential that you report the details of your model if you are going to use it for prediction or to compare models.

10.3.6 **Literature link: nodules**

online resource centre

Voisin *et al.* (2003) J. Exp. Bot. 54: 2733–2744 (OUP)

Plants need nitrogen to make proteins from the carbohydrates they produce by photosynthesis. Most plants get this nitrogen from the soil, but some plants can also get it from the air by a process known as nitrogen fixation. These plants can do this because they have special symbiotic nitrogen-fixing bacteria living in little lumps, called nodules, on their roots. Voisin *et al.* (2003) investigated this symbiotic nitrogen-fixation (SNF) process in the pea plant *(Pisum sativum)*.

As part of their analyses, Voisin *et al.* (2003) looked at the rate of SNF in relation to the mass of nodules on the plant. They measured SNF as milligrams of nitrogen fixed per plant per day, and nodule mass as grams per plant. Both these variables are continuous scale and one, SNF, can be thought of as potentially dependent on the other, nodule mass. In other words, SNF was their dependent variable and nodule mass their independent variable. They looked at the relationship between these two variables for three different stages of plant phenological development: the vegetative stage (when the plant is just growing), the flowering stage (when the plant is growing and producing flowers), and the seed-filling stage (when the plant is making seeds). Some plants were grown in a soil enriched with nitrogen and others were not, giving two different nitrogen (N) treatments.

In their results Voisin *et al.* (2003) tell us:

> 'SNF was then related to nodule biomass. SNF activity increased with biomass for both N treatments at the vegetative and at the flowering stages (Figure 4) whereas it remained low and similar regardless of nodule biomass at seed filling. A linear relationship could be established between SNF activity and nodule biomass at each phenological stage regardless of the N treatment (Figure 4).'

Voisin *et al.*'s Figure 4 is a scatterplot with SNF on the *y*-axis and nodule biomass on the *x*-axis, with a straight line drawn through the points for each stage labelled as follows:

Vegetative stage $y = 72.6x - 0.04,\ R^2 = 0.66$

Flowering stage $y = 58.8x - 1.4,\ R^2 = 0.86$

Seed-filling stage $y = 3.8x - 0.2,\ R^2 = 0.13$

Notice that we are not actually given the result of the regression test. We are given the models and the coefficients of determination (R^2). However, we are not told the F-statistic details and so do not know for sure if the models explained a significant amount of variation in SNF. In other words we are given the answer to our first and third regression questions, but not our second (Section 10.1.3). It is therefore ambiguous as to whether the slopes of the lines are significantly different from zero. If any were in fact non-significant then the coefficients of determination would become irrelevant and the data could not be said to show a linear relationship.

Tracking backwards and looking at the statistics section of the Methods we are told: 'Analysis of variance was performed with the GLM procedure of SAS...'. Analysis of variance (Anova) was dealt with in the previous chapter. SAS is an alternative statistical package to SPSS and GLM in this example stands for general, or generalized, linear model (Chapter 12). We are not told explicitly how the regressions were performed. They could have been carried out with the GLM procedure in SAS, or through a regression-specific option.

10.4 Model I and model II regression

Section 9.5 introduced the idea of model I and model II Anovas. These concepts can also be applied to regression. We have only looked at Model I regression, and for most of the situations you are likely to find yourself in this will serve you fine. However, if you want to explore this further there is a full discussion of model I and II regression in Sokal and Rohlf (2012; pp. 535–544) and Grafen and Hails (2002; pp. 43–45) also has a good summary. We will just touch on the idea briefly here.

If the independent variable is determined by the investigator (**fixed**) then **model I regression** is appropriate. If it is not (that is, it is **random**) then a **model II regression** is appropriate. If an independent variable is interpreted as being measured by the investigator, then model I applies and this means that model I covers most circumstances.

Model I regression is appropriate when the independent variable is **fixed** and **model II regression** when it is **random**.

In Section 9.5 we defined a fixed factor as one that was determined by the investigator and a random factor as one that was not. Saying an independent variable is fixed is to assume that it is measured completely accurately every time, without error. It is usually perfectly reasonable to assume that an independent variable in an Anova is measured without error. This is because the independent variable tends to divide the data up into clearly defined treatments such as the early and late stage of infection. However, such an assumption may be more problematic in the case of regression, where the independent variable is measured at the scale level. For example, although weight (Table 10.1) or reserve size (Section 10.2) can be determined by the investigator in the sense of being deliberately selected, both are subject to measurement error. Just think of your bathroom scales at home: stand on them twice within a few seconds and you often get a different reading for your weight, due to the vagaries of the equipment.

Model II-type analyses are free from the assumption that the independent variable is determined by the investigator and measured without error. The bad news is that model II is a hornet's nest mathematically. The good news is that we don't have to be really strict about the assumption for model I regression to be reliable. However, when interpreting the results of a model I regression, what we need to do is to remember that the independent variable is as measured in that particular case. For the example data in this chapter, species richness was related to reserve size as measured by Justin Brashares and his colleagues (Section 10.2).

Checklist of key points

- A regression test is performed as part of a regression analysis. A regression test is used to test for a relationship between two variables.
- Regression analysis involves answering three questions.
  1. What is the best linear model? To do this you need to find the values of the slope, b, and the intercept, c, in the equation for a line. (Answered with information on SPSS output labelled Coefficients.)
  2. Is this model significant? To do this you need to use a regression test involving F. If the answer is yes, you can use your answer to question 1 for prediction and it is worth asking question 3. (Answered with information on SPSS output labelled Anova.)
  3. How much does the model explain? To do this you calculate a coefficient of determination (R^2). But remember it is only worth asking this question if the answer to question 2 is yes. (Answered with information on SPSS output labelled Model Summary.)
- Multiple regression involves several independent variables. In addition to doing a regression test (involving F) to see if the model as a whole is significant, you can use a test involving *t* to see if each independent variable explains a significant amount of variation.

This test assesses if the slope of the line, b, is significantly different from zero for a line drawn for each independent variable. For bivariate regression, the regression test on the whole model using F, and the test of the slope of the line using *t*, achieve the same thing; sometimes *t* is reported instead of F in answer to question 2 above.

- You should consider a regression test when:
  - the numbers you see when looking at the data are measured at the scale level;
  - the numbers in one sample come from the dependent variable and numbers in the other sample come from the independent variable.
- The regression test is parametric. The parametric criteria for the regression tests are that residuals of the dependent variable must be normally distributed around each value of the independent variable, and that variances of the dependent variable must be similar across all values of the independent variable. These criteria are generally assumed to hold if both variables are at the scale level and there is no reason to think otherwise.
- In addition to meeting the parametric criteria, data must also potentially fit a linear model for a regression test to work. This can be checked by visual inspection of a scatterplot.
- The four steps of a regression test are:
  1. State H_0: there is no relationship between the populations.
  2. Choose a critical significance level: typically $\alpha = 0.05$.
  3. Calculate your statistic: calculate F and both degrees of freedom according to Box 10.2.
  4. Reject or accept H_0: reject H_0 if $F \geq F_{critical}$ (Box 10.3) or if $P \leq \alpha$ (Box 10.4).
- If your regression test is significant, you can use your model for prediction and your coefficient of determination to tell you how much has been explained.

Self-help questions

1. For each of the following features say which alternative after the colon applies to a regression test.
   (a) Category of test: parametric/nonparametric. *Link back: Section 10.1*
   (b) Numbers in the formula for the statistic: frequencies/measurements. *Link back: Sections 10.1.1 and 10.3.2*
   (c) Test to assess: difference/relationship. *Link back: Section 10.1*
   (d) If the calculated statistic is greater than the critical value of the statistic looked up in a table: accept/reject. *Link back: Section 10.3.2*
   (e) If *P* on the SPSS printout is less than your critical significance level: Accept/Reject. *Link back: Section 10.3.2*
2. (a) What are the three regression questions? *Link back: Section 10.1.3*
   (b) Which question does a regression test address? *Link back: Section 10.1.3*
   (c) In addition to a regression test, what else does regression analysis involve? *Link back: Section 10.1.4*
   (d) What statistic does the regression test use? *Link back: Section 10.3.2*
   (e) What is the symbol for the coefficient of determination? *Link back: Section 10.1.3*
3. A researcher working in a zoo recorded the number of prairie dogs out of their burrows and the number of visitors present at noon on 22 different days. She thought that the behaviour of the prairie dogs might depend on how many visitors were around the enclosure. She used SPSS to perform a regression analysis to investigate. The output she got is shown in Figure 10.7.
   (a) What is the key information she would need to report? *Link back: Section 10.3.5*
   (b) Comment on the observation that both variables are discrete scale rather than continuous scale. *Link back: Section 10.1*

Regression

Variables Entered/Removed[b]

Model	Variables Entered	Variables Removed	Method
1	Number of visitors[a]		Enter

[a]All requested variables entered
[b]Dependent Variable: Number of prairie dogs out of burrows

Model Summary

Model	R	R Square	Adjusted R Square	Std. Error of the Estimate
1	.559[a]	.312	.278	9.37314

[a]Predictors: (Constant), Number of visitors

ANOVA[b]

Model		Sum of Squares	df	Mean Square	F	Sig.
1	Regression	798.338	1	798.338	9.087	.007[a]
	Residual	1757.117	20	87.856		
	Total	2555.455	21			

[a]Predictors: (Constant), Number of visitors
[b]Dependent Variable: Number of prairie dogs out of burrows

Coefficients[a]

Model		Unstandardized Coefficients		Standardized Coefficients		
		B	Std. Error	Beta	*t*	Sig.
1	(Constant)	29.406	3.029		9.709	.000
	Number of visitors	–.478	.158	–.559	–3.014	.007

[a]Dependent Variable: Number of prairie dogs out of burrows

Figure 10.7 SPSS output referred to in self-help question 10.3.

11 Tests of relationship: correlation

Chapter Aims

The subject of this chapter is tests of relationship for two interdependent variables. The parametric Pearson correlation and nonparametric Spearman correlation are presented. The chapter starts by comparing these tests and dealing with general points relating to them. The four steps of null hypothesis significance testing introduced in Chapter 5 are applied to both the Pearson correlation and the Spearman correlation. This is first done in a general context and then using example data worked through using SPSS. An example of the use of each test in the literature is also presented.

11.1 Introduction to the Pearson and Spearman correlation tests

This chapter is all about testing whether two different variables vary together in a linear fashion due to chance alone, or to chance plus something of potential biological interest. In other words, we will be looking at bivariate linear correlation. There are parametric and nonparametric ways of doing this: we will be looking at the parametric Pearson product-moment correlation and the nonparametric Spearman rank-order correlation: the Pearson and Spearman correlations for short.

You should note from the outset that although two variables are involved, it is *not* the case that one is the dependent variable and the other is the independent variable. Correlation is about investigating the covariation between variables, and not the variation of a dependent variable in relation to an independent variable. Examples of potential pairs of variables on which a bivariate correlation could be performed include:

- Tusk weight and body weight in elephants.
- Mammalian diversity and insect diversity in nature reserves.
- Swimming speed and body length in brine shrimps.
- Aspirin consumption and paracetamol consumption in humans.

Bivariate linear correlation is a way of assessing if two variables covary in a linear fashion. As for regression, a linear model (Section 10.1.2) underlies the calculations. We do not usually try to calculate the details of the model for a correlation, as doing so is much trickier than for regression. However, before proceeding with either the Pearson or Spearman correlation, you need to generate scatterplots in order to assess your data visually for linearity. As long as the data points do not fall around an obviously curvy line, it is reasonable to proceed with a linear correlation. For example, it would be reasonable to perform a linear correlation on all five data sets portrayed in Figure 11.1.

A correlation can be either positive or negative. If one variable increases as the other increases this is called a **positive correlation** (Figure 11.1a). If one variable increases and the other variable decreases, this is a **negative correlation** (Figure 11.1e). The scatterplots in Figure 11.1 show patterns ranging from an apparent strong positive correlation (Figure 11.1a), through no apparent correlation (Figure 11.1c), to an apparent strong negative correlation (Figure 11.1e). The null hypothesis significance tests presented in this chapter give us an objective way of deciding how scattered the points can be, and it still be reasonable for us to regard the variables as correlated. Consider, for example, Figure 11.1b; the data points appear to fall in a line, but this pattern might be produced by chance alone.

When two variables increase together it is a **positive correlation**. If one variable increases when the other one decreases it is a **negative correlation**.

In this introduction I am going to emphasize some general points about bivariate linear correlation:

- The type of data on which bivariate linear correlation tests work.
- Similarities and differences between Pearson's and Spearman's techniques.
- The parametric criteria as applied to a Pearson correlation.
- The statistics used in correlation: correlation coefficients.
- Extensions of bivariate linear correlation.

11.1.1 **Variables and levels of measurement needed**

When you look at a data set that is appropriate for bivariate linear correlation there will be two samples: one from each of two variables. The numbers in these samples will be at least ordinal level and, since the data are related, sample sizes will be the same. This is similar to data on which you would do regression (Section 10.1.1). The key difference is that the two variables are not considered as a dependent/independent variable pair.

Look at Table 11.1 to see some data on great tit anatomy suitable for analyis using correlation. The two variables are wing span and body length, both measured in centimetres and both scale level. If we performed a correlation analysis on these data, we would be asking if wing span and body length vary together. If two variables are correlated they lack independence (Section 2.2.4).

11.1.2 **Comparison of Pearson's and Spearman's tests**

Both a Pearson's test and a Spearman's test are used to look for a correlational relationship between two variables. In both cases you are testing for a linear relationship.

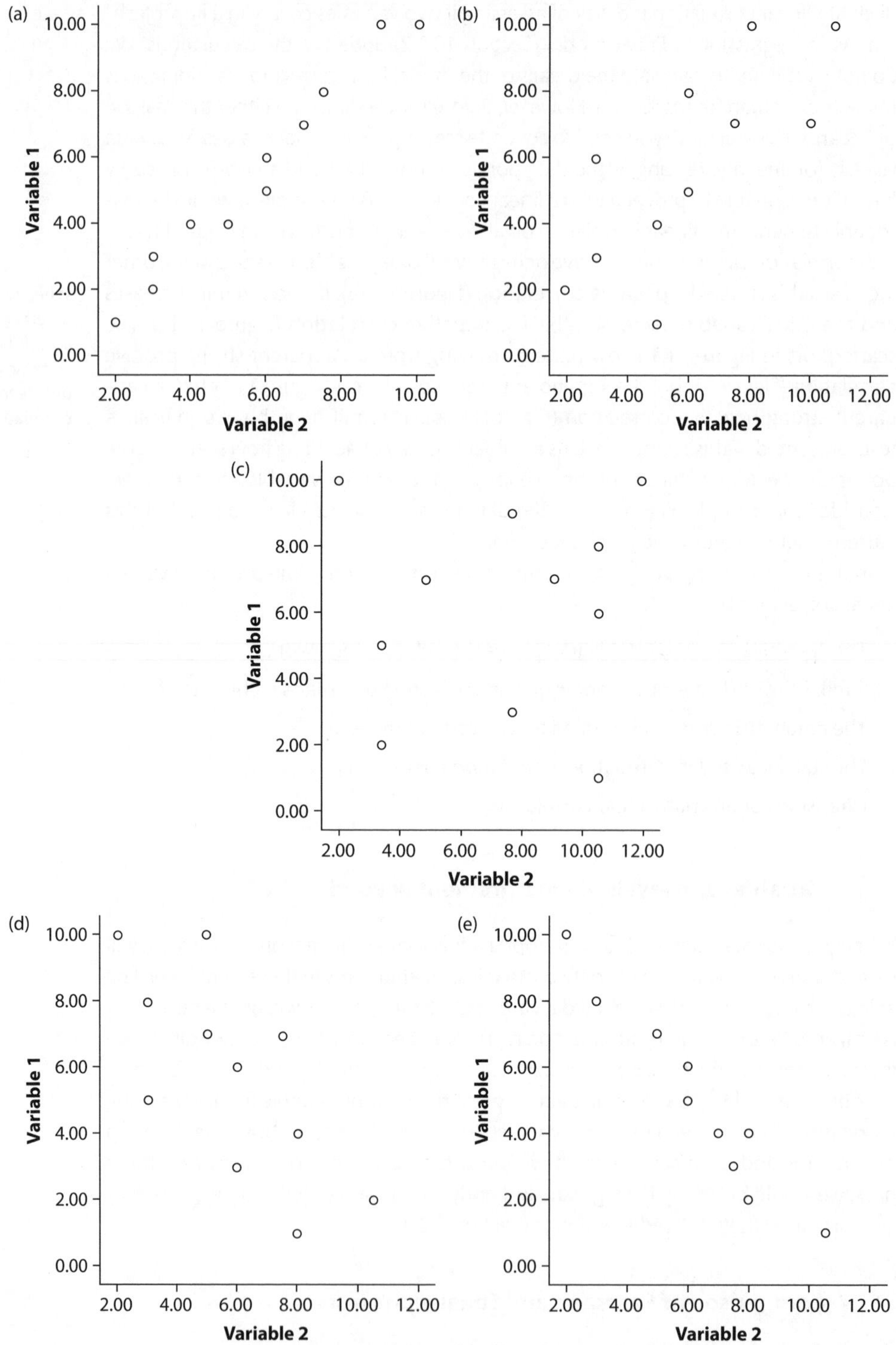

Figure 11.1 Variation of variable 1 with variable 2. (a) Strong positive linear relationship. (b) Weak positive linear relationship. (c) No linear relationship. (d) Weak negative linear relationship. (e) Strong negative linear relationship.

Bird ID number	Wing span (cm)	Body length (cm)
209	22.2	12.5
215	22.4	13.2
231	23.1	13.8
322	23.2	14.8
234	24.0	14.9
256	24.1	15.2
276	24.5	15.3
299	24.9	15.5

Table 11.1 Wing span and body length in great tits.

The data must be measured at a minimum of ordinal level, and data points in the two samples related.

The Pearson's test differs from the Spearman's test in that Pearson's is parametric and Spearman's is nonparametric. Since Pearson's test is parametric, the data must fulfil special criteria if the test is to be valid (Sections 5.3 and 11.1.3), and data can only possibly fulfil parametric criteria if they are measured at scale level. On the other hand, Spearman's test can work when the data have been measured at either the ordinal or scale levels.

The similarities and differences between these two tests are summarized in Table 11.2.

11.1.3 **Pearson and the parametric criteria**

A Pearson test is a parametric test and therefore you must be satisfied that the data meet the parametric criteria (Section 5.3). In the case of the Pearson test, this means that the two samples come from a bivariate normal distribution. This means that for each value from one of the variables, the corresponding values of the other variable should be normally distributed and vice versa. For this to be the case, both variables should technically be measured at the continuous, rather than discrete, scale level. However, this is a grey area (Section 3.3.4) and going with just scale is generally okay.

In terms of checking this assumption, the situation is similar to that for simple linear regression (Section 10.1.5). You will tend to have only a single value of one variable for each value of the other, and therefore not actually be able to check your data for this feature. The precedent is that, providing you have no reason to think that the data might not conform to these criteria, you can assume that they do (Section 5.4).

	Pearson's test	Spearman's test
Similarities	Tests of linear relationship Two samples from each of two variables (not dependent/independent) Data in samples related	
Differences	Parametric Scale data only	Nonparametric Scale or ordinal data

Table 11.2 Comparison of Pearson's and Spearman's correlation tests.

In practice, this means that you should at least check to make sure each sample is normally distributed.

11.1.4 The correlation coefficient

Correlation coefficient is the statistic used in correlations.

The statistic used in both a Pearson's and a Spearman's test is a **correlation coefficient**. For Pearson's the symbol is r and for Spearman's it is r_s. In addition to being used in a null hypothesis significance test to determine if the linear relationship between two variables is significant, correlation coefficients indicate:

1. The strength of the relationship (from 0, no correlation, to 1, a perfect correlation).
2. The direction of the relationship (positive versus negative).

The strength of the relationship is indicated by the numerical size of the correlation coefficient. The direction is indicated by the sign of the correlation coefficient. For example a correlation coefficient of:

- 0.9 indicates a strong positive correlation, for example as shown in Figure 11.1(a).
- −0.9 is a strong negative correlation, for example as shown in Figure 11.1(e).
- 0.3 is a weak positive correlation, for example as shown in Figure 11.1(b).
- 0.0 is no correlation, for example as shown in Figure 11.1(c).

The weaker the correlation the more likely it is to be non-significant, that is, any apparent pattern being merely a chance happening. However, specific values regarded as significant or non-significant depend on the critical significance level chosen and the sample size.

For Pearson's correlation, the correlation coefficient can be squared to give a **coefficient of determination** (r^2) for correlation. If $r = 0.9$ then $r^2 = 0.81$; in other words, the two variables covary for 81% of their variability. Put another way, 81% of the variability in one variable is matched by variability in the other variable.

Coefficient of determination (r^2) for correlation is a measure of how much variables vary together. For example, a coefficient of determination of 0.25 means that they vary together for 25% of their variation. It is the square of the Pearson correlation coefficient. It does not apply to nonparametric situations.

11.1.5 Partial, multiple, and multivariate correlation

Pearson's correlation analysis can be extended into a technique that measures the relationship between two variables when one or more other variables have been controlled. For example, we might want to look at the correlation between wing span and body length in great tits, while controlling for a third variable, say body mass. This would tell us how wing span and body length would be correlated if mass were constant. This is called partial correlation and we will not be dealing with it in this book. Spearman's correlation cannot be extended in this way, but an alternative non-parametric correlation, called Kendall rank-order correlation, can be. Correlations involving more than two variables are called multiple or multivariate correlations.

We are not going to deal with partial, multiple, or multivariate correlation in this book. If you want to learn more about these you can turn to one of the books that are listed in the Selected further reading section, for example, Zar (1999) and Siegel and Castellan (1988).

11.2 Example data: eyeballs

In this section I am going to introduce the data set that I will be using in the worked examples in Sections 11.3 and 11.4, for a Pearson correlation and a Spearman correlation. I have chosen a data set that meets the parametric assumptions, so that we can use it for both parametric and nonparametric tests. Using the same data set for both tests emphasizes that the basic data structure required for a Pearson and a Spearman correlation is the same. You need to remember that if parametric assumptions are not met, then you must pursue the nonparametric alternative. Furthermore, if you have the option of either parametric or nonparametric tests, you should do the parametric test because it is more powerful (Section 5.3); there is no need to do both.

Glaucoma is a leading cause of preventable sight loss worldwide. Damage to the nerve carrying signals from the eye to the brain (the optic nerve) is untreatable and causes blindness. However, factors causing this damage are potentially treatable and it is this that makes the glaucoma preventable. The problem is that the disease can progress without symptoms until it is too late, and the causal factors are not fully understood. Scientists used to think that high pressure inside eyeballs (intraocular pressure, IOP) was the main cause of this optic nerve damage. Since sufferers of glaucoma include people with normal IOP, they have now recognized that other factors must also be involved. Helping to identify these factors was the motivation behind the PhD work of Pinakin Gunvant (Pinakin, 2002). As part of a wide-ranging study of the measurement of and patterns shown by variables connected with the biology of the eye, he found that corneal thickness and corneal curvature are both important predictors of glaucoma. Furthermore, he found that these factors are related to the rate of blood flow to the eye (known as pulsatile ocular blood flow, POBF).

We have already seen some data on IOP and corneal thickness from Pinakin's work in Chapter 8 (Table 8.1). In this chapter, we are going to use an extract from his data set on pressure in an eyeball (IOP) and blood flow to that eyeball (POBF), to demonstrate correlation. These data are presented in Table 11.3. They show the IOP and POBF of one eyeball from each of 20 people. Whether the eyeball was the left or the right was a random choice.

There are two samples in Table 11.3, one from each of two variables. One is from the pressure in the eyeball, IOP, measured in terms of pressure, in units of millimetres of mercury (mmHg). The other variable is the rate of blood flow to the eye, POBF, measured as microlitres per minute (μl/min). Both variables are measured at the scale level. Histograms for each sample are presented in Figure 11.2 and a scatterplot of their relationship is shown in Figure 11.3. The former indicates that the data are normally distributed and that the parametric criteria can therefore be assumed to apply (Sections 5.3 and 11.1.3). The latter indicates that an assumption that the relationship is linear is reasonable.

Subject or ID	IOP (mmHg)	POBF (μl/min)
1	15.3	858
2	11.3	1192
3	17.3	987
4	14.4	1264
5	15.1	1372
6	18.6	649
7	18.1	979
8	21.7	523
9	13.8	1177
10	9.2	1060
11	7.5	1761
12	22.7	618
13	15.8	1484
14	15.6	1062
15	15.7	1038
16	12.3	1116
17	22.3	1066
18	12.3	1267
19	18.2	465
20	10.1	948

Source of data: courtesy of Pinakin Gunvant, Russell Watkins, and Dan O'Leary.

Table 11.3 Example data: eyeballs ($n = 20$). Intraocular pressure (IOP) and pulsatile ocular blood flow (POBF) pressure in 20 humans.

11.3 Pearson correlation test

The topic of this section is the Pearson correlation test. It covers:

- When to use and when not to use a Pearson correlation.
- The four steps as applied to a Pearson correlation test.
- A worked example using SPSS, including how to find the coefficient of determination for correlation.
- An example of the use of a Pearson correlation from the literature.

The data used in the worked example come from Pinakin Gunvant's work and were introduced in Section 11.2.

11.3.1 When to use

A Pearson correlation is a parametric test for assessing whether the linear relationship between two samples can be accounted for by sample error alone. Since it is a parametric test your data must fulfil the parametric criteria (Sections 5.3 and 11.1.3) before proceeding. You also need to look at a scatterplot of the data to check that a linear model might be reasonable. Issues relating to when and when not to use this test were covered in the introduction to this chapter (Section 11.1). They are summarized in Box 11.1.

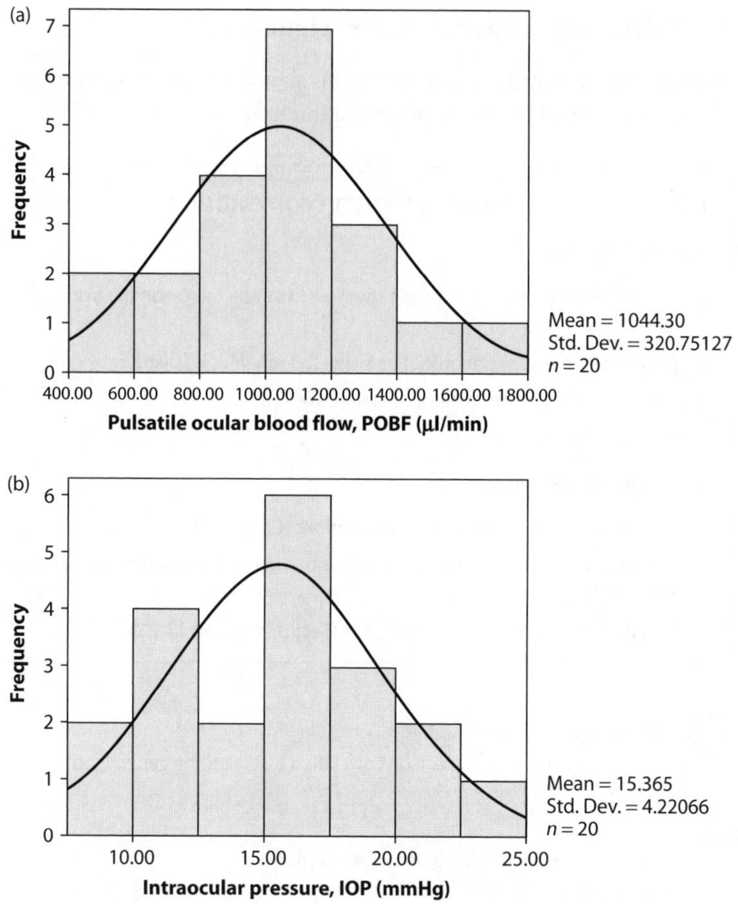

Figure 11.2 Histograms of example data: eyeballs. Normal distributions are indicated by the curved lines. (a) Pulsatile ocular blood flow ($n = 20$). (b) Intraocular pressure ($n = 20$).

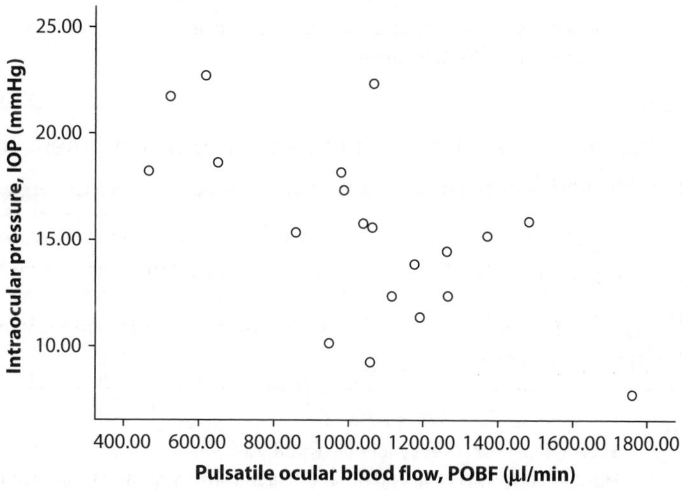

Figure 11.3 Variation of pulsatile ocular blood flow with intraocular pressure ($n = 20$).

11.3.2 Four steps of a Pearson correlation test

Here are the four steps of null hypothesis significance testing outlined in Chapter 5, specifically applied to a Pearson correlation test.

Box 11.1 When to use a Pearson correlation test

Use this test when:

- You are looking for a *relationship* between *two* samples, one sample from each of *two variables*.
- You can assume any relationship between the variables is *linear*.
- Data in the samples are measured at the *scale* level.
- Data are *related*.

Do not use this test when:

- You want to compare frequency distributions (Chapter 6).
- You want to interpret your results in terms of a dependent and an independent variable (Chapter 10).
- Data do not fulfil the parametric criteria (Sections 5.3 and 11.1.3).

STEP 1 State the null hypothesis (H_0)
The null hypothesis for a Pearson correlation test takes the general form:

H_0: The two variables do not covary in a linear fashion.

STEP 2 Choose a critical significance level (α)
Typically this is 5% (0.05).

STEP 3 Calculate the test statistic
For a Pearson test the statistic is *r*. This statistic was introduced in Section 11.1.4. It is called a correlation coefficient. The formula for Pearson correlation coefficient, *r*, shown in Box 11.2. The degrees of freedom equals the number of pairs of data minus two. The number of pairs will be the size of either sample.

Box 11.2 The formula for Pearson's product-moment correlation coefficient

$$r = \frac{\sum (y-\bar{y})(z-\bar{z})}{\sqrt{\sum (y-\bar{y})^2 \sum (z-\bar{z})^2}}$$

Where: $\bar{y}$ = mean of sample from variable 1;
$\bar{z}$ = mean of sample from variable 2.

Degrees of freedom = $n - 2$, where n = size of first sample = size of second sample = number of pairs of data points.

STEP 4 Reject or accept your null hypothesis
How you do this will depend on whether you are doing the calculations by hand or using a statistical computing package.

Using critical-value tables

First you need to look up the critical value of r given the critical significance level you have chosen in step 2 and the degrees of freedom calculated in step 3 (Appendix II, Table AII.7). Then you need to compare your value for r calculated in step 3 with this critical value. Finally, you should decide to reject or accept your null hypothesis according to the rule in Box 11.3 . Appendix I has notes on how to enter data into SPSS.

> **Box 11.3** Decision using critical values for a Pearson correlation test
>
> If $r \geq r_{critical} \rightarrow$ reject $H_0 \rightarrow$ significant result.
> If $r < r_{critical} \rightarrow$ accept $H_0 \rightarrow$ non-significant result.

Using *P* values on computer output

Find *P* on the computer output and make your decision according to the rule in Box 11.4.

> **Box 11.4** Decision using *P* values for a Pearson correlation test
>
> If $P \leq \alpha \rightarrow$ reject $H_0 \rightarrow$ significant result.
> If $P > \alpha \rightarrow$ accept $H_0 \rightarrow$ non-significant result.

11.3.3 **Worked example: using SPSS**

online resource centre

We are going to do a Pearson correlation test on the eyeball data (Section 11.2) using SPSS. To do this the data must be entered into SPSS as shown in Figure 11.4. In newer versions of SPSS the **Data Editor** window has a **Variable View** tab and a **Data View** tab. Figure 11.4a shows the former and Figure 11.4b the latter.

STEP 1 State the null hypothesis (H_0)

H_0: Intraocular pressure (IOP) and pulsatile ocular blood flow (POBF) do not covary in a linear fashion.

STEP 2 Choose a critical significance level (α)

We will use 5% (0.05).

STEP 3 Calculate the test statistic

To get SPSS to conduct a Pearson correlation on your data you must first open the data file. Then you must make the following selections:

Analyze
→ Correlate
  → Bivariate...

A window like that shown in Figure 11.5 will appear. You need to select the variables that you want to include in your test and send them over to the **Variables** box. For our

(a)

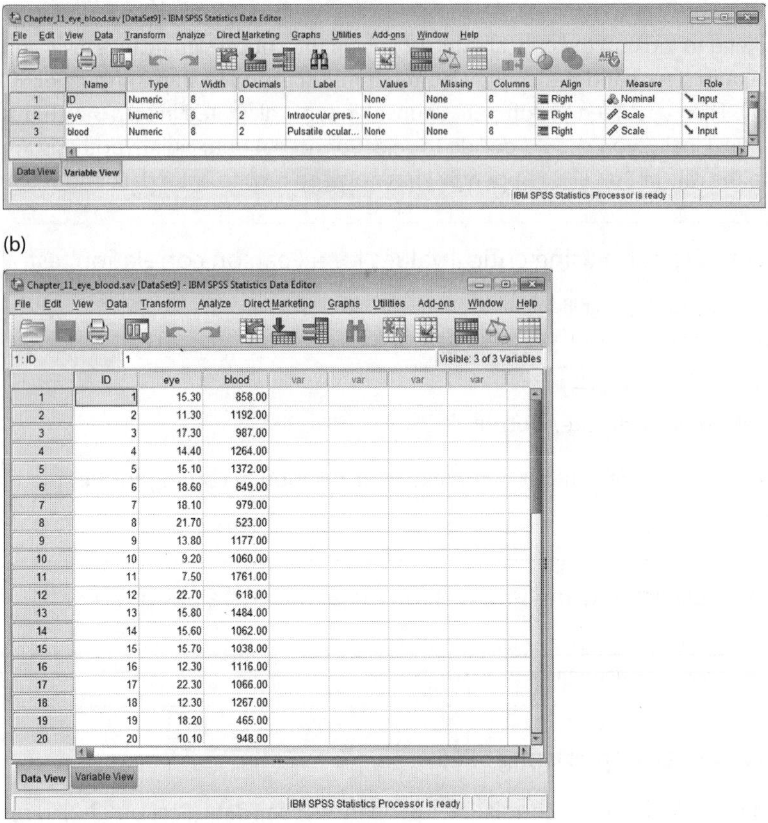

Figure 11.4 Example data in SPSS: eyeballs. (a) Variable View. (b) Data View.

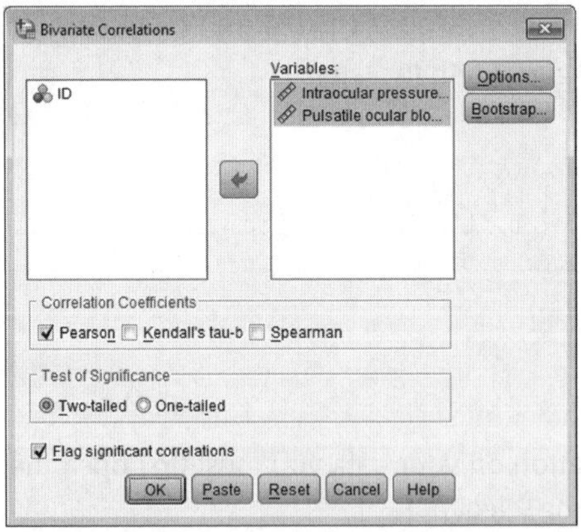

Figure 11.5 Conducting a Pearson correlation using SPSS. Main dialogue window.

example data these are *Intraocular pressure* and *Pulsatile ocular blood flow*. You can just ignore the *ID* column. Check that **Pearson** is selected under **Correlation Coefficients**. Once you have done this you can click **OK** and immediately you will get output like Figure 11.6. The key information is: $r_{18} = -0.642$, $P = 0.002$, remembering that the degrees of freedom is two less than the number of pairs of data points (Box 11.2).

Correlations

Correlations

		Intraocular pressure, IOP (mmHg)	Pulsatile ocular blood flow, POBF (microlitres/min)
Intraocular pressure, IOP (mmHg)	Pearson Correlation	1.000	−.642**
	Sig. (2-tailed)		.002
	N	20.000	20
Pulsatile ocular blood flow, POBF (microlitres/min)	Pearson Correlation	−.642**	1.000
	Sig. (2-tailed)	.002	
	N	20	20.000

**Correlation is significant at the 0.01 level (2-tailed).

Statistic (*r*)

P

Size of samples (N)

Figure 11.6 SPSS output for a Pearson correlation with key information annotated.

STEP 4 Reject or accept your null hypothesis using *P* values on the computer output

Since the critical significance level of 0.05 is larger than the value 0.002 of *P*, we must reject the null hypothesis according to Box 11.4.

In short:

$P(0.002) < \alpha(0.05) \rightarrow$ reject $H_0 \rightarrow$ significant result.

We can conclude that intraocular pressure (IOP) and pulsatile ocular blood flow (POBF) are negatively correlated. We know that it is a negative correlation because the sign of the correlation coefficient, *r*, is negative. This means that as IOP increases, POBF decreases. (Pearson correlation $r_{18} = -0.642$, $P = 0.002$).

The coefficient of determination for correlation in this example is the square of -0.642, which is 0.412. In other words, 41.2% of the variation in one of the variables is related to the other.

11.3.4 **Literature link: male sacrifice**

online resource centre

Snow and Andrade (2004) Behav. Ecol. 15: 785-792 (OUP).

Redback spiders are delightful little creatures that lurk, among other places, in Australian toilets. The female redback spider is particularly engaging because she usually cannibalizes the male of her species during copulation. By giving the female something to munch on he prolongs copulation and increases his chances of paternity compared with non-cannibalized suitors. An assumption has been that longer copulation allows the male to transfer more sperm, thus giving him an advantage if his sperm have to compete for access to eggs with the sperm of other males who have copulated with the same female.

Snow and Andrade (2004) decided to test this assumption. One way in which they did this was to see if the duration of copulation in redback spiders correlated with the number of sperm transferred. They found that the 'absolute number of sperm transferred to the female's spermatheca was not correlated with copulation duration ($R^2 = 0.012$, $p = 0.449$, $n = 36$) (Figure 3)'. A spermatheca is a sac-like bit of a female spider's anatomy where she stores sperm.

The authors had to conclude that the amount of sperm transferred to the spermatheca and the duration of copulation were not correlated because of the high *P* value, much higher than a typical critical significance level of 0.05. The authors chose to report the coefficient of determination rather than the correlation coefficient (Section 11.1.4), which they reported as R^2. However, we can deduce that the correlation coefficient would have been ± 0.110, that is, the square root of 0.012. The figure they cite to further support this finding, their Figure 3, is a scatterplot. The result that the amount of sperm transferred and copulation duration were not correlated suggests that the assumption that redback males get a paternity advantage from longer copulation, because it allows them to transfer more sperm to the female, is false.

Earlier in the paper, under analyses in the methods section, we are told: 'All statistical and power analyses were completed by using Systat 10.2 (SPSS 2002) with consultation from Zar (1984).' Systat is a statistical software package whose makers are connected commercially with SPSS. The number 10.2 tells us what version of Systat they used. Zar (1984) is a heavyweight statistical textbook, the latest edition of which was published in 1999 and is listed in the Selected further reading section at the end of this book.

The distributions of the two variables involved in the correlation are not mentioned, but it is implied that they were normal and did not require transformation for use in a parametric test. We are also not given any specific information about the correlation procedure used, but can tell that it must be parametric from the use of the coefficient of determination.

11.4 Spearman correlation test

The topic of this section is the Spearman correlation test. It covers:

- When to use and when not to use a Spearman correlation.
- The four steps as applied to a Spearman correlation test.
- A worked example using SPSS.
- An example of the use of a Pearson correlation in the literature.

The data used in the worked example come from Pinakin Gunvant's work and were introduced in Section 11.2.

11.4.1 **When to use**

A Spearman correlation is a nonparametric test for assessing whether the linear relationship between two variables can be accounted for by sample error alone. You need to look at a scatterplot of the data to check that a linear model might be reasonable.

Issues relating to when and when not to use this test were covered in the introduction to this chapter (Section 11.1). They are summarized in Box 11.5.

Box 11.5 When to use a Spearman correlation test

Use this test when:

- You are looking for a *relationship* between *two* samples, one sample from each of *two variables*.
- You can assume that any relationship between the variables is *linear*.
- Data in the samples are measured at the *ordinal* or *scale* level.
- Data are *related*.

Do not use this test when:

- You want to compare frequency distributions (Chapter 6).
- You want to interpret your results in terms of a dependent and an independent variable (Chapter 10).

11.4.2 **Four steps of a Spearman correlation test**

Here are the four steps of null hypothesis significance outlined in Chapter 5, specifically applied to a Spearman correlation test.

STEP 1 State the null hypothesis (H_0)
The null hypothesis for a Spearman correlation test takes the general form:

H_0: The two variables do not covary in a linear fashion.

STEP 2 Choose a critical significance level (α)
Typically this is 5% (0.05).

STEP 3 Calculate the test statistic
For a Spearman test the statistic is Spearman's rho (r_s). This statistic was introduced in Section 11.1.4. It is called a correlation coefficient. The formula for the Spearman correlation coefficient is shown in Box 11.6. The formula is exactly the same as for the Pearson correlation coefficient (Box 11.2), except that data are converted to ranks as for the Mann-Whitney U test and the Wilcoxon signed-rank test in Chapters 7 and 8. The other piece of important information is the number of pairs of data points (N).

Box 11.6 The formula for the Spearman rank-order correlation coefficient

$$r_s = \frac{\sum(y-\bar{y})(z-\bar{z})}{\sqrt{\sum(y-\bar{y})\sum(z-\bar{z})}}$$

Where: $\bar{y}$ = mean rank of the sample from variable 1; $\bar{z}$ = mean rank of the sample from variable 2.

N = number of pairs of data points.

STEP 4 Reject or accept your null hypothesis
How you do this will depend on whether you are doing the calculations by hand or using a statistical computing package.

Using critical-value tables

First you need to look up the critical value of r_s given the critical significance level you have chosen in step 2 and the degrees of freedom calculated in step 3 (Appendix II, Table AII.8). Then you need to compare the value of r_s calculated in step 3 with this critical value. Finally, you should decide to reject or accept your null hypothesis according to the rule in Box 11.7.

Box 11.7 Decision using critical values for a Spearman rank-order correlation test

If $r_s \geq r_{s\,critical} \rightarrow$ reject $H_0 \rightarrow$ significant result.
If $r_s < r_{s\,critical} \rightarrow$ accept $H_0 \rightarrow$ non-significant result.

Using *P* values on computer output

Find *P* on the computer output and make your decision according to the rule in Box 11.8.

Box 11.8 Decision using *P* values for a Spearman rank-order correlation test

If $P \leq \alpha \rightarrow$ reject $H_0 \rightarrow$ significant result.
If $P > \alpha \rightarrow$ accept $H_0 \rightarrow$ non-significant result.

11.4.3 **Worked example: using SPSS**

online resource centre

We are going to do a Spearman correlation test on the eyeball data (Section 11.2) using SPSS. To do this the data must be entered into SPSS as for the Pearson test (Section 11.3.3, Figures 11.4a and 11.4b).

STEP 1 State the null hypothesis (H_0)

H_0: Intraocular pressure (IOP) and pulsatile ocular blood flow (POBF) do not covary in a linear fashion.

STEP 2 Choose a critical significance level (α)
We will use 5% (0.05).

STEP 3 Calculate the test statistic

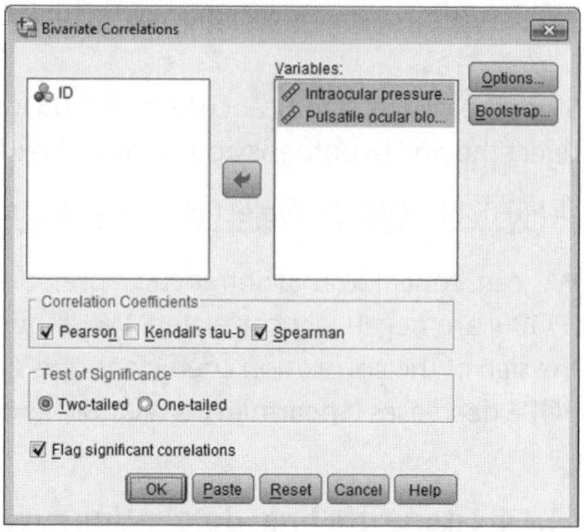

Figure 11.7 Conducting a Spearman correlation using SPSS. Main dialogue window.

To get SPSS to conduct a Spearman correlation on your data you must first open the data file. Then you must make the following selections:

Analyze
→ Correlate
→ Bivariate

A window like that shown in Figure 11.7 will appear. You need to select the variables that you want to include in your test and send them over to the **Variables** box on the window together. For our example data these are *Intraocular pressure* and *Pulsatile ocular blood flow*. You can just ignore the ID column. Check that the Spearman option is selected under **Correlation Coefficients** (you can either click Pearson to deselect it or do both tests at the same time). Once you have done this you can click **OK** and immediately you will get output like Figure 11.8. The key information is: $r_s = -0.576$, N = 20, $P = 0.008$.

Figure 11.8 SPSS output for a Spearman correlation with key information annotated.

Nonparametric Correlations

Correlations

			Intraocular pressure, IOP (mmHg)	Pulsatile ocular blood flow, POBF (microlitres/min)	
Spearman's rho	Intraocular pressure, IOP (mmHg)	Correlation Coefficient	1.000	−.576**	← **Statistic (r_s)**
		Sig. (2-tailed)	.	.008	← ***P***
		N	20	20	← **Size of sample (N)**
	Pulsatile ocular blood flow, POBF (microlitres/min)	Correlation Coefficient	−.576**	1.000	
		Sig. (2-tailed)	.008		
		N	20	20	

**Correlation is significant at the 0.01 level (2-tailed).

STEP 4 Reject or accept your null hypothesis using the *P* values on the computer output

Since the critical significance level of 0.05 is larger than the value 0.008 of *P*, we must reject the null hypothesis according to Box 11.8. In short:

$P(0.008) < (0.05) \rightarrow$ reject $H_0 \rightarrow$ significant result.

We can conclude that intraocular pressure (IOP) and pulsatile ocular blood flow (POBF) are negatively correlated. We know that it is a negative correlation because the sign of the correlation coefficient, r_s, is negative. This means that as IOP increases, POBF decreases (Spearman correlation: $r_s = -0.576$, $N = 20$, $P = 0.008$).

11.4.4 **Literature link: defoliating ryegrass**

online resource centre

Morvan-Bertrand *et al.* (1999) J. Exp. Bot. 50:1817–1826 (OUP).

Perennial ryegrass *(Lolium perenne)* is common on disturbed ground, such as roadsides and cultivated fields, in Europe and North America. Morvan-Bertrand *et al.* (1999) investigated the influence of various biochemical factors on the speed of regrowth of ryegrass after cutting it down to near ground level. The act of cutting defoliates the grass, just as a large grazing animal or a lawn mower would do, and a plant which can show speedy regrowth after such treatment will be at an adaptive advantage.

In their methods, Morvan-Bertrand *et al.* (1999) tell us that 'Significant differences are reported for $P < 0.05$'. In other words, they used a critical significance level of 0.05. They go on to say, 'The correlation between initial parameters and regrowth after defoliation were analysed by Spearman correlations (StatView, 4.02). A nonparametric correlation was used because the data were not normally distributed.' StatView is another statistical package, like SPSS; the number 4.02 indicates the version of this software that they used.

The initial parameters they measured were the contents of various biochemicals in the roots, leaves, and leaf bases per tiller (grass plant) at defoliation. One of these was the sugar fructans. To assess regrowth they used production of leaf dry matter per tiller, in units of milligrams of dry weight after 2, 6, and 28 days. They presented the results of their correlations between regrowth and initial parameters in one big table (Morvan-Bertrand *et al.* 1999, Table 6). The following are examples of Spearman correlation coefficients they found between production of leaf dry matter per tiller after 2 days and initial fructan per tiller from this table:

Initial fructans per tiller in roots, $r_s = 0.274$ n.s.

Initial fructans per tiller in leaf sheaths, $r_s = 0.604$*

Initial fructans per tiller in elongating leaf bases, $r_s = 0.808$***

They used n.s. to indicate non significance and asterisks to indicate significance, specifically, * for $P <$ or $= 0.05$, ** for $P <$ or $= 0.01$, *** for $P <$ or $= 0.001$). They tell us what these codes mean and that the value of N is 18 in the table legend (although they use the symbol n rather than N). These results indicate that initial fructan levels and regrowth were not correlated in the roots but were positively correlated in the leaf sheaths and elongating leaf bases.

11.5 Comparison of correlation and regression

Regression and correlation are both used to investigate relationships between variables. Regression is used to investigate the dependence of one variable on another. Correlation is used to investigate the interdependence between two variables.

Mathematically, the connection between the parametric regression analysis we dealt with in Chapter 10 and the parametric correlation we have looked at in this chapter is close. In fact, in practice, for these particular techniques the coefficient of determination for regression and correlation are the same. Just for fun, you could try doing a Pearson's correlation analysis for the species richness and reserve area example data from Chapter 10. You will find $r = 0.884$ and, therefore, $r^2 = 0.78$ (to two decimal places) for the correlation compared with $R = 0.884$ and $R^2 = 0.782$ in the Model Summary table in Figure 10.5.

Nevertheless, it is very important that you apply regression and correlation appropriately even in situations where in practice it would make no difference to the number produced, for two main reasons. Firstly, it will assist you in making relevant interpretations of your findings. Secondly, it will make you better placed to learn extension techniques in which regression and correlation give different results in calculation as well as concept. Unfortunately you will not always find good practice on the correct application of regression and correlation in the primary literature.

The regression and correlation techniques we have looked at are compared in Table 11.4 according to a range of features discussed in Chapters 9, 10, and 11.

Checklist of key points

- The Pearson correlation and the Spearman correlation are used to test for a relationship between two variables.
- You should consider these tests when:
  - the numbers you see when looking at the data are measured at the ordinal or scale level;
  - the numbers in one sample come from one variable and numbers in the other sample come from a different variable.
- The minimum level of measurement needed for the Pearson correlation is scale and for a Spearman correlation it is ordinal.
- For either a Pearson or a Spearman test to work the data must potentially fall along a straight line. This can be checked by visual inspection of a scatterplot.
- The Pearson correlation is parametric and the Spearman correlation is nonparametric. The parametric criterion for the Pearson test is that the two variables come from a bivariate normal distribution.

Table 11.4 Comparison of simple bivariate linear regression and correlation.

	Regression	Correlation
Variables involved	Dependent (*x*) and independent (*y*)	Variable one and variable two (*y* and *z*)
Nature of variables: random or fixed (Sections 9.5 and 10.5)	**For model I:** *y* should be random and *x* fixed. **For model II:** Both *x* and *y* should be random. (NB: only model I is considered in detail in this book.)	For correlation both *y* and *z* should be random.
Nature of variables: levels of measurement (Section 2.3.2)	**For parametric:** Both *x* and *y* should be scale. Technically *y* should be continuous scale but you can usually get away with it being discrete. **For nonparametric:** both *x* and *y* must be at least ordinal. (NB: nonparametric alternatives are not commonly used and not dealt with in detail in this book.)	**For Parametric:** technically both *y* and *z* should be continuous scale but you can usually get away with either or both being discrete. **For nonparametric:** both *y* and *z* must be at least ordinal.
Relationship investigated	The extent that variation in the dependent variable, *y*, depends on the independent variable, *x*.	Interdependence; the extent two variables, *y* and *z*, covary.
Model used	Linear model. The values of b and c in the equation for a line can be calculated, for example, based on ordinary least squares method.	Linear model. Model calculations are not straightforward and not dealt with in this book.
Used in prediction?	Yes	No
NHST assesses	If the model based on *x* explains a significant amount of the variation in *y*. Involves the calculation of F statistic (parametric).	If *y* and *z* covary significantly. Involves the calculation of the correlation coefficient, *r* (parametric) or r_s (nonparametric)
Coefficient of determination	Can be calculated for parametric situations. Tells us how much of the variation in the dependent variable is explained by variation of the independent variable.	Can be calculated for parametric situations. Tells us to what extent the two variables covary.

- The four steps of a correlation test are as follows:
  1. State H_0: H_0 = there is no relationship between the populations.
  2. Choose a critical significance level: typically $\alpha = 0.05$.
  3. Calculate your statistic: the statistic is called a correlation coefficient. For Pearson correlation, calculate *r* and the degrees of freedom according to Box 11.2. For Spearman, calculate r_s and number of pairs (N) according to Box 11.6.
  4. Reject or accept H_0: For Pearson correlation, reject H_0 if $r \geq r_{critical}$ (Box 11.3) or if $P \leq 0.05$ (Box 11.4). For Spearman correlation, reject H_0 if $r_s \geq r_{s\ critical}$ (Box 11.7) or if $P \leq 0.05$ (Box 11.8).
- If you use Pearson correlation you can square the correlation coefficient to give a coefficient of determination.

Self-help questions

1. For each of the following features say whether it applies to a Pearson correlation, a Spearman correlation, both, or neither.
   (a) Parametric test. *Link back: Section 11.1*
   (b) Nonparametric test. *Link back: Section 11.1*
   (c) Numbers plugged into the formula for the statistic are frequencies. *Link back: Section 11.1*

(d) Numbers plugged into the formula for the statistic are ranks. *Link back: Section 11.1*
(e) Used on unrelated data. *Link back: Section 11.1.1*
(f) Used on related data. *Link back: Section 11.1.1*
(g) Uses a statistic called a correlation coefficient. *Link back: Section 11.1.4*
(h) Uses the *r* statistic. *Link back: Section 11.3.3*
(i) Uses the r_s statistic. *Link back: Section 11.4.2*
(j) Reject the null hypothesis if the calculated statistic is greater than or equal to the critical value of the statistic looked up in a table. *Link back: Sections 11.3.2 and 11.4.2*
(k) Reject the null hypothesis if the calculated statistic is less than or equal to the critical value of the statistic looked up in a table. *Link back: Sections 11.3.2 and 11.4.2*
(l) Reject the null hypothesis if *P* on an SPSS printout is less than 0.05. *Link back: Sections 11.3.2 and 11.4.2*

2. A researcher recording the vocalizations of squirrel monkeys in the South American jungle wanted to know if the peak frequency of a call and the duration of that call were correlated. He used SPSS to perform a Spearman correlation. The output he got is shown in Figure 11.9.
(a) What is the key information he would need to report? *Link back: Section 11.4.3*
(b) What would he have concluded from this result? *Link back: Section 11.4.3*
(c) The data were scale level so it is not this that stopped him doing a Pearson correlation. Comment on why he might have chosen to do a nonparametric test and how he might have overcome this. *Link back: Sections 5.3 and 11.1.3*

Nonparametric Correlations

Correlations

			Peak Frequency (Hz)	Duration (ms)
Spearman's rho	Peak Frequency (Hz)	Correlation Coefficient	1.000	−.295
		Sig. (2-tailed)	.	.352
		N	12	12
	Duration (ms)	Correlation Coefficient	−.295	1.000
		Sig. (2-tailed)	.352	.
		N	12	12

Correlations

Correlations

		Peak Frequency (Hz)	Duration (ms)
Peak Frequency (Hz)	Pearson Correlation	1.000	−.314
	Sig. (2-tailed)		.321
	N	12.000	12
Duration (ms)	Pearson Correlation	−.314	1.000
	Sig. (2-tailed)	.321	
	N	12	12.000

Figure 11.9 SPSS output referred to in self-help question 2.

12 Introducing the Generalized Linear Model: General Linear Model

Chapter aims

The subject of this chapter is the generalized linear model. In particular, the sub-framework known as the general linear model, and how it can be used to assess the relative effects of multiple explanatory (independent) variables on a single response (dependent) variable, which is measured at scale level. We start by dealing with general points relating to the use of the general linear model and consider some example data before looking in more detail at when and how to use the general linear model. This includes reviewing lessons learned from the four-step null hypothesis significance testing procedure introduced in Chapter 5, and introducing you to a modified approach more appropriate for use with the general linear model. We work through a basic general linear model analysis using SPSS and look at an example of the use of the general linear model from the literature. Finally, we will look at some additional features of the general linear model: its use in a multiple model approach and its extension into the generalized linear model.

The **Generalized Linear Model (GLZM)** is an extremely useful, and increasingly popular, framework approach to analysing data. The **General Linear Model (GLM)** is a sub-framework of the generalized linear model which uses the normal distribution. (There are other models within the generalized linear model that assume other distributions.) The abbreviation **GLM** can be used for either, although in this textbook it is only used to refer to the general linear model.

Notes on terminology

The three letters **GLM** are commonly used as an abbreviation for both the **general linear model** and the **generalized linear model**. In this book I will only use **GLZM** to refer specifically to the generalized linear model and **GLM** just for the general linear model. You will have to use the context to judge which one applies in other books and papers.

In Section 2.2.2 I mentioned that a variety of terms can be used instead of the words dependent variable and independent variable. In GLM the alternatives

response variable, for a **dependent variable**, and **explanatory variable**, for an **independent variable**, are more typically used and I'll be changing to this convention for this chapter.

Response variable, test variable, outcome variable, or **data variable** are alternative names for a **dependent variable**. **Explanatory variable, grouping variable, predictor variable, factor**, or **covariate** are alternative terms for an **independent variable**.

12.1 Introduction to the general linear model

This chapter introduces the **generalized linear model**. The generalized linear model is an extremely useful, and increasingly popular, framework approach to analysing data. Since it relies on making assumptions about the distribution of data it is parametric. In particular, we are going to use the **general linear model (GLM** – see notes on terminology). GLM is a sub-framework of the generalized linear model which uses the normal distribution.

Procedures that fall under the umbrella of the generalized linear model but not GLM, make assumptions involving other types of distribution (Section 3.3.5). We will consider two of these in more detail in Chapter 13: the binary logistic model, which uses the binomial distribution, and the loglinear model, which uses the Poisson distribution.

Historically, there has been much more work done by statisticians to develop the use of GLM compared with the broader generalized linear model. Combined with the fact that the normal distribution is so widely occurring in biological data (Section 3.3.3), the GLM framework will cover much of your needs. However, once you have mastered this you will be well placed to go on to learn how to use other sub-frameworks within the generalized linear model (Section 12.7 and Chapter 13).

GLM incorporates all the most commonly used parametric analyses, such as *t*-tests, Anova, and linear regression, in a single framework. In other words, GLM can be used to test for differences between two or more samples (as covered in Chapters 9 and 10) and for relationships between two samples (as covered in Chapter 11). GLM can also be used to do both at the same time (which is also known as analysis of covariance or Ancova: see Section 9.1.4), and it can handle more than two explanatory variables (either multi-way Anova, see Section 9.1.4 and/or multiple regression, see Section 10.1.6).

A great advantage of GLM is that you can carry out all the analyses we have looked at in previous chapters, including extensions that involve multiple explanatory variables, under a single framework; thus it radically reduces what you have to remember in order to carry out your tests. Moreover, once you have mastered the procedure, you will be in an excellent position to take a more sophisticated modelling approach to data analyses (Section 12.6). A disadvantage of using the GLM is that it requires computer access and knowledge of how to use a statistical package, such as SPSS and R, and is not easily done using a calculator or Excel spreadsheet. Moreover, you still have to know about the traditional tests, because you will read about them in the literature and you will need to be able to understand and critically evaluate their use. Also, GLMs are not reliable with small samples.

In this chapter we are going to use the GLM to analyse data sets where there is one scale level response variable (univariate GLM) and where data are not related (Section 2.3.1). The GLM can handle multiple response and related data and the basics you learn here will help you to learn these techniques should you need them (see Selected further reading).

In the rest of this introduction I am going to emphasize some general points about the GLM as listed below.

- The type of data on which GLM works.
- The idea of a word equation.
- The idea of GLM.
- The language of GLM.
- The things you can find out by performing GLM analysis.
- The assumptions of GLM.

12.1.1 **Variables and levels of measurement**

Data sets appropriate for analyses with the GLM framework are varied in their structure, but in all cases the response variable must be measured at scale level and can reasonably be assumed to be normally distributed (Section 3.3.3). There will then be one or more explanatory variables measured at nominal, ordinal, or scale level. Data sets suitable for analyses using GLM include those on which Anovas (Section 9.1.1) and bivariate linear regression (Section 10.1.1) can be conducted.

Data from a **categorical variable**, or **qualitative variable**, are measured using one of a limited number of categories. Data from a **continuous variable**, or **quantitative variable**, can fall anywhere along a continuum.

When considering a data set for GLM, it is important to be able to tell whether the way explanatory variables are measured leads them to be best treated as **categorical** (**qualitative**) or **continuous** (**quantitative**) (Section 2.3.3). For *t*-tests and Anova, the explanatory variable data can be measured at any level, provided that this can be used to distinguish a small number of categories (two in the case of a *t*-test). An example here is the variable colour when measured as red, white, or blue. For linear regression, the explanatory variable must be measured at continuous scale level. An example here is the dosage of a medicine measured in micrograms. In a GLM you can have one or more categorical or continuous explanatory variables, or indeed a mixture.

Here are some examples. If you had the yield of wheat per acre from fifty farms, and information on which of three alternative varieties were grown, and whether the regime they were grown under was organic or not, you could use a GLM to assess the effect of the two categorical explanatory variables, *variety* and *regime*, on the response variable, *yield*. You could also analyse this using two-way Anova (Section 9.1.4). If you had the flight speed of fifty blue tits, their weight, and the ambient temperature, you could use a GLM to assess the effect of the two continuous explanatory variables, *weight* and *temperature*, on the response variable, *flight speed*. You could also analyse this using multiple regression (Section 10.1.6). If you had the waist measurement of fifty adult humans, their sex, and the amount of time they spent exercising per week, you could use a GLM to assess the effect of the two explanatory variables, *sex* (categorical: male or female) and *time spent exercising* (continuous) on the response variable, *waist size*. You could also analyse this using Ancova (Section 9.1.4).

12.1.2 **The linear model revisited**

We have already come across the idea of a **linear model** in Chapter 10. In particular you should consider (re)reading Section 10.1.2, to refresh your memory. A model is nothing scary, simply a summary of data in word, equation, or graphical form. A model provides, or attempts to provide, a summary description of patterns in your data, so that you can get a better understanding of what is going on biologically. **Linear** simply means that the model is based on the equation for a line.

A **model** seeks to summarize patterns in data so that you can get a better understanding of what is going on biologically. **Linear** model simply means the model is based on the equation for a straight line.

When using univariate GLM, you are dealing with models that are seeking to describe the influence of one or more explanatory variables on a single response variable. Specifically, a GLM uses these explanatory variables to model the mean of the response variable. It is useful to get into the habit of writing out equations for any model you are considering in words. We are going to call these **word equations**. Writing word equations is very straightforward, and just means writing out what you want to look at in words using an equals sign after the response variable and plus signs between explanatory variables.

A **word equation** is a representation of a model using words and simple mathematical symbols. For a GLM, the response variable goes to the left of an equals sign and the explanatory variable(s) to the right. Explanatory variables are separated by plus signs.

The general form of a word formula for a model involving two explanatory variables would be

Response variable = Explanatory variable 1 + Explanatory variable 2.

It is also worth noting whether you are going to treat each explanatory variable as categorical or continuous. Look at Table 12.1 for examples, some of which link back to previous chapters.

As I have already mentioned, one of the special features of GLM is that it brings together a range of traditional analyses under one umbrella procedure. How does this work? Let us look at traditional bivariate linear regression and one-way Anova to get some insight. In Chapter 10, I explained that we can think of a bivariate linear regression test as a way of assessing whether our data are best modelled by a single horizontal straight line (as defined by the null hypothesis), or a single sloping straight line. Well, we can actually think of a *t*-test or a one-way Anova in a similar way: it allows us to assess whether our data are best modelled by a single horizontal straight line (as defined by the null hypothesis), or a series of separate horizontal straight lines, one for each of the categories of the explanatory variable. Have a look at Figures 12.1 and 12.2: they show the data on brine shrimp from Chapter 7 and weight and fat from Chapter 10, presented in such a way as to illustrate these ideas.

You should not expect your data to fall exactly on the line or lines defined by a model. The differences between data points and the line or lines of a model are known as **residuals** or **errors**. We have, in fact, already come across the idea of residuals in Section 10.1.2. and Figure 10.1, in the context of regression. Take a look at Figures 12.1 and 12.2 to compare residuals in regression with residuals in a *t*-test/one-way Anova and see that they represent the same thing in both.

The differences between a data point and the line, or lines, of a model are known as **residuals** or **errors**. A GLM models the mean of the response variable and assumes that residuals around the model are normally distributed. All GLMs have **error terms** which represent this, even though they are not always explicitly written down in equations.

As I said earlier, a GLM models the mean of the response variable using one or more explanatory variables. What a GLM does is assume that the residuals around this model are normally distributed. This makes sense if the response variable itself is normally distributed. Strictly speaking all our equations, both the word equations

Example	Word equation	Traditional analysis
Does bone density differ between males and females? (Section 7.2)	*Bone density* = *Sex* Where *Sex* is categorical	*t*-test
Does the nitrogen content of reeds differ between sites in a reed bed? (Section 9.2)	*Nitrogen content* = *Site* Where Site is categorical	One-way analysis of variance (Anova)
Is species richness related to game reserve size? (Section 10.2)	*Species richness* = *Reserve size* Where *Reserve size* is continuous	Bivariate linear regression
Does wheat yield differ depending on variety and/or regime (organic or not)? (Section 12.1.1)	*Yield* = *Variety* + *Regime* Where *Variety* and *Regime* are categorical	Two-way Anova
Is flight speed in blue tits related to their weight and/or the ambient temperature? (Section 12.1.1)	*Flight speed* = *Weight* + *Temperature* Where *Weight* and *Temperature* are continuous	Multiple regression
Does waist size differ between males and females and/or is it related to time spent exercising? (Section 12.1.1)	*Waist size* = *Sex* + *Time exercising* Where *Sex* is categorical and *Time exercising* is continuous	Analysis of covariance (Ancova)

Table 12.1 Example word equations with the names of the traditional test that could be used to analyse the data.

below and the mathematical ones we are about to come on to, should include an **error term** that represents this, but I have omitted this for the sake of simplicity.

The model, in mathematical equation form, for a linear relationship between a response variable and a continuous explanatory variable is the equation for a straight line. This is typically written as

$$y = \mathrm{b}x + \mathrm{c}$$

Coefficient is the general name given to the values of the intercept, the slopes for continuous variables, and the relative positions of the means for categorical variables, in a GLM model. They are often represented by the letters b and c. Since they are estimates, they have associated confidence intervals.

where *y* and *x* are constants and b and c are called **coefficients.**

The coefficient, b is the slope of the line. The coefficient, c is the value of *y* when *x* equals zero, and is called the intercept. Written in full, the equation would look like this:

Response variable = Slope × Explanatory variable + Intercept

When more than one continuous explanatory variable is involved there is a 'b*x*' term for each one. For example, with two continuous explanatory variables the equation would be

$$y = \mathrm{b}_1x_1 + \mathrm{b}_2x_2 + \mathrm{c}$$

When an explanatory variable is continuous, its coefficient in the model will be a single number indicating the slope of a line. When the explanatory variable is categorical, there will be a set of coefficients indicating the position of the means of

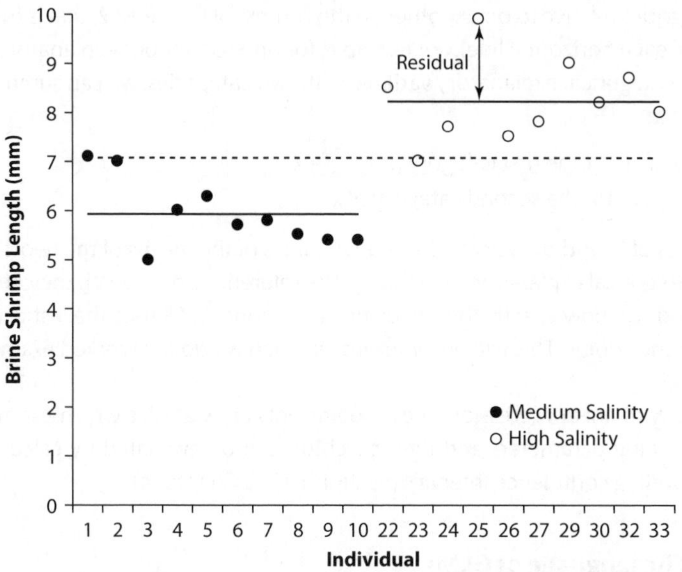

Figure 12.1 The length of brine shrimp and whether they come from a medium or high salinity environment ($n_1 = 10$, $n_2 = 10$). The horizontal dashed line is a model of no difference between medium and high salinity. The two separate solid lines are an alternative model where there is difference. The vertical distance between a point and a model is called a residual; an example is marked with a double ended arrow.

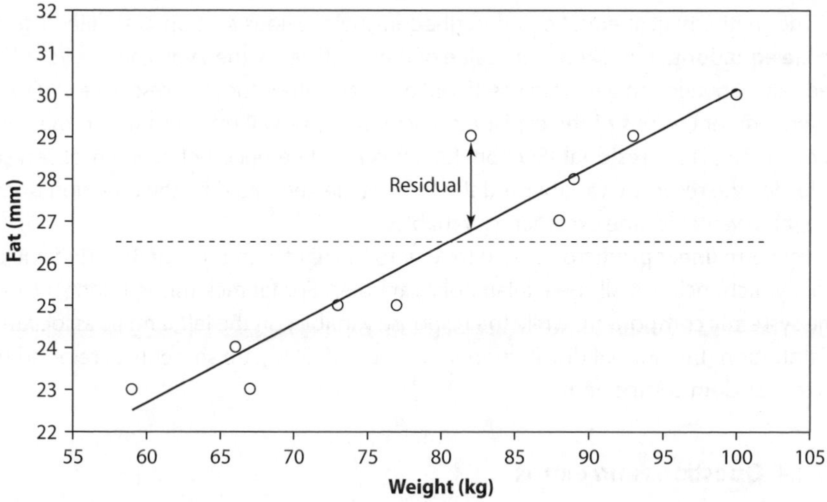

Figure 12.2 Fat against weight in humans ($n = 10$). The horizontal dashed line is a model of no relationship between fat and weight. The sloping line is an alternative model where there is a relationship. The vertical distance between a point and a model is called a residual; an example is marked with a double ended arrow.

the categories relative to one another (in the context of Figure 12.2, that is the relative height of each horizontal line). For example, for one continuous explanatory variable and one categorical explanatory variable with two categories, we can summarize the equation as

$$y = b_1x_1 + \begin{cases} b_2 \text{ for the first category of } x_2 \\ b_3 \text{ for the second category of } x_2 \end{cases} + c$$

The values of b_2 and b_3 indicate the relative values of the means of the two categories of the categorical explanatory variable x_2. The reference point which they are relative to depends on how exactly the calculations are done. SPSS uses the last category as the reference point. This will become clearer when we do the worked example (Section 12.3.3).

Just as with linear regression, these coefficients are statistics which estimate their corresponding parameters and their precision can be evaluated by calculating the corresponding confidence intervals (Section 10.1.2; Chapter 4).

12.1.3 **The language of GLMs**

There are some alternative terms used when talking about GLMs, which might be confusing if you don't realize that they are referring to the same concepts that you have come across in other tests. I have already mentioned the convention of using the alternatives **response variable**, for a dependent variable, and **explanatory variable**, for an independent variable, in the notes on terminology at the start of this chapter. Continuous explanatory variables are often called **covariates**. Categorical explanatory variables are often called **factors,** and the categories within factors are called levels.

The mathematical equations described in the previous section are called **fitted value equations**. If you know the value of the coefficients, then you can use your fitted value equation to generate **predicted** or **fitted values** for your response variable, given different values of the explanatory variables (we will return to this in the next section, 12.3.3). A **residual** (Section 12.1.2) is the difference between an observed value for the response variable and the fitted value generated by the equation using the same values for the explanatory variables.

The term **linear predictor** is used to refer to the right-hand side of the fitted equation, which includes all the explanatory variables. Sometimes this is referred to as the **systemic component**, while the response variable, on the left, and its associated distribution (the normal distribution in the case of GLM) are sometimes referred to as the **random component**.

Response variable, **test variable**, **outcome variable**, or **data variable** are alternative names for a **dependent** variable. **Explanatory variable**, **grouping variable**, **predictor variable**, **factor**, or **covariate** are alternative terms for an **independent variable**.

A **fitted value equation** is a mathematical equation representing a model. Fitted value equations can be used to generate **predicted** or **fitted values** for different sets of values of the explanatory variables.

A **residual** is the difference between an observed value for the response value and the fitted value generated by the equation using the same values for the explanatory variable.

The **linear predictor** includes all the explanatory variables. This is also referred to as the **systemic component**. The response variable and its associated distribution are referred to as the **random component**.

12.1.4 **Questions and extras**

In Chapter 10 we looked at three different questions regression could be used to answer, and how the answers could be used for prediction and for figuring out how much variation is explained by the model (Sections 10.1.3 and 10.1.4). We can use GLMs in the same way. We can ask:

1. What is the linear model that best describes changes in the response based on the explanatory variables? This question is answered by finding all the b values and c (i.e. the coefficients) in the fitted value equation, which can then be used for prediction if the answer for question 2 is yes.
2. Does this model describe a significant amount of the variation in the response variable? For models with more than one explanatory variable this question has two parts:
   2a. Does the overall model, including all the explanatory variables, explain a significant amount of the variation in the response variable?
   2b. Does a particular explanatory variable explain a significant amount of variation in the response variable?

Both questions can be answered by calculating F statistics.

3. How much variation does the model explain? This question is answered by looking at the **coefficient of determination (R^2)**. It is only generally useful to ask this question if the answer to question 2 is yes.

Coefficient of determination (R^2) is a measure of how much the response variable varies with the linear predictor (all the explanatory variables combined). For example, a coefficient of determination of 0.25 means the model explains 25% of the variation of the response variable.

Predicted, or fitted (Section 12.1.3) values are calculated by substituting values into the equation just as for bivariate linear regression (see Sections 10.1.2 and 10.3.4).

12.1.5 **Types of sums of squares**

Question 2b needs a little bit more discussion. In this chapter we will be focusing on whether a particular explanatory variable explains a significant amount of variation in the response variable, once all the other explanatory variables in the model have been taken into account. This last bit 'once all the other explanatory variables in the model have been taken into account' means that the **sum of squares** (Section 3.2.2, variance and Section 9.1.5) that goes into determining the value of F has to be calculated in a particular way. The sum of squares must be determined for each explanatory variable after all other explanatory variables have been included. This gives you a **type III** or **adjusted sum of squares**. However, there are alternatives. For example, a **type I** or **sequential sum of squares** is calculated for an explanatory variable if only those listed before it in the model have been included. Clearly, with a type I sum of squares, the order in which the explanatory variables appear in the model will be important. Types I and III sums of squares will be the same for the last explanatory variable in your model. Therefore, if there is only one explanatory variable in your model, Types I and III sums of squares will be the same.

Type III (adjusted) sums of squares for an explanatory variable are calculated after all other explanatory variables in the model have been taken into account. **Type I (sequential) sums of squares** for an explanatory variable are calculated after any explanatory variables listed before it in the model have been taken into account.

Comparing type I and type III sums of squares, within and between models, can give you important insights that can help you choose which explanatory variables to include in your model (Section 12.6). For example, it is one way of assessing whether two continuous explanatory variables are correlated, and therefore should not both be included in the model because they share information. Detailed discussion of this is beyond the scope of this book, but try Selected further reading: Grafen and Hails (2002) if you want to find out more. We are going to stick to using just the type III sum of squares.

12.1.6 **Model assumptions and the parametric criteria**

The idea that you should check special criteria when doing parametric tests was introduced in Section 5.3, and you were given basic rules of thumb to follow using skills covered in Chapter 3 to check for normality and homogeneity of variances. In Chapters 10 and 11, you heard about regression and correlation techniques which also assume that the relationship between variables is linear and you were recommended to check for this by eyeballing scatter plots of the data. When you do a GLM you assume that all three—homogeneity, normality, and linearity—apply. If one or more of these assumptions is not met, then your analysis may not reliable. The rest of this section gives a little more detail on each of these three model assumptions, in the context of a GLM. For each assumption a simple way to assess it is described. There are other ways, but these are sufficient.

Previously we have looked at the data in advance of doing our tests, and not proceeded if we have been worried that the criteria have not been met. In contrast, when doing a GLM it is usual to check the assumptions after, or as part of, the analysis.

Normality of error

This means that the residuals (Section 12.1.2) around the model must be normally distributed. You can check this by drawing a histogram of the residuals and looking to see that it approximates the normal distribution. Look at Figure 12.3: in (a) the residuals have an approximately bell-shaped symmetrical shape while in (b) the residuals are skewed to the left.

Homogeneity of variance

This refers to the variances of the residuals which need to be similar, regardless of the values of the explanatory variables. You can check this by plotting residuals against fitted values (otherwise known as predicted values: Section 12.1.2) and look for even scatter above and below line for all parts of the model. Look at Figure 12.4: in (a) the residuals have an even scatter across all fitted values, whereas in (b) the scatter is much larger for higher fitted values, giving the plot a funnel shape.

Linearity

This means that the model is assumed to take a linear form. You can check for this by looking for linear patterns in a plot of residuals against fitted values and plots of residuals against each continuous variable in the model. A problem of nonlinearity can be caused by failing to include significant interaction terms, which are discussed more in Section 12.4. Look at Figure 12.5: in (a) the points don't seem to form any sort of shape and could be reasonably said to fall in a straight line, whereas in (b) the points form an upside down U.

Like almost all types of statistical analysis, a GLM also requires independence of response data points, and consequentially of residuals (Section 2.1.1). This will follow from good experimental design and a good understanding of your data.

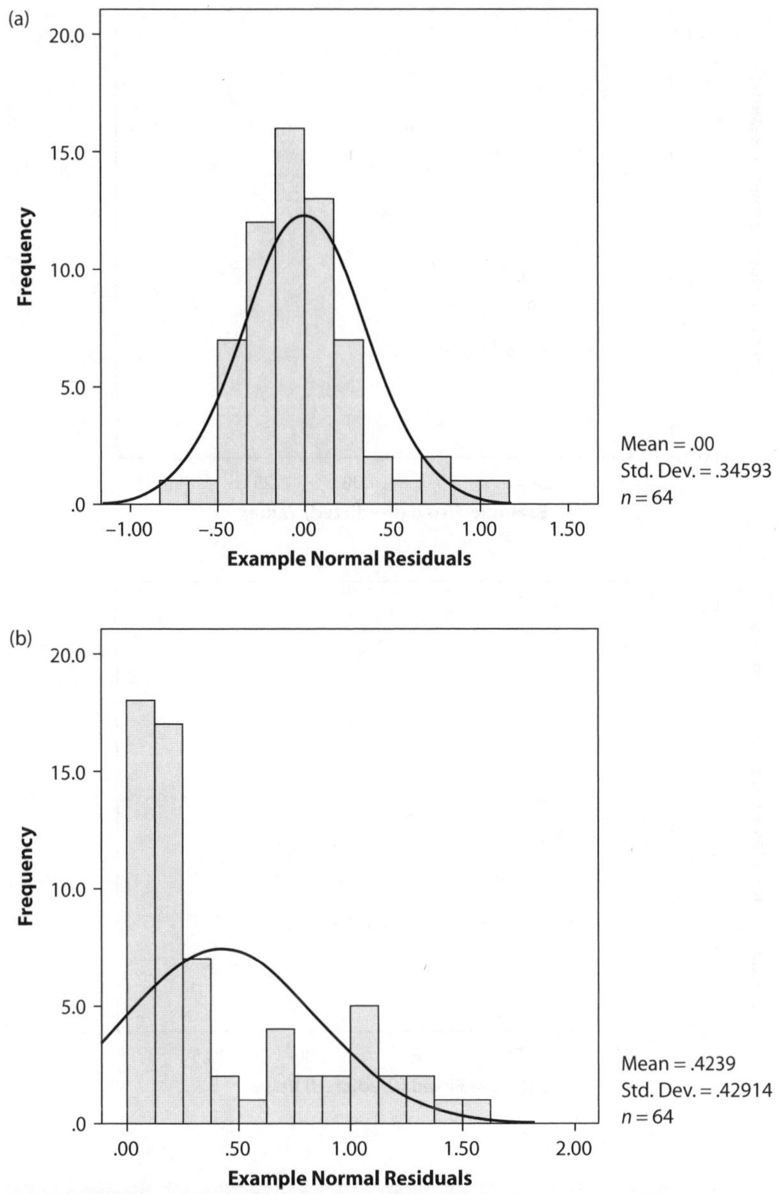

Figure 12.3 Example histograms of residuals (a) Normal; (b) Non-normal. The solid black lines indicate what normal distribution should look like for these data.

12.2 **Example data: watered willows**

In this section I am going to introduce the data set that I will be using in the worked examples in Section 12.3. The data were collected as part of a big European Union funded project called FLOBAR2, short for, Floodplain Biodiversity and Restoration. Humans manage water in the world around them in order to satisfy their domestic,

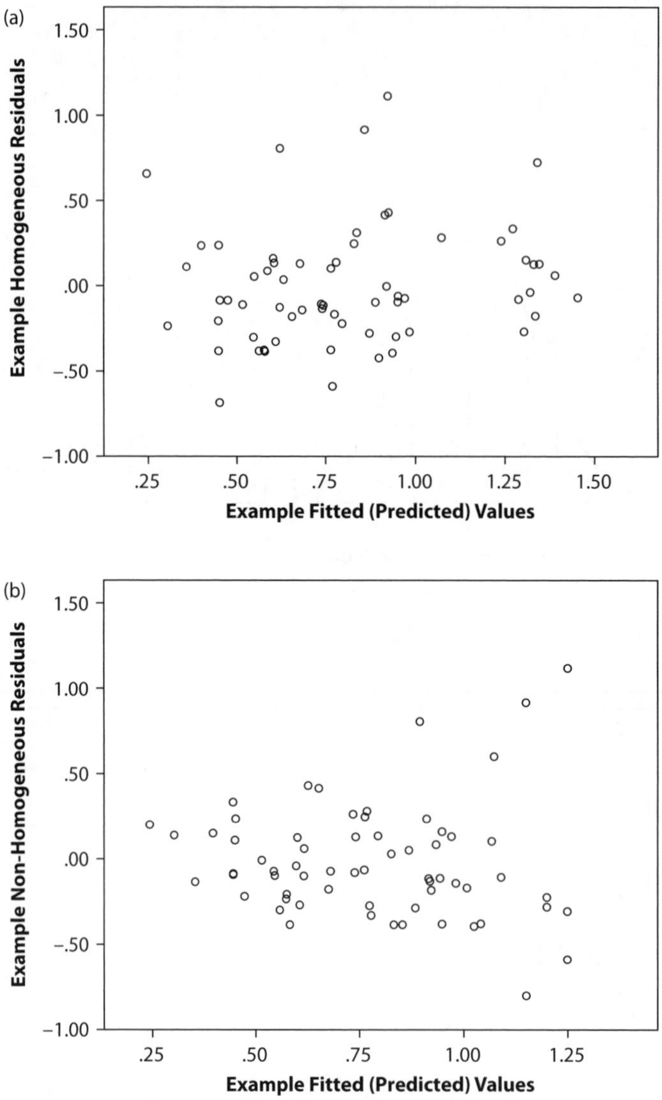

Figure 12.4 Examples of plots of residuals against fitted values (a) Homogeneous; (b) Non-homogeneous.

industrial, and agricultural needs. A scientific understanding of the ecosystems that supply this water, and an ability to restore and protect them, is vital to maintaining these supplies and protecting biodiversity. The FLOBAR2 project addressed key gaps in this understanding, and integrated this information with water management policies. You can find out more about FLOBAR2 from its website:
http://www.geog.cam.ac.uk/research/projects/flobar2/

The example data for this come from part of the FLOBAR2 project, looking at the influence of hydrological regimes on the relative success of male and

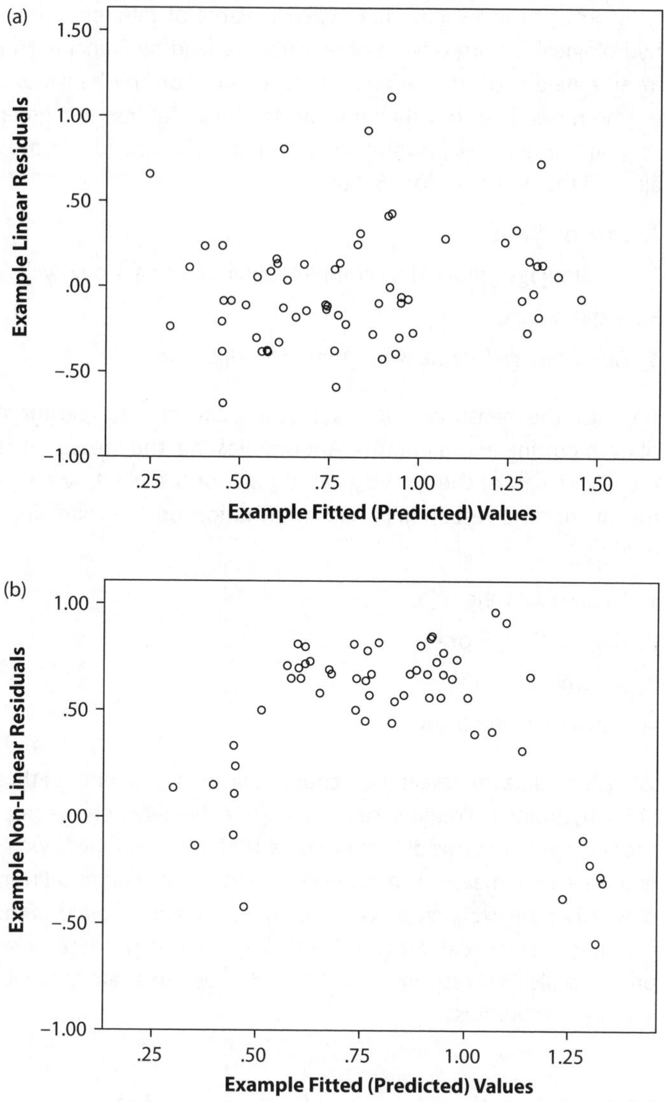

Figure 12.5 Example of plots of residuals against fitted values (a) Linear; (b) Non-linear.

female willow plants that live along rivers in northern Sweden (Hughes *et al.* (2009) in preparation). If male and female plants respond differently to environmental factors such as soil moisture and drainage, then this could lead to males and females growing in different areas (spatial sexual segregation), which would consequently have an impact on their ability to reproduce sexually. This is one way in which changes in the hydrological environment, like changes to rainfall amount or dam building, could lead to fundamental changes in plant communities on river floodplains.

In order to investigate how willow plants of different sexes are influenced by hydrological factors, a team of researchers, lead by Francine Hughes, took cuttings from female and male willow plants growing on the banks of the Vindel River in northern Sweden. They then transported these cuttings to a greenhouse where they randomly applied each cutting to one of the following four hydrological regimes and allowed them to grow for 85 days:

1. Slow drainage.
2. No drainage followed by complete drainage on a 4-day cycle.
3. Fast drainage.
4. Fast drainage for one week, then slow drainage.

Amongst the measurements taken at the start of the experiment was the diameter of each cutting in millimetres. After harvesting, the measurements taken included the dry weight of the above ground parts of the plant, known as the shoots. This meant that the researchers had information on the following variables for each cutting:

1. Final shoot weight (g).
2. Regime (1, 2, 3, or 4).
3. Sex (male or female).
4. Initial diameter (mm).

With these data the researchers could explore the influence of three explanatory variables, hydrological regime, sex, and cutting diameter, on the response variable, final shoot weight. In terms of levels of measurement, final shoot weight and initial diameter were both measured at scale level, and regime and sex at nominal level. In terms of whether the explanatory variables were categorical or continuous, cut diameter is continuous and sex and regime both categorical. In short, this example data set comprises a scale level response variable and three explanatory variables, two categorical and one continuous.

12.3 Using the general linear model

12.3.1 When to use

You can use a GLM when you have a response variable that is measured at scale level and one or more explanatory variables that are either continuous and/or categorical. GLM is a parametric modelling technique and before proceeding, you should think whether there is any reason that your data will definitely not conform to the criteria of normality, homogeneity, and linearity (Section 12.1.6). If you do not come up with a reason, then you can proceed, but you will need to check the assumptions you have made as part of the procedure. Issues relating to when and when not to use a GLM were covered in the introduction to this chapter (Section 12.1). They are summarized in Box 12.1.

Box 12.1 When to use a univariate GLM

Use this test when:

- You want to assess the influence of *one or more explanatory variables* (independent variables) on a *single response variable* (dependent variable).
- The explanatory variable is measured at *any level* but the response variable is *scale*.
- It is reasonable to assume that a *linear* model could summarize the data.
- You want to take a *modelling* approach and/or *prediction* would be useful.

Do not use this test when:

- The response variable is not scale.
- Small sizes are very small.
- The model is unlikely to, or definitely does not, conform sufficiently to the criteria of normality, homogeneity, and linearity (Section 12.1.6).

12.3.2 **GLM and the four steps**

Up until now we have applied the four steps of null hypothesis significance outlined in Chapter 5 to each test. We are not going to do this for GLMs, but there are some lessons to remember. Step 1, stating your null hypothesis, has helped you to be clear on the context of your test—exactly what are you testing to see is different or related? When using GLMs, you have your model to keep you on the straight and narrow. The first thing to do when using a GLM is to write your word equation. There are also important lessons to carry forward from Step 2. This involves choosing a critical significance level. This has helped you to realize that the decision-making process is not set in stone, but determined by you, the scientist. You've also learned that 0.05 is an often, but not universally used, critical significance level, and a good scientist needs to be clear at what level they are working. The critical significance level sets the probability of making a type I error (Section 5.2), and this reminds us that we still might be getting it wrong, even if we find a significant result.

We also calculate statistics (step 3) and make decisions (step 4) when doing GLMs. However, it's time to take some shortcuts. What you are on the look out for are *P* values that are less than a critical significance level chosen by you. When the *P* value is associated with a model, then this indicates that this model explains more variation in the response variable than can be accounted for by chance alone. When this *P* value is associated with a specific explanatory variable, it indicates that this explanatory variable, once other explanatory variables in the model have been taken into account, explains more variation in the response variable than can be accounted for by chance alone. This is summarized in Box 12.2. The critical significance level most often chosen is 0.05. This is the level we are going to use in our worked example.

Box 12.2 Decision using *P* values for GLM

If $P \leq \alpha \rightarrow$ significant contribution to variation in response variable.
If $P > \alpha \rightarrow$ non significant result.

There are two main sets of statistics with GLM:

1. F statistics, found in an Anova table. (In SPSS this table is labelled **Tests of Between-Subjects Effects.**)
2. Model coefficients, found in a coefficients table. (In SPSS this table is labelled **Parameter Estimates.**) Model coefficients have *t*-statistics and confidence intervals associated with them.

The F statistic is the same statistic used in Anova (Chapter 9) and regression (Chapter 10). An F statistic is a ratio of mean squares (otherwise known as variances), which are themselves calculated by dividing a sum of squares (Section 12.1.5) by its associated degrees of freedom. An F statistic is calculated in relation to each explanatory variable in the model, as well as for the model overall. Each F statistics has two different degrees of freedom. The first degrees of freedom is associated with the overall model or explanatory variable in question. For a categorical explanatory variable this will be one less than the number of categories, and for a continuous variable this will be one. The second degrees of freedom is the 'error' degrees of freedom.

12.3.3 Worked example: using SPSS

online resource centre

We are going to conduct a GLM on the willow data (Section 12.2) using SPSS. To do this, the data must be entered into SPSS, as shown in Figure 12.6. In newer versions of SPSS (version 10 onwards), the **Data Editor** window has a **Variable View** tab and a **Data View** tab. Figure 12.6a shows the former and Figure 12.6b the latter. Appendix I has notes on how to enter data into SPSS.

You should start by writing out the model you are considering. In this example, we are going to look at the model:

Shoot Weight = *Regime* + *Sex* + *Cut Diameter*

where *Regime* and *Sex* are categorical and *Cut Diameter* is continuous.

Looking at the model overall (answering question 2a, Section 12.1.4)

In this section we are going to focus on the overall model, answering our question 2a: Does the overall model, including all the terms defined in the word equation, explain a significant amount of the variation in *Shoot Weight*?

To do this using SPSS, you must first open the data file. Then you must make the following selections:

Analyze
→General Linear Model
→Univariate

A window like that shown in Figure 12.7 will appear. You need to select *Shoot Weight* from the list on the left, and send it to the box labelled **Dependent Variable** by pressing the top arrow. Then you need to send *Regime* and *Sex* to the box labelled

(a)

(b)

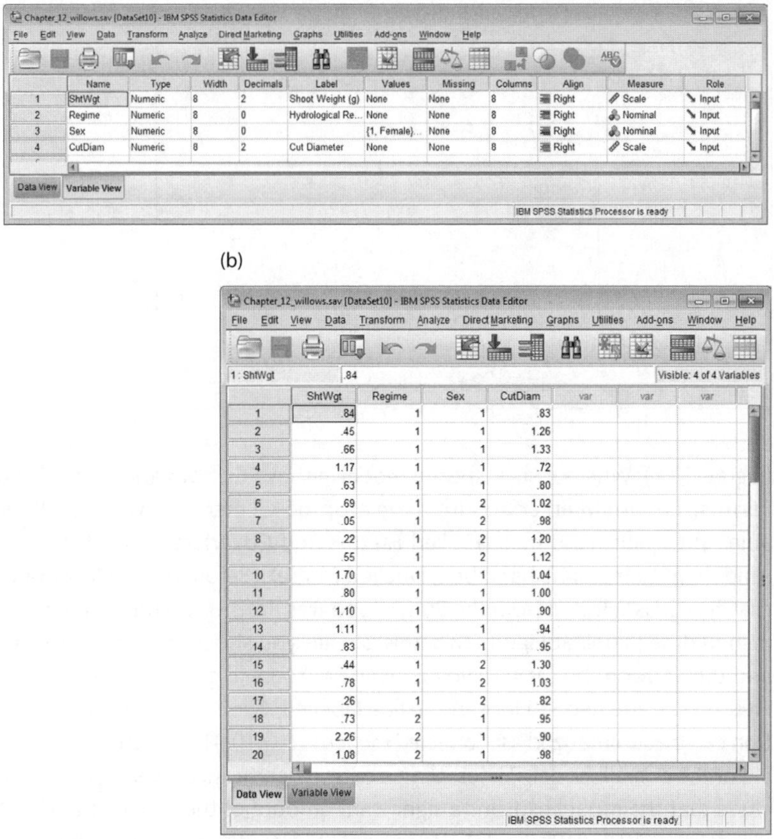

Figure 12.6 Example data in SPSS, willows. (a) Variable view. (b) Data view.

Fixed Factors(s), by pressing the second arrow down. Then you need to send *Cut Diameter* to the box labelled **Covariate(s)**, by pressing the fourth arrow down. Look at Figure 12.7, this is what you should now see. Once you have done this, you need to press the button that says **Model.** This will bring up the **Univariate: Model** window

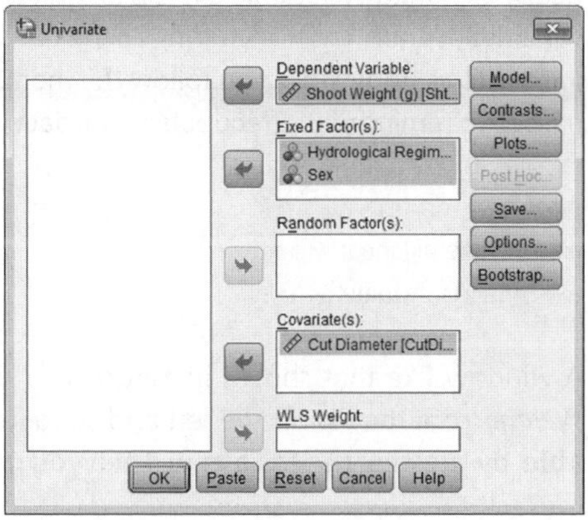

Figure 12.7 Conducting a univariate GLM using SPSS. Main dialogue window.

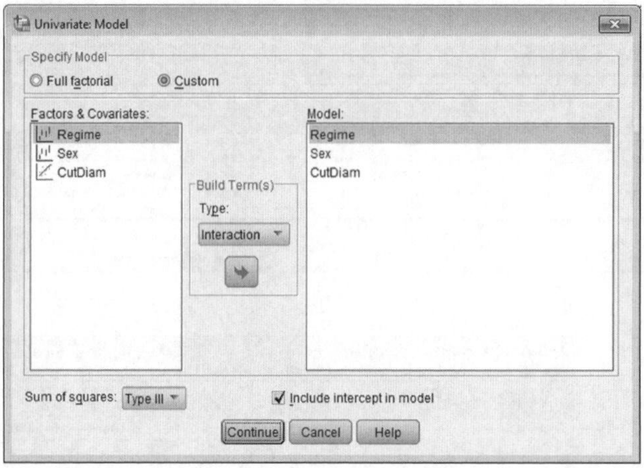

Figure 12.8 Conducting a univariate GLM using SPSS. Model window.

(Figure 12.8). Where it says, specify model, choose **Custom** and where it says **Build Term(s),** choose **Main effects** from the drop-down menu. Now select all three variables in the left-hand box, labelled **Factors and Covariates,** and click the arrow to send them to the right-hand box, labelled **Model**. (Before proceeding double check that *Type III* is selected from the **Sum of squares** drop-down menu, it's the default so it should be.) Look at Figure 12.8, this is what you should now see. Press **Continue** and then **OK** on the main dialogue window. You will get output like that shown in Figure 12.9. You need to look at the line labelled, *Corrected Model*. The key information can be summarized as: *Corrected model*: $F_{5,57} = 10.016$, $P = 0.000$

For the overall model, *P* is less than a critical significance level of 0.05. We can conclude that the model explains a significant amount of the variation in shoot weight (GLM: $F_{5,57} = 10.016$, $P = 0.000$). We can not tell from this which explanatory variables are associated with the variation in shoot weight.

Looking at individual explanatory variables (answering question 2b, Section 12.1.4)

In this section we are going to focus on answering our question 2b: Does a particular explanatory variable explain a significant amount of variation in the response variable?

To do this using SPSS, you follow the menu choices as when looking at the overall model. To remind you, after opening the data, you must make the following selections:

Analyze
→General Linear Model
→Univariate

A window like that shown in Figure 12.7 will appear. You need to select *Shoot Weight* from the list on the left and send it to the box labelled **Dependent Variable**, by pressing the top arrow. Then you need to send *Regime* and *Sex* to the box

Univariate Analysis of Variance

Between-Subjects Factors

		Value Label	N
Hydrological Regime	1		17
	2		16
	3		15
	4		15
Sex	1	Female	38
	2	Male	25

Tests of Between-Subjects Effects

Dependent Variable: Shoot Weight (g)

Degrees of freedom **Statistic (F)** ***P***

Source	Type III Sum of Squares	df	Mean Square	F	Sig.
Corrected Model	6.031[a]	5	1.206	10.016	.000
Intercept	5.411	1	5.411	44.934	.000
Regime	1.876	3	.625	5.193	.003
Sex	1.081	1	1.081	8.980	.004
CutDiam	1.866	1	1.866	15.498	.000
Error	6.864	57	.120		
Total	52.943	63			
Corrected Total	12.895	62			

[a]R Squared = .468 (Adjusted R Squared = .421)

Coefficient of determination (R^2) **Adjusted coefficient of determination (R^2_{adj})**

Figure 12.9 SPSS output for a univariate GLM. Tests of Between-Subject Effects (Anova Table) annotated for key information for overall model.

labelled **Fixed Factors(s)**, by pressing the second arrow down. Then you need to send *Cut Diameter* to the box labelled **Covariate(s)**, by pressing the fourth arrow down. Look at Figure 12.7, this is what you should now see. Once you have done this, you need to press the button that says **Model**. This will bring up the window shown in Figure 12.8. Where it says, **specify model**, choose **Custom** and where it says **Build Term(s)** choose **Main effects** from the drop-down menu. Now select all three variables in the left-hand box labelled **Factors and Covariates,** and click the arrow to send them to the right-hand box labelled **Model**. (Before proceeding double check that *Type III* is selected from the **Sum of squares** drop-down menu, it's the default so it should be.) Look at Figure 12.8, this is what you should now see. Press **Continue** and then **OK** on the main dialogue window. You will get output like that shown in Figure 12.10. The key information can be summarized as

Regime: $F_{3,57} = 5.193$, $P = 0.003$

Sex: $F_{1,57} = 8.980$, $P = 0.004$

Cut Diameter: $F_{1,57} = 15.498$, $P = 0.000$

Univariate Analysis of Variance

Between-Subjects Factors

		Value Label	N
Hydrological Regime	1		17
	2		16
	3		15
	4		15
Sex	1	Female	38
	2	Male	25

Tests of Between-Subjects Effects

Dependent Variable: Shoot Weight (g)

Degrees of freedom **Statistic (F)** ***P***

Source	Type III Sum of Squares	df	Mean Square	F	Sig.
Corrected Model	6.031[a]	5	1.206	10.016	.000
Intercept	5.411	1	5.411	44.934	.000
Regime	1.876	3	.625	5.193	.003
Sex	1.081	1	1.081	8.980	.004
CutDiam	1.866	1	1.866	15.498	.000
Error	6.864	57	.120		
Total	52.943	63			
Corrected Total	12.895	62			

[a]R Squared = .468 (Adjusted R Squared = .421)

Figure 12.10 SPSS output for a univarate GLM. Tests of Between-Subject Effects (Anova Table) annotated for key information for the individual explanatory variable *Regime*.

In each case, the *P* value is less than a critical significance level of 0.05. We can conclude that hydrological regime (GLM: $F_{3,57} = 5.193$, $P = 0.003$), sex (GLM: $F_{1,57} = 8.890$, $P = 0.004$), and cut diameter (GLM: $F_{1,57} = 15.498$, $P = 0.000$) all make a significant contribution to the model, and thus account for a significant amount of variation in shoot weight. Remember that since we are using a type III sum of squares (Section 12.1.5), the effect for each variable is, with the other variables, being controlled.

Looking at R^2 (answering question 3, Section 12.1.4)

In this section we are going to focus on answering our question 3: How much variation does the model explain? This question is answered by looking at the coefficient of determination (R^2), found at the bottom of the **Tests of Between-Subjects Effects** table (Figure 12.9). The key information is: $R^2 = 0.468$.

This is saying that our model explains 46.8% of the variation in shoot weight. For our example analyses it is valid to look at this value, because we have established that our model of willow shoot weight is significant. However, it is important to remember that it is usually only useful to use the R^2 if the answer to question 2a is yes.

It is also worth noting the value of the adjusted R^2, which is in brackets right next to the R^2 value in the **Tests of Between-Subjects Effects** table (Figure 12.9). The adjusted R^2 takes into account the number of coefficients in the model, which can be useful when comparing between models (Section 12.6). The key information is: $R_{adj}^2 = 0.421$.

Using coefficients (answering question 1, Section 12.1.4)

In this section, we are going to focus on answering our question 1: What is the model? This question is answered by finding all the b values and c, the coefficients, in the fitted value equation. We find these in a table labelled **Parameter Estimates**, which is just like the table labelled **Coefficients** in regression (Figure 10.6). However, unlike for regression, SPSS does not produce a table with coefficients by default. To tell SPSS to do so you must click on the **Options** button in the main dialogue window (Figure 12.7). Look at Figure 12.11 to see the **Options** window that should appear. You need to select **Parameter estimates** from the items listed under **Display** in the bottom half of the window. When you are done, click **Continue**. If you run the analyses with this option activated, then the additional output will look like that shown in Figure 12.12.

By substituting the coefficients into the fitted value equation which looks like:

$$\begin{aligned} \textit{Shoot Weight} = {} & 0.036 \text{ for } \textit{Regime } 1 + 0.269 \text{ for } \textit{Sex } 1 - 1.045 \times \textit{Cut Diameter} + 5.411 \\ & 0.400 \text{ for } \textit{Regime } 2 + 0.000 \text{ for } \textit{Sex } 2 \\ & -0.053 \text{ for } \textit{Regime } 3 \\ & 0.000 \text{ for } \textit{Regime } 4 \end{aligned}$$

Notice that the coefficients for the categorical explanatory variables are all relative to the final category, which is set to zero. This is the way SPSS does it by default, but other packages may have different default reference points.

Look at Figure 12.2 and notice that each coefficient is reported along with a confidence interval and a t statistic. The confidence interval gives you an idea of how

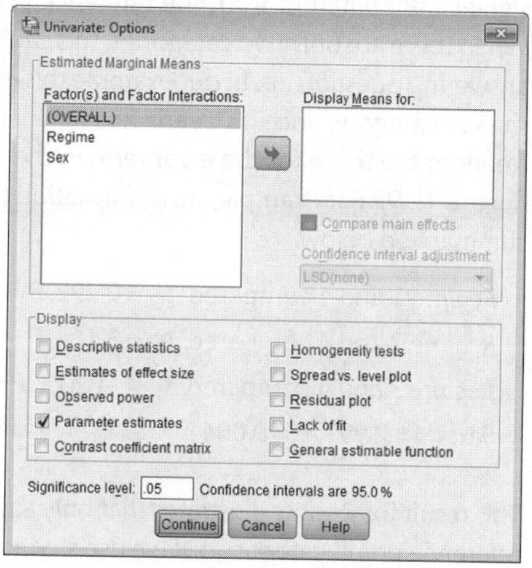

Figure 12.11 Conducting a univariate GLM using SPSS. Option window.

Parameter Estimates

Dependent Variable: Shoot Weight (g)

Parameter	B	Std. Error	t	Sig.	95% Confidence Interval Lower Bound	95% Confidence Interval Upper Bound
Intercept	1.588	.289	5.500	.000	1.010	2.166
[Regime = 1]	.036	.123	.292	.771	−.211	.283
[Regime = 2]	.400	.126	3.164	.002	.147	.653
[Regime = 3]	−.053	.128	−.414	.680	−.310	.203
[Regime = 4]	0[a]	.	.	.	.	.
[Sex = 1]	.269	.090	2.997	.004	.089	.448
[Sex = 2]	0[a]	.	.	.	.	.
CutDiam	−1.045	.265	−3.937	.000	−1.576	−.513

[a]This parameter is set to zero because it is redundant.

Intercept (c)

Coefficients for Regime

Coefficient for Cut Diameter

t statistic for Cut Diameter

P value for *t* statistic for Cut Diameter

95% Confidence interval for Cut Diameter

Figure 12.12 Additional SPSS output for a univarate GLM Parameter Estimates table with model intercept and key information for model coefficients annotated for *Regime* and *Cut Diameter.*

precise is the coefficient (Section 4.2.2). For *Cut Diameter,* the coefficient is −1.045, with a 95% confidence interval from −1.576 to −0.513 (Figure 12.12).

For continuous variables, the *t* statistic can be used as an alternative to the F statistic for answering question 2b, since they will give the same result. The degrees of freedom are those for the error term in the **Tests of Between-Subjects Effects** table. For our example: *Cut Diameter:* $t_{57} = -3.937$, $P = 0.000$. From this we can conclude that *Cut Diameter* makes a significant contribution to the model.

For categorical variables, the *t* statistic can be used to assess if each category is significantly different from the reference point which, for SPSS is the last category. When there are only two categories, this can again be used as an alternative means of answering question 2b. In our example this applies to *Sex*. No *t* statistic is reported for the last category, since it's clearly not significantly different from itself. The degrees of freedom are those for the error term in the **Tests of Between-Subjects Effects** table (Figure 12.9). For example, the information for the *t* statistics for *Regime* and *Sex* are summarized below.

Regime 1 and 4 compared: $t_{57} = 0.292$, $P = 0.771$

Regime 2 and 4 compared: $t_{57} = 3.164$, $P = 0.002$

Regime 3 and 4 compared: $t_{57} = -0.414$, $P = 0.680$

Sex: $t_{57} = 2.997$, $P = 0.004$

This result for *Regime* illustrates that only category 2 is significantly different from the reference point, category 4, if we use a critical significance level of 0.05.

Using coefficients (making predictions)

Once you have your fitted value equation you can generate fitted, or predicted, values:

For example, if the *Regime* is 1 and the *Sex* is male (2) and *Cut Diameter* is 1.02, then the predicted value is 0.56, to two decimal places, given by:

Fitted value for *Shoot Weight* = 0.036 + 0.000 + (–1.045 × 1.02) + 1.588

If you look at the data (Figure 12.14b) you can see that we actually have an observed value for shoot weight with this set of explanatory variables values and it is 0.69. This gives a residual of 0.13 given by:

Residual = Observed value (0.69) – Predicted value (0.56)

For our example analyses, it's valid to use the fitted value equation for generating fitted values for combinations of explanatory variables not in our observed data set i.e. for prediction, because we have established that the model is significant. However, it is important to remember that it is only meaningful to use the fitted value equation for prediction if the answer to question 2a is yes. Also, do not forget that you should not predict outside the range of values in your data.

Checking model assumptions

One way to check for the three model assumptions of normality, homogeneity, and linearity (Section 12.1.6) is to get SPSS to calculate two additional columns of data when it does the analysis:

1. **Fitted/Predicted values:** the value that the model generates for each observed value of the response variable. This is generated by substituting the values of the explanatory variables into the model as explained in the previous section.
2. **Residuals/Errors:** the distances between the observed and fitted/predicted values (Section 12.1.2).

To tell SPSS to do this you must click on the **Save** button in the main dialogue window (Figure 12.7). Look at Figure 12.13 to see the **Save** window that should appear. You need to select **Unstandardized** from the items listed under **Predicted Values** in the box on the left, and from the items listed under **Residuals** in the box on the right. When you are done, click **Continue**. If you run the analyses with these choices activated, SPSS will add the two columns of data you need to **Data View**—look at Figure 12.14. The column labelled **PRE_1** is the predicted value for that set of values of the explanatory variables, and the column labelled **RES_1** is the residual. Notice that the residual is the column labelled **shoot weight** (observed value) minus the column labelled **PRE_1** (predicted value). Also take a look at the predicted value in Figure 14b, when the example *Regime* is 1 and the *Sex* is male (2) and the *Cut Diameter* is 1.02, and reassure yourself that it's 0.56, the same as when we calculated it by hand in the previous section.

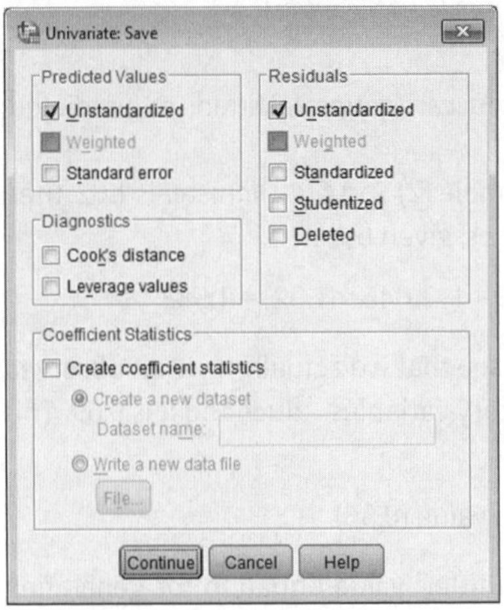

Figure 12.13 Conducting a univariate GLM using SPSS. Save window.

Before proceeding there is one more step that you need to take. You need to make sure that the two new variables, PRE_1 and RES_1, are defined as scale level in variable view (Figure 12.14b).

Now that you have these data, you can check for normality of error by drawing a histogram of the residuals and looking to see that it approximates to the normal distribution (Section 12.1.6). You should return to Chapter 3, Section 3.6 for guidance on how to produce histograms using SPSS. The results for the example data are in Figure 12.15. Looking at Figure 12.15, you should agree that our model fulfills this assumption.

You can check for homogeneity of variances by plotting residuals against predicted values and look for even scatter above and below line for all parts of the model (Section 12.1.6). There is guidance in Chapter 13, Section 13.4 on how to make scatterplots using SPSS. The results for the example data are in Figure 12.16. Looking at Figure 12.16 you should agree that our model fulfills this assumption.

You can check for linearity by looking for a linear pattern in the plot of residuals against predicted values that you used to check for homogeneity, and a plot of residuals against the continuous variable, *Cut Diameter*. The results for the example data for the latter are shown in Figure 12.17. Again, there is guidance in Chapter 13, Section 13.4 on how to make scatterplots using SPSS. Looking at Figure 12.15, Figure 12.16, and Figure 12.17, you should agree that our model fulfills this assumption.

12.3.4 **Reporting results from GLM**

When reporting a result involving the overall model, you should report both the F statistic and the R^2 value, as this gives an indication of the size of the effect. For example, you might report that shoot weight varied with hydrological regime, and the sex and initial diameter of the cutting (GLM: $F_{5,57} = 10.016$, $P = 0.000$; $R^2 = 0.468$).

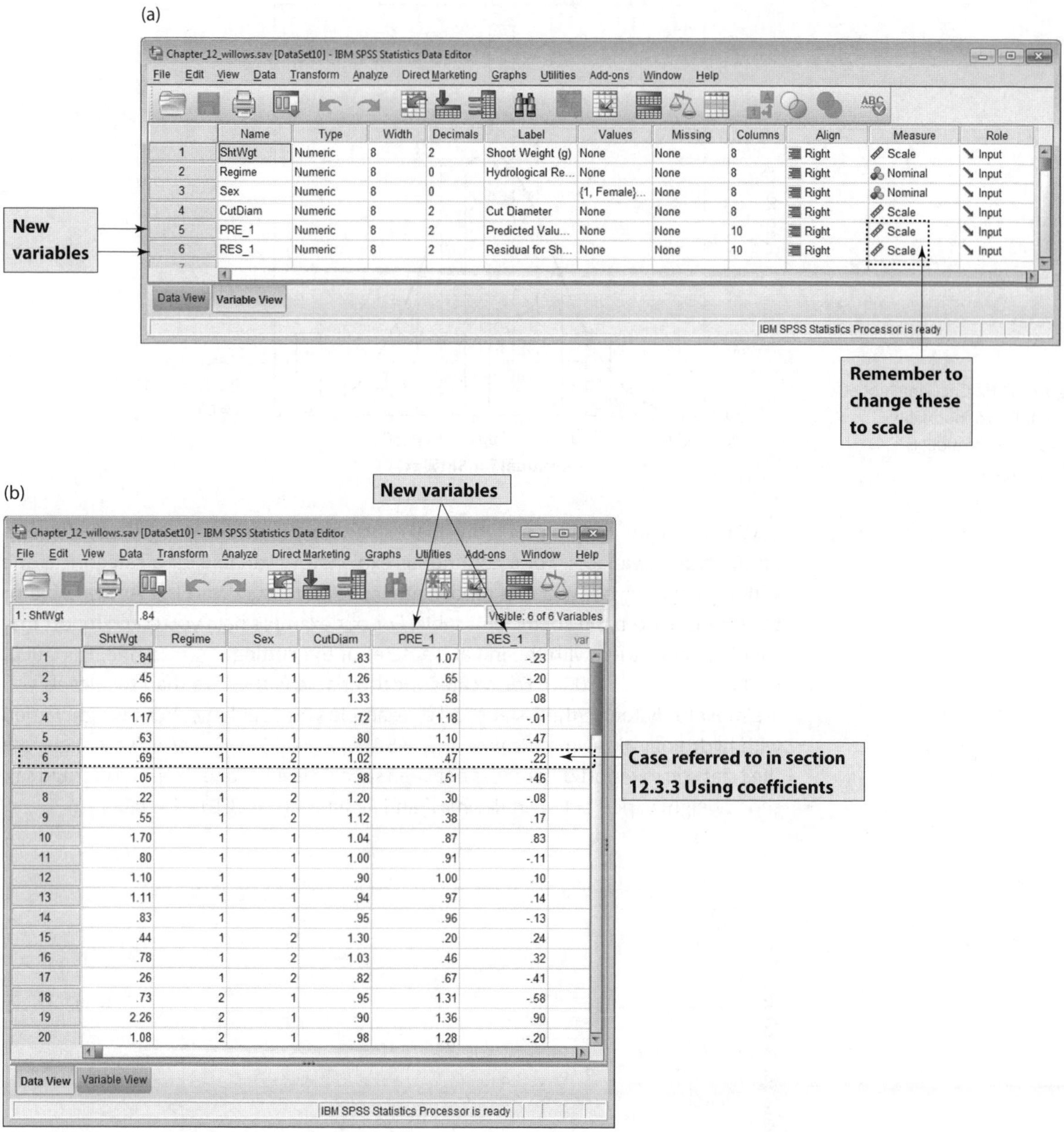

Figure 12.14 Example data in SPSS with residuals and predicted (fitted) values saved. (a) Variable view. (b) Data view. Marked case referred to in text.

Reporting the details of a model with all its coefficients is rather cumbersome and probably not necessary, unless you are dealing with prediction or model comparison (Section 12.6). However, if you have one continuous explanatory variable it is easy enough to report this detail as for regression (Section 10.3.5).

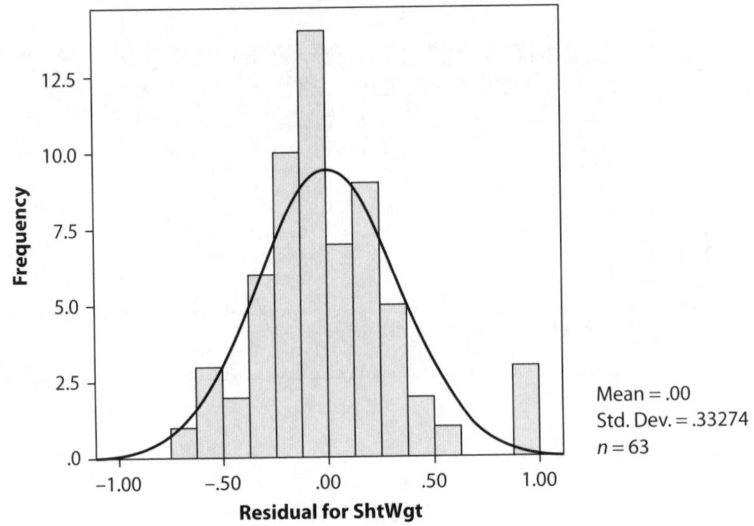

Figure 12.15 Checking model assumption of normality. Histogram of residuals (SPSS output).

When you are focusing on individual explanatory variables your choices depend on the type of variable. For a continuous explanatory variable, or a categorical one with only two levels, you can report either the F statistic from the Anova table or the *t* statistic from the coefficients table. For our example data you could report that shoot weight varied with the initial diameter of the cutting $F_{1,57} = 15.498$, $P = 0.000$) or ($t_{57} = 3.937$, $P = 0.000$). For a categorical variable with more than two levels you are most likely to use the F statistic. For example you could say, shoot weight varied with hydrological regime (*Regime*: $F_{3,57} = 5.193$, $P = 0.003$). However, you might use the *t* statistic to report differences between specific levels. For example you could say, shoot weight varied between *Regime 2* and *Regime 4* ($t_{57} = 3.164$, $P = 0.002$).

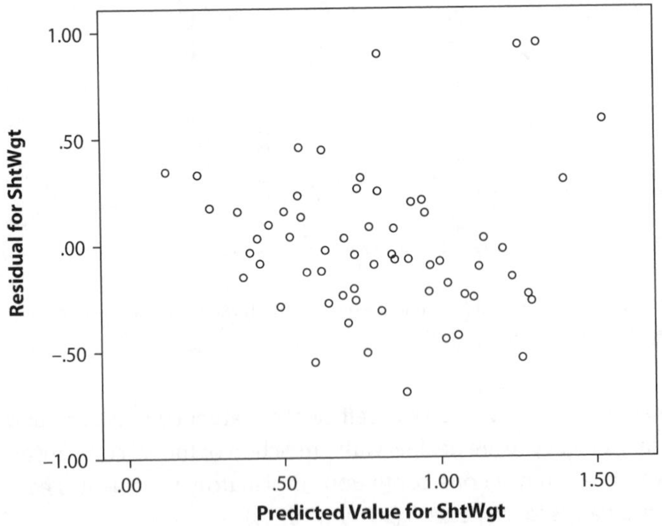

Figure 12.16 Checking model assumptions of homogeneity and linearity. Plot of residuals against fitted (predicted) values (SPSS output).

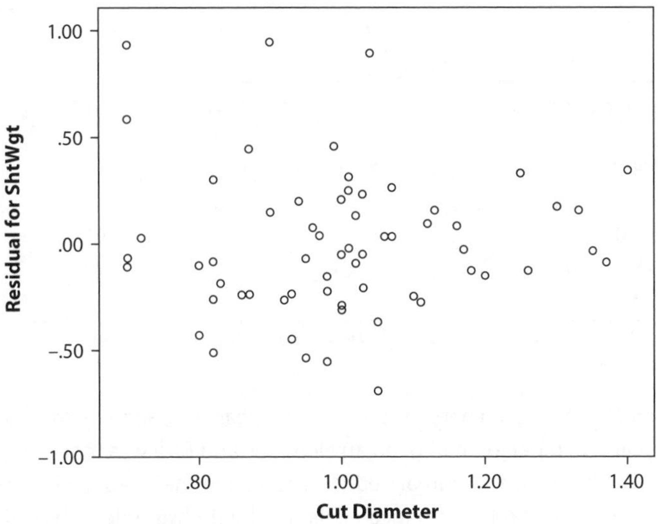

Figure 12.17 Checking model assumption of linearity. Continuous variable plotted against residuals (SPSS output).

online resource centre

12.3.5 **Literature link: Brains and Booze**

Wilhelm *et al.* (2008) Alcohol & Alcoholism 43: 296–299 (OUP).

It's well established that excessive booze is bad for the brain, but how do beer, wine, and spirits compare in their effect? Julia Wilhelm and her colleagues carried out research aimed at investigating the influence of these different types of alcoholic beverage. In particular, they investigated this factor in relation to the hippocampal volume. Humans have two hippocampi in their brains: one on the left, one on the right. They play important roles in short-term memory and spatial navigation.

Wilhelm *et al.* used an MRI machine to measure the hippocampal volume in 52 alcoholics. In the statistical analysis section of their methods, they tell us, 'To compare the impact of different covariates and factors on hippocampal volume, we used a general linear model...'. They also tell us, 'For all statistical tests we applied a significance level of $\alpha = 0.05$ (two-sided)' (by significance level they meant critical significance level), and that they used SPSS 14.0.

In their results section, they tell us, 'we applied a general linear model with hippocampus as dependent variable and preferred beverage and gender as fixed factors and homocysteine levels, years of drinking, age, and BMI as covariates ($R^2_{corr} = 0.560$).' This is telling us what variables were included in the model and whether they were treated as continuous or categorical, and that the model overall had an adjusted R^2 of 0.560. If written out, the word equation would have been:

Hippocampal volume = *Beverage type* + *Gender* + *Homocysteine levels* + *Years drinking* + *Age* + *BMI*

where *Beverage type* and *Gender* are categorical and *Homocysteine levels*, *Years drinking*, *Age*, and *BMI* are continuous.

Beverage type was measured as beer, wine, or spirits, and was determined from interviewing each person. *Gender* was male or female. Homocysteine is an amino acid

	Df	F	*P*-value	% Variance explained by the variable
Corrected Model	7	10.260	0.000	0.620
Gender	1	12.411	0.001	0.220
Preferred beverage	2	5.341	0.008	0.195
Homocysteine (µmol/l)	1	8.313	0.006	0.59
BMI	1	5.746	0.021	0.116
Years of drinking	1	3.690	0.061	0.077
Age (years)	1	0.060	0.808	0.001

BMI, body mass index. Results of the general linear model. Statistical details are summarized in the Results section.

Table 12.2 Contents of Table 2 from Wilhelm et al. (2008) entitled "Impact of different variables on hippocampal volume".

and levels in each person's blood were measured. *Years of drinking* was the number of years each person had been an alcoholic and *BMI* was their body mass index.

Wilhelm *et al.* summarized their results in their Table 2 (Table 12.2). It includes a column labelled '% Variance explained by the variable' (which they obtained by selecting the '**Estimates of effect size**' on the **Option** window in SPSS: Figure 12.11). As the title of this column suggests, it indicates the relative influence of the different explanatory variables on the response variable. Based on this, the authors conclude that 'gender had the largest influence on hippocampal volume, followed by the kind of preferred beverage, and homocysteine levels. With the exception of 'age' and 'years of drinking', all other variables studied also had a significant impact on hippocampal volume (Table 2). No interaction between the fixed factors gender and preferred beverage was observed.' Using post hoc analyses and graphical inspection of data (Section 9.1.6), they determined that beer drinking alcoholics have larger hippocampal volume than spirit or wine drinking alcoholics.

The authors do not give us the error degrees of freedom but we can work it out, if we assume that there are no missing data points, as 52 subjects minus the degrees of freedom for the model, minus another one for the intercept, gives 44. They don't tell us what types of sum of squares were used, but we can assume it was Type III, as the default in SPSS. This means that, for example, the influence of beverage type was significant once the influence of all other explanatory variables had been taken into account.

12.4 Interaction

The **main effects** in a model are those of the explanatory variables. Models can also include **interaction** terms. If two variables interact, the effect one has on the response variable depends on the other. If main effects are significant but interaction terms are not, the main effects are said to affect the response variable in an **additive** manner.

So far we have only considered the **main effects** of the explanatory variables, which are their effects individually, controlling for the effect of the other explanatory variables in the model if we use Type III sum of squares. Something we have not yet considered is that things can get more complicated when you have more than one explanatory variable, because they can have an interactive effect on the response variable. Returning to the wheat yield example mentioned in Section 12.1.1, it is possible that the *variety* of wheat interacts with the *regime* used (organic or not). It could be that farms that use organic regimes have lower yields, but the difference is not the same for each regime. A more extreme example of **interaction** would be if yield in one variety were lower with an organic regime, but for another variety, yield were higher with an organic regime. If main effects are significant but interaction terms are not, people say that the main effects are affecting the response variable in an **additive** manner.

A strategy of including all possible interaction terms is not the way to go. Inclusion of interaction terms in your model complicates the interpretation of *P* values. For example, when using Type III sum of squares (Section 12.1.5), you would be asking: does variety contribute to wheat yield when regime and the interactive effect of regime and variety have been taken into account? How this should be handled is debated by statisticians, and in any event the question is not very useful biologically. You could use Type I sum of squares instead, but then the order in which you have your variables in the model is an issue (see Selected further reading: Grafen and Hails (2002).

A better strategy is to only include an interaction term when you have good reason to do so. There are two main reasons for including an interaction term in your model. Firstly, the interaction may be what you are interested in rather than the main effects. For example, your question could be: is there an interaction between the *variety* and *regime* (organic or not) on wheat *yield*? For this reason, your decision to include interaction terms should be guided by theory and previous research. Secondly, failing to include a significant interaction is likely to mess things up for you. If a significant interaction only has a small effect, then you might be okay if you leave it out—otherwise leaving out a significant interaction means you won't be able to rely on your results. This is because the criteria for linearity will definitely not be met (Section 12.1.6). For example, if the interaction between *variety* and *regime* is significant but you leave it out, your model will fail the test of linearity. Conversely, if a model passes inspection in terms of linearity, without the inclusion of an interaction term, and you have no other reason to include it, then your decision is made: there is no need to add it.

So, in summary, when should you include interaction terms? As a general rule of thumb, if you were not interested in interaction as such and the interaction term is not likely to be significant, then leave it out. Only revisit this decision if you have a problem with linearity.

The maximum number of interaction terms you can have in your model increases rapidly if you add more explanatory variables. If you have just two explanatory variables, there can only be one interaction between them. However, if you have three explanatory variables, the maximum number is four: all the possible pairwise combinations, 1 & 2, 1 & 3, 2 & 3, plus all three could be interacting 1, 2, & 3.

If you want to use a model with interaction terms in it, then you can use a * sign to indicate the variables involved. For example, the general form of word equation for a model involving two explanatory variables and their interaction would be:

Response variable = Explanatory variable 1 + Explanatory variable 2 + Explanatory variable 1 * Explanatory variable 2

And the mathematical equation if they were both continuous would be:

$$y = b_1x_1 + b_2x_2 + b_3x_1x_2 + c$$

Something that you have to keep in mind with interactions is that they can purely be a result from your experimental design (Section 1.4; Appendix IV), and have nothing to do with the biology of the system. For example, the more unbalanced your design (Section 2.3.1), the more likely it is that any interaction you detect will not be of biological significance. In our wheat example, you would detect an interaction between *variety* and *regime* (organic or not) if most farms in the study grew a particular variety non-organically, whether or not something of biological interest was happening. With continuous categorical variables it is practically impossible to eliminate the possibility of purely design-generated interaction all together. If the effect is slight, then this doesn't cause problems worth worrying about and can be safely ignored. If the effect is more than slight, or you suspect it might be, you will need to take action (see Selected further reading).

12.4.1 **Worked example: Using SPSS, interaction**

We are going repeat the basic GLM on the willow data (Section 12.2), using SPSS to look at the interaction between *Regime* and *Sex*. Check back to Section 12.3.3 and Figure 12.6 to see how the data look.

We are now going to include an interaction term in our model:

$$Shoot\ Weight = Regime + Sex + Cut\ Diameter + Regime * Sex$$

where *Regime* and *Sex* are categorical and *Cut Diameter* is continuous.

As in Section 12.3.3, to get SPSS to conduct a univariate GLM on your data you must first open the data file. Then you must make the following selections:

Analyze
→General Linear Model
→Univariate

A window like that shown in Figure 12.7 will appear. You need to select *Shoot Weight* from the list on the left and send it to the box labelled **Dependent Variable**, by pressing the top arrow. Then you need to send *Regime* and *Sex* to the box labelled **Fixed Factors(s)**, by pressing the second arrow down. Then you need to send *Cut Diameter* to the box labelled **Covariate(s)**, by pressing the fourth arrow down. Once you have done this you need to press the button that says **Model**. This will bring up the window shown in Figure 12.18. Where it says, specify model, choose **Custom** and where it says **Build Term(s)** choose **Main effects** from the drop-down menu. Now select all three variables in the left-hand box labelled **Factors and Covariates**, and click the arrow to send them to the right-hand box labelled **Model**. (Before proceeding, double check that *Type III* is selected from the **Sum of squares** drop-down menu, it's the default so it should be.)

So far the procedure has been the same as when no interaction was included, but at this point you need to take an additional step in order to add the interaction. Where it says **Build Term(s)** choose **Interaction** from the drop-down menu. Now select both *Regime* and *Sex* in the left-hand box labelled **Factors and Covariates** and click the arrow to send them both to the right-hand box labelled **Model**. Look at Figure 12.18 to see what your **SPSS Model window** should now look like.

Press **Continue** and then **OK** on the main dialogue window. You will get output like that shown in Figure 12.19. The key information can be summarized as *Regime* × *Sex*: $F_{3,54} = 0.766$, $P = 0.518$.

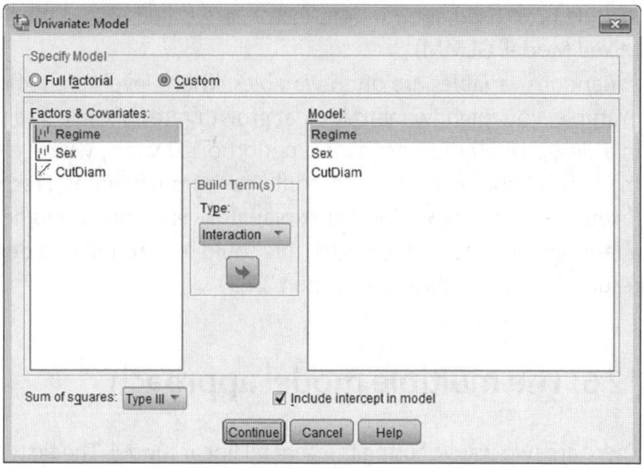

Figure 12.18 Conducting a univariate GLM using SPSS. Model window.

The *P* value is more than a critical significance level of 0.05 (Box 12.2). We can conclude that regime and sex do not interact (GLM: $F_{3,54} = 0.766$, $P = 0.518$).

A **fixed factor** is a categorical explanatory variable whose levels measure something that's likely to cause variation in the response variable, such as habitat type. A **random factor** is a categorical explanatory variable whose levels are random entities and likely to contribute only to background variation. A **mixed model** (or General Linear Mixed Model, **GLMM**) is one that includes both fixed and random factors.

12.5 **Random factors and mixed models**

We have dealt with only one type of categorical explanatory variable, or factor: **fixed factors**. There is another type: **random factors**. Random variables contribute to variation in the response variable, but not to its mean. Usually you include random variables in models for reasons of experimental design and not because you are interested in assessing their effects as such. I will only briefly review random factors here. You should consult Selected further reading for more information. Models that

Degrees of freedom

Tests of Between-Subjects Effects

Statistic (F) **P**

Dependent Variable: Shoot Weight (g)

Source	Type III Sum of Squares	df	Mean Square	F	Sig.
Corrected Model	6.311[a]	8	.789	6.470	.000
Intercept	4.883	1	4.883	40.051	.000
Regime	1.764	3	.588	4.823	.006
Sex	1.008	1	1.008	8.270	.006
CutDiam	1.577	1	1.577	12.985	.001
Regime*Sex	.280	3	.093	.766	.518
Error	6.584	54	.122		
Total	52.943	63			
Corrected Total	12.895	62			

[a]R Squared = .489 (Adjusted R Squared = .414)

Figure 12.19 SPSS output for a univarate GLM. Between-Subjects Effects table only with key information for interaction between *Regime* and *Sex* annotated.

include both fixed and random factors are called **mixed models** (or General Linear Mixed Model, **GLMM**).

Random variables are often variables whose levels are biological individuals. For example, you might want to look at growth rates of male and female baby baboons in a group of 20 mothers over a period of 20 years, where each mother has more than one offspring. Your model could have *growth rate* as a response variable and *sex* (fixed) and *mother* (random) as explanatory variables. If *mother* was not included as a random variable in the model, this could lead to inflated degrees of freedom as a result of pseudoreplication (section 2.1.1).

12.6 The multiple model approach

There are two ways of using the general linear model. The first, as already mentioned, is that it is a framework encompassing a range of traditional techniques such as *t*-tests, Anova, and regression. As such it simplifies life greatly. The second feature relates to the way it can be used to generate and compare multiple models. This feature isn't evident from the way we have used GLM in this chapter. We have focused on how to carry out a single GLM analysis. However, GLMs are often used to generate and compare more than one model, with the aim of finding the best one. This is using GLMs in a different way, not just using as an alternative to a traditional test. I shall give you a taste of this here. If, after reading the following sections, you want to use this approach for analysing your data, you should read up a bit more before proceeding (see Selected further reading: Burnham and Anderson (2002); Crawley (2005); Grafen and Hails (2002)).

The multiple model approach is particularly useful if you have a lot of potential explanatory variables and you want to explore which are most important in telling you what's going on with, or can best predict, the response variable. You will often be in this position if you are working with observational data sets (Section 2.2.3). For example, you might be looking at the *area of crops damaged by elephants* (response variable) and have data on the following potential explanatory variables: *distance from nearest protected area*, such as a National Park (km); *distance to nearest permanent water* (km); *density of human population* (people per km^2); season (wet or dry); *moon phase* (new, waning, full, or waxing); *rainfall in previous month* (mm); *rainfall in previous two months* (mm). You could then use a multiple model approach to help you understand which combination of explanatory variables best explains, or predicts, the area of crops damaged.

The process of generating a model is called **model formulation**. Any model which is significant and fits the criteria for normality, homogeneity, and linearity is called an **adequate model**. The process of deciding which model is best is called **model choice** or **model selection**. A model which includes all available explanatory variables is called a **maximal model**. The process of reducing the number of variables and factor levels in a model is called **model simplification**.

12.6.1 Finding the best model

The process of generating models is called **model formulation**. Any model which you decide is significant and fits the criteria for normality, homogeneity, and linearity is called an **adequate model**. You may well find that you generate more than one adequate model. The process of deciding which one of your adequate models is best to use, is called **model choice** or **model selection**. The process of model formation and model selection is somewhat of an art. This sounds a bit odd in the world of science, but there are no hard and fast rules and experience and context are very important.

There are ways to get computers to select the best model for you on purely statistical criteria, but experienced users of GLM generally do not advocate a solely automatic procedure. Keeping control of the model building process yourself allows you, for example, to give preference to models containing variables that are easy and/or cheap to measure, reflect what you know about the mechanics of what's going on, or follow from important design principles.

The most widely used principle in model selection is that of seeking simplicity. A common approach to finding the best model is to throw all the explanatory variables in at the start, which is the **maximal model**, and then to go through a process of **model simplification**. An obvious way to simplify a model is to remove non-significant terms and combine factor levels that are not significantly different. Removing variables that share information is also good practice. These can be identified by either preliminary correlational analyses (Chapter 11), or by comparing types of sums of square (Section 12.1.5). However, you should not exclude insignificant main effects if the variable is part of a significant interaction, and you should always refer back to your experimental design and the aim of your study before leaving anything out.

There are some numbers that can help you. R^2 is useful for comparing models with the same number of coefficients, but it will inevitably increase when you add variables or factor levels. Adjusted R^2 is more useful in this context. Other indices that can be used to compare models include Akaike's Information Criteria (AIC). All have their strengths and weaknesses.

Another thing to keep in mind is **type I errors** (Section 5.2). If you use a critical significance level of 0.05, then you can expect that, on average, one in every twenty *P* values will be significant regardless. It is easy to get carried away generating multiple models using your computer, forget this and then get very excited by a rare significant result that is likely to be a type I error.

A **type I error** is rejecting a true null hypothesis. The critical significance level (α) sets the probability of making a type I error.

12.6.2 **Report results from multiple models**

No standard way has yet emerged for reporting of results when using the multiple model approach for GLMs. However, in order to understand and evaluate what's going on your reader will need at least the following information:

- What explanatory variables were available and whether they were treated as categorical (fixed or random) or continuous.
- What process was used to find the best model, specifying the statistical and other criteria used.
- What the best model is with statistical details from the Anova and coefficients tables.

Ideally, the F statistic for the overall model for all, or at least the most relevant, models should be presented along with any statistical criteria in a **model comparison table**. This may take up space but a piece of work can not be critically evaluated without this information.

A **model comparison table** is a way of presenting criteria used to compare models.

12.7 The general and generalized linear models compared

A **generalized linear model** has three characteristics:

1. A **response variable** and its **probability distribution** (the **random component**);
2. A set of explanatory variables called the **linear predictor** and an **error** term (the **systematic component**);
3. A **linking** (or **link**) **function** connecting the above two.

As I said at the start of this chapter, the general linear model is a sub-framework of the **generalized linear model**. If you have a grasp of the former you will be well placed to expand your skills to the latter. The main advantage of the generalized linear model is that it can cope with non-scale **response variables**, where the assumptions of normality and/or homogeneity can not be met (see Table 12.3 for examples). The generalized linear model is able to do this by getting you to use a **link** or **linking function** (such as identity, log, or logit) and making you specify a **probability distribution** (such as normal, Poisson, binomial). The link function transforms the fitted value so that it equates to the **linear predictor** in a linear fashion:

Link function (fitted value) = Linear predictor

In Table 12.3 you will find some combinations of link functions and probability distributions that work well with different types of response variables. The generalized

When the response variable is a: - examples	**The distribution is typically:**	**The link function is usually:**	**Sub-framework**
Measure e.g., – Birth weight of a ground squirrel (g) – Area of a field damaged by elephants (km^2) – Mass of a tusk (kg) – Bone density (g/cm^2) – Shoot weight (g) – Tumour size (mm)	Normal	Identity	General linear model
Count/Frequency e.g. – Number of dead individuals – Number of pea plants – Number of cells on a microscope slide – Number of days crops raided by elephants	Poisson	Log	Loglinear model
One of two alternatives (binary) e.g. – Yes/No – Male/Female – Live/Dead – Healthy/Diseased	Binomial (with a trial size of 1)	Logit	(Binary) Logistic model
Proportion e.g. – Percentage male (sex ratio data) – Proportion of individuals surviving to age 5 – Proportion of fields experiencing crop raiding – Proportion of abnormal cells	Binomial	Logit	Logistic model

Table 12.3 Examples of the range of types of response variable that the generalized linear model can handle, with their typical probability distributions and link function that usually works.

	Generalized linear model	General linear model
Similarities	Can handle one or more explanatory (independent) variables. Relies on assumptions about the distribution of data (parametric). Can be used in single or multiple model approaches.	
Differences	Response (dependent) variable can be measured at any level.	Response (dependent) variable must be measured at scale level.
	Can use distributions other than the normal distribution.	Uses the normal distribution.
	A link function, which transforms the linear predictor into the fitted value, must be specified.	A link function is not used (or alternatively you can say that the link function is identity i.e. the fitted value equals the linear predictor).
	Model can only reliably be estimated using ML.	Model can be estimated using OLS or ML.
	Parameter estimates (coefficients) more reliable than model fit (analysis of variance).	Both model fit (analysis of variance) and parameter estimates (coefficients) reliable and useful.

Table 12.4 Comparison of generalized and general linear model. The general linear model is a sub-framework of the generalized linear model. ML stands for maximum likelihood and OLS stands for ordinary least squares (Section 10.1.2).

linear model with an identity link (which doesn't actually do anything to the response variable) and a normal distribution is a general linear model.

Look at Table 12.4 to see a summary of how the general and generalized models compare.

Checklist of key points

- The generalized linear model is a broad and flexible parametric framework for analysing data. The general linear model is a sub-framework of the generalized linear model which uses the normal distribution. The abbreviation GLM may be used to refer to either the generalized linear model or more specifically to the general linear model. (In this book it is only used for the latter.)
- You should consider using a GLM when you are interested in determining the influence of one or more explanatory (independent) variables on a single response (dependent) variable. The response variable needs to be scale level. Explanatory variables can be at any level of measurement, but need to be identified as either continuous or categorical.
- Advantages of GLM:
  - Simplifies your life because procedures equivalent to a range of parametric tests, including those dealing with multiple explanatory variables, are covered by a single framework.
  - Allows you to take an alternative approach to data analyses because it lends itself to generating and comparing multiple models.
- Disadvantages of GLM:
  - Needs computing facilities and specialist software.
  - You still need to know about traditional tests.
  - Not reliable with small samples.

- Models should always be clearly stated. This can be done using word equations in the general form:

  Response variable = Explanatory variable 1 + Explanatory variable 2 + ...

  Whether an explanatory variable is categorical or continuous should also be stated.
- Do not expect your data to fit your model exactly. The differences between observed data points and their corresponding place on a model are called errors or residuals. Error around a model is represented, when included in equations, by an error term.
- GLM involves answering three questions.
  1. What is the best linear model? To do this you need to find the values of the model coefficients in the equation for a line. Coefficients have confidence intervals and *t*-statistics associated with them. (In SPSS this information is produced in a table labelled the Parameter Estimates.)
  2. Does the model overall explain a significant amount of variation in the response variable? And, does each explanatory variable make a significant contribution to the model? To do this you need to use the F statistic. Where *P* values are less than your critical significant level (typically 0.05), this indicates that the model, or one of its component explanatory variables, explains a significant amount of variation in the response variable. If the model is significant, you can use your answer to question 1 for prediction and it is worth asking question 3.
  3. How much does the model explain? To do this you calculate a coefficient of determination (R^2). Remember it is only worth asking this question if the answer to question 2 is yes.
- There are two main sets of statistics with GLM:
  1. F statistics found in the Anova table used to answer questions 2 and 3. In SPSS this table is labelled **Tests of Between-Subjects Effects**.
  2. Model coefficients found in the coefficients table used to answer question 1. They have *t*-statistics and confidence intervals associated with them. In SPSS this table is labelled **Parameter Estimates**.
- For a GLM analysis to be valid it must conform to the criteria of normality, homogeneity, and linearity. These can be assessed using plots of residuals and fitted values. You should disregard a GLM analysis that does not conform to these criteria.
- When you have more than one explanatory variable they may interact. You should include an interaction term in your model if it is special interest and/or without it your model does not fulfil the criteria of linearity.
- The GLM can be used in data analysis in two ways:
  1. Instead of using traditional parametric test procedures, such as *t*-test, one-way Anova, two-way Anova, bivariate linear regression, multiple linear regression, and Ancova.
  2. In an approach where multiple models are generated and compared. This is often a particularly informative approach to take with observational data.
- There are no fixed rules for choosing the best model when you take a multiple model approach, but in general the simpler the better. Statistical tools for model choice include the use of adjusted R^2 values and fully automated procedures. Non-statistical factors that can have an influence on model choice include the goal of the study, study design, results of exploratory analyses, budgetary constraints, and ethical considerations.
- The generalized linear model uses a link function to cope with response variables which are non-normal and/or do not have homogenous variance.

Self-help questions

1. What advantages does using a GLM have over using the tests covered in previous chapters? *Link back: Section 12.1*

2. Name the traditional test that could be done instead of using a GLM on the following models: *Link back: Section 12.1.1 and Table 12.1*
   (a) *Force need to withdraw spinal catheter* = *Position of patient* (sitting, or lying) where *Position of patient* is categorical.
   (b) *Body mass index* = *Sex* (male/Female) + *Age* where *Sex* is categorical and *Age* is continuous.
   (c) *Time spent feeding* = *Rainfall* (mm previous month) + *Temperature* (mean previous month)

3. (a) What are the three GLM questions? *Link back: Section 12.1.4*
   (b) What are the two main sets of statistics in a GLM and how are they used? *Link back: Section 12.3.2*
   (c) What are R^2 and R_{adj}^2? *Link back: Sections 12.1.4 and 12.3.3*
   (d) What does it indicate about a model if its associated *P* value is less than your critical significance level? *Link back: Section 12.3.2*
   (e) What does it indicate about an explanatory variable if its associated *P* value is less than your critical significance level? *Link back: Section 12.3.2*

4. A researcher wanted to investigate what factors affect the levels of stress hormone in dogs. In particular she was interested whether the presence of a toy affected stress levels. Suggest some explanatory variables on which she might want to collect data. Give your answer by writing out the model (without any interaction). *Link back: Section 12.1.2*

5. What are the three criteria models have to meet in order for the results of a GLM analysis to be valid? What two variables do you have to generate in order to check these criteria have been met? *Link back: Section 12.1.6*

6. Use the GLM option in SPSS to analyse the example data in Chapters 9 and 10 (assume the criteria for normality, homogeniety, and linearity are met).
   (a) Write out the key information for the overall models in each case and compare it to the key information from a GLM.
   (b) For the Chapter 9 example data, state and justify whether you think the nitrogen content is different between sites 1 and 3 using your GLM output. *Link back: Sections 12.3.3 and 12.3.4, Chapters 9 and 10.*

13 More on the generalized linear model: logistic and loglinear models

Chapter aims

Chapter 12 introduced the generalized linear model (GLZM) and led you through the sub-framework of the GLZM, known as the general linear model (GLM), designed for use with response variables measured at scale level data that can be assumed to be normally distributed. In this chapter we look at sub-frameworks of the GLZM that can be used for nominal level data with two categories (binary logistic model) and scale data in the form of counts (loglinear model) and how they can be used to assess the relative effects of multiple explanatory (independent) variables on a single response (dependent) variable. We start by dealing with general points relating to the use of the these two models and consider some example data, before looking in more detail at when and how to use, first the binary logistic model, and then the loglinear model. For each model, we work through a basic analysis using SPSS and look at an example of its use from the literature. Finally, we will compare the models and briefly consider what alternatives and extensions are available.

It is worth reading Chapter 12 before reading this chapter in order to introduce yourself to, or refresh your memory about, the generalized linear model. At the very least you should review the notes on terminology used in this book, covered in the Chapter aims in Chapter 12.

13.1 Introduction to the logistic and loglinear models

Like the **general linear model (GLM)** covered in Chapter 12, the **binary logistic model** and the **loglinear model** are sub-frameworks of the **generalized linear model** (GLZM).

This chapter extends using the **generalized linear model** to situations where response variable data cannot be assumed to be normally distributed and therefore do not fulfil the criteria needed to make the results of a **general linear model** reliable. In particular, we are going to look at the **binary logistic model**, which can be used where a **response variable** is measured at nominal level with two categories (such as yes or no, male or female, diseased or healthy) and the **loglinear model** which can be used where a response variable is measured at scale level, but in the form of counts.

In the rest of this introduction I am going to emphasize some general points about the GLZM as listed below.

- Variables and levels of measurement needed.
- What the link functions are and what they do.
- The things you can find out by performing binary logistic and loglinear model analyses.
- The assumptions of binary logistic and loglinear model analyses.

Response variable, **test variable**, **outcome variable**, or **data variable** are alternative names for a **dependent** variable. **Explanatory variable, grouping variable, predictor variable, factor**, or **covariate** are alternative terms for an **independent variable**.

13.1.1 **Variables and levels of measurement**

It is the response variable that determines which sub-framework you should use within a GLZM (Chapter 12, Table 12.3). If the response variable is measured at nominal level as one of two categories (otherwise known as **binary data**), then this indicates that you should use a **binary logistic model**. If the response variable is scale level but in the form of **count data**, then this indicates that you should use a **loglinear model**.

Let us think a little bit more about these two types of data. Biologists often collect binary data, for example by recording presence or absence of a species, whether a specimen is male or female, if tissue is diseased or healthy, or if an individual is alive or dead. Since binary data are nominal, it is definitely not appropriate to assume that they are normally distributed. Only scale data can be normally distributed. The good news is that there is an alternative theoretical distribution that can be used with such data: the **binomial distribution.**

Biologists also count things a lot too, for example, the size of a clutch of eggs in a bird's nest, the richness of species of vertebrates in a forest patch, or the number of cells on an agar plate. Count data is discrete rather than continuous scale data and only comes as 'non-negative integers' (that is positive whole numbers: Section 2.3.2), and as such cannot, technically, ever be normally distributed. Again there is good news in that there is an alternative theoretical distribution that generally works well with count data: the **Poisson distribution**.

The Binary Logistic Model uses the **binomial distribution** and is useful if your response variable is **binary**: nominal level data with just two categories such as alive/dead or presence/absence. The **Loglinear Model** uses the **Poisson** and is useful if your response variable is **count data:** scale level but only positive whole numbers thus not normally distributed.

As mentioned in Section 3.3.4, there is a bit of leeway here, especially if the range your count data can take is reasonably large. For example, the maximum number of eggs a gull might lay is, say, three, whereas for a pheasant it could be 15 or more. Data on clutch sizes in pheasants are more likely to approximate the normal distribution than data on clutch sizes in seagulls. Nevertheless in both cases your results will typically be more reliable if you use a technique that assumes the Poisson distribution rather than the normal distribution.

For explanatory variables, it is the same for the binary logistic and loglinear models as it was for the general linear model covered in Chapter 12: you can have one or more **categorical** or **continuous explanatory variables**, or indeed a mixture.

Data from a **categorical variable**, or **qualitative variable**, are measured using one of a limited number of categories. Data from a **continuous variable,** or **quantitative variable**, can fall anywhere along a continuum.

Here are some examples: If you had data on whether 100 people had contracted flu or not during the previous winter and information on their sex and age, you could use a binary logistic model to assess the effect of the two explanatory variables, *sex (categorical: male/female)* and *age (continuous: years)*, on the response variable, *flu (binary: yes/no)*. If you had information on the number of times they had had flu in the last 10 years rather than if they had had it or not during the previous winter, you could use a loglinear model to assess the effect of the two explanatory variables, *sex (categorical: male/female)* and *age (continuous: years)*, on the response variable, *flu (counts: number of times)*. Just for comparison, if you had information on the maximum body temperature each person had had during their last bout of flu, rather than if they had had it or not during the previous winter or the number of times they had flu in the last 10 years, then you could use a general linear model to assess the effect of the two explanatory variables, *sex (categorical: male/female)* and *age (continuous: years)*, on the response variable, *flu (scale, continuous: maximum temperature in degrees Celsius)*.

13.1.2 **Link functions revisited**

A **GLZM** has three characteristics:

1. A **response variable** and its **probability distribution** (the **random component**).
2. A set of explanatory variables called the **linear predictor** and an **error** term (the **systematic component**).
3. A **linking** (or **link**) **function** connecting the above two.

The **random component** and **systematic component** of a generalized linear model are connected by a **link function** (see Section 12.7).

The link function builds in a transformation (Section 5.3.4) process. What happens is that the **linear predictor** generates a fitted value which is transformed by the **link function:**

Link function (Fitted value) = Linear predictor

With the general linear model the link function is identity and in fact does not actually change anything; it is often not even mentioned, but with other sub-frameworks of the GLZM the link function does do something, so it needs to be both mentioned and understood.

We are going to start by looking at the link function for the binary logistic model which is called **logit**. If a number falls within the range zero to one, then its logit is given by the formula

$$\text{Logit}\,(x) = \log\,(x/1 - x)$$

Logs, or logarithms, can be based on any number, but typically the so-called natural log, which is based on the number known as 'e' (2.718, to three decimal places), is most often used.

Since the response variable in a binary logistic model is one of two categories, often represented by 0 and 1, the linear predictor cannot be predicting directly comparable numbers. What the linear predictor actually generates in a binary logistic model is not the 'raw' response variable, which would be one of the two binary categories, but the probability of getting one outcome compared to the other. For example, using our flu scenario, the model would predict the probability, ranging from zero to one, of a person of a particular age and sex getting flu.

Now let us turn to the link function for the loglinear model which is called **log** and it is simply the log of the number under consideration, as shown in the following formula:

Log (x)

Again, logs based on any number can be used but the natural log is most common.

The binary logistic model uses the **logit** link function and the loglinear model uses the **log** link function.

13.1.3 **Questions and extras**

There are three basic questions to answer about a model: What is it? Is it significant? How good is it? We considered these questions in the context of regression (Section 10.1.3) and GLM (Section 12.1.4), and you might want to look back at these sections before reading on here but in summary they are:

1. What is the model? This question is answered by finding all the b values and c (i.e. the coefficients) in the fitted value equation, which can then be used for prediction if the answer to question 2 is yes.
2. Is the model significant? For models with more than one explanatory variable this question has two parts:

   2a. Does the overall model, including all the explanatory variables, explain a significant amount of the variation in the response variable?

   2b. Does a particular explanatory variable explain a significant amount of variation in the response variable?

   For both the binary logistic and loglinear models, these questions can be answered by calculating likelihood or Wald chi-square statistics.
3. How good is the model? There are various ways of assessing this, but the **Akaike information criterion (AIC)** is widely quoted and often used. The AIC is not an absolute measure of how good the model is, but a relative one, so, strictly speaking, the question it answers is: how good is this model compared to other models based on the same data? The calculation and use of AIC come from what is known as **information theory** or **information-theoretic approach**. AIC is really useful if you are using a model selection approach (Section 12.6). When comparing models the one with the highest AIC is the best quality.

AIC stands for **Akaike information criterion** and comes from **information theory** or the **information-theoretic approach**. It can be used to compare the quality of different models based on the same data where a bigger AIC indicates a better quality model.

Predicted, or fitted (Section 12.1.3) values are calculated by substituting values into the equation, just as for bivariate linear regression (Sections 10.1.2 and 10.3.4) and the general linear model (Sections 12.1.4 and 12.3.3).

13.1.4 **Model assumptions and overdispersion**

In a comparable way to the general linear model, both the binary logistic and loglinear models assume a linear relationship between the link function transformed response variable and the explanatory variables. However, it's not clear how often people check this assumption and certainly it is not universal practice to do so, and at this introductory level we are not going to worry about it any further. However, unlike the general linear model, the binary logistic model and the loglinear model do not assume that residuals are normally distributed, or that the variance of the residuals is the same regardless of the values of explanatory variables.

Overdispersion is when the variability in your data is greater than would be expected, based on the theoretical distribution used in your model.

There is one assumption-related problem that you do need to look out for with the binary logistic and loglinear models and that's **overdispersion.** This is when the variability in your data is greater than would be expected, based on the theoretical distribution used in your model, in our case binary and Poisson respectively. One common cause of this is having excessive zeros in your data. As an example, if you collected count data on the number of cigarettes smoked per day by people in the UK, you would be likely to get excessive zeros because for all non-smokers the value would be zero.

One way of checking for overdispersion is to examinine the deviance (or Pearson chi-square) results that appear in the **Goodness-of-Fit Test** in SPSS. In particular, you are looking for the ratio of these values divided by the degrees of freedom to be close to 1. If these ratios are not close to 1, say nearer 2 or greater, then you may need to address the problem of overdispersion. One solution would be to include more explanatory variables in the analyses, but this may not be an option, in which case you can try using an alternative model type, for example one with count data that uses the *negative binomial* distribution.

For count data, the reverse of too many zeros is another potential issue. Imagine that you collected count data on numbers of cigarettes smoked per day, just from smokers, then you would have no zeros at all. There are *zero-truncated* models that can deal with such situations.

An awareness of whether your data are related or not (Section 2.3) is also crucial to checking that you are using GLZMs inappropriately. If your design includes, for example repeated measures, you will need take this into account (Section 13.6).

13.2 Example data: urban birds

In this section I am going to introduce the data set that I will be using in the worked examples in Section 13.3. The example comes from long-term research conducted by Shelley Hinsley, from Centre for Ecology and Hydrology, and Nancy Harrison and Julia Mackenzie, from Anglia Ruskin University. The example data we are going to use is part of a subset of the data which Julia used for her PhD. She looked at the reproductive success of great tits and blue tits in different habitats, to give insights into the impact of urbanization on British wildlife (Mackenzie, 2010). Julia's findings highlighted the importance of native trees and shrubs to the success of both species, but especially great tits, and demonstrated that we need to understand both the foraging strategies of animals, as well as the characteristics of their habitats, in order to identify and mitigate impacts of urbanization on British wildlife.

The subset of data from this project that we are going to use contains information of fledging success from bird boxes for just two of the sites that Julia studied: Brampton Wood, a patch of native woodland just outside Cambridge, and the Cambridge University Botanic Garden (CUBG), a patch of wooded parkland within the city itself, containing many non-native plant species. In addition to site, the other explanatory variable in the data set gives the date at which eggs were laid relative to the first of

April, which is often the earliest egg laying date in the breeding season. All other things being equal, it is known that birds born earlier in the breeding season tend to be better quality and more likely to fledge, so putting egg date in the model will control for this phenomenon.

The response variable in this example, fledging success, is measured in two ways: a record of whether fledging has occurred or not (binary data) and the number of nestlings that successfully fledge (count data). The binary variable can be derived from the count variable. In this case it is somewhat redundant to look at both, since the count data are reliable and they are both good quality variables. We are just looking at both for illustrative purposes. However, there are circumstances where you might have count data, but decide it is more reliable to look at a derived binary variable. If, for example, your observational conditions made you think that you were sure about whether fledging occurred, or not so confident about the number of nestlings that actually fledged, then a decision to go with the binary version of the variable would lose you precision, but gain you accuracy.

In summary, the example data set for this chapter contains the following variables.

1. Fledging success measured as yes, at least one chick fledged from this box, or no, there weren't any chicks that fledged from this box.
2. Fledging success measured as number of chicks fledged.
3. Site (Brampton Wood or CUBG).
4. Egg date (when the eggs were laid relative to the first of April).

With these data we can explore the influence of the explanatory variables, site and egg date, on the response variable, fledging success, either in its binary (which we will use in a binary logistic model) or count (which we will use in a loglinear model) form. While the characteristics of the response variable are important for determining which model we use, we need to decide if each of our explanatory variables is categorical or continuous. Site is nominal and clearly categorical. Egg date is discrete scale data and can be treated as continuous, because of the wide range of values it can take (Section 2.3.2).

13.3 Using the binary logistic model

13.3.1 When to use

You can use a binary logistic model when you have a response variable that is measured at nominal level with just two categories, and one or more explanatory variables that are either continuous and/or categorical. Before proceeding you should think whether there is any reason that your data are likely to be overdispersed (Section 13.1.4). If you do not come up with a reason, then you can proceed, but it is still good to check for overdispersion as part of the procedure. Issues relating to when and when not to use a binary logistic model were covered in the introduction to this chapter (Section 13.1). They are summarized in Box 13.1.

Box 13.1 When to use a binary logistic model

Use this test when:

- You want to assess the influence of *one or more explanatory variables* (independent variables) on a *single response variable* (dependent variable).
- The explanatory variable(s) is(are) measured at *any level* but the response variable is *nominal level with only two categories* (i.e. *binary*).
- You want to take a *modelling* approach and/or *prediction* would be useful.

Do not use this technique when:

- The response variable is not binary.
- Sample sizes are very small.
- Your data are overdispersed (Section 13.1.4).

13.3.2 Binary logistic models and the four steps

As explained in Section 13.3.2, we are not going to explicitly use our four-step approach now that we have graduated to GLZMs. For the equivalent of step 3, the statistics calculated are chi-squares, but the likelihood or Wald versions, rather than the Pearson version are used in chi-square null hypothesis significance tests (Chapter 6), and we shall still be using 0.05 as our guide when judging statistical significance (Box 13.2).

Box 13.2 Decision using *P* values for the binary logistic model

If $P \leq \alpha \rightarrow$ significant contribution to variation in response variable.
If $P > \alpha \rightarrow$ non significant result.

13.3.3 Worked example: using SPSS

online resource centre

We are going to conduct a binary logistic model analysis on the bird box data (Section 13.2) using SPSS. To do this the data must be entered into SPSS as shown in Figure 13.1. In newer versions of SPSS (version 10 onwards) the **Data Editor** window has a **Variable View** tab and a **Data View** tab. Figure. 13.1a shows the former and Figure. 13.1b the latter. Appendix I has notes on how to enter data into SPSS.

You should start by writing out the model you are considering. In this example, we are going to look at the model:

Fledging Success (yes/no) = Site + Egg Date

where *Fledging Success* is binary
and where *Site* is categorical and *Egg Date* is continuous.

(a)

(b)

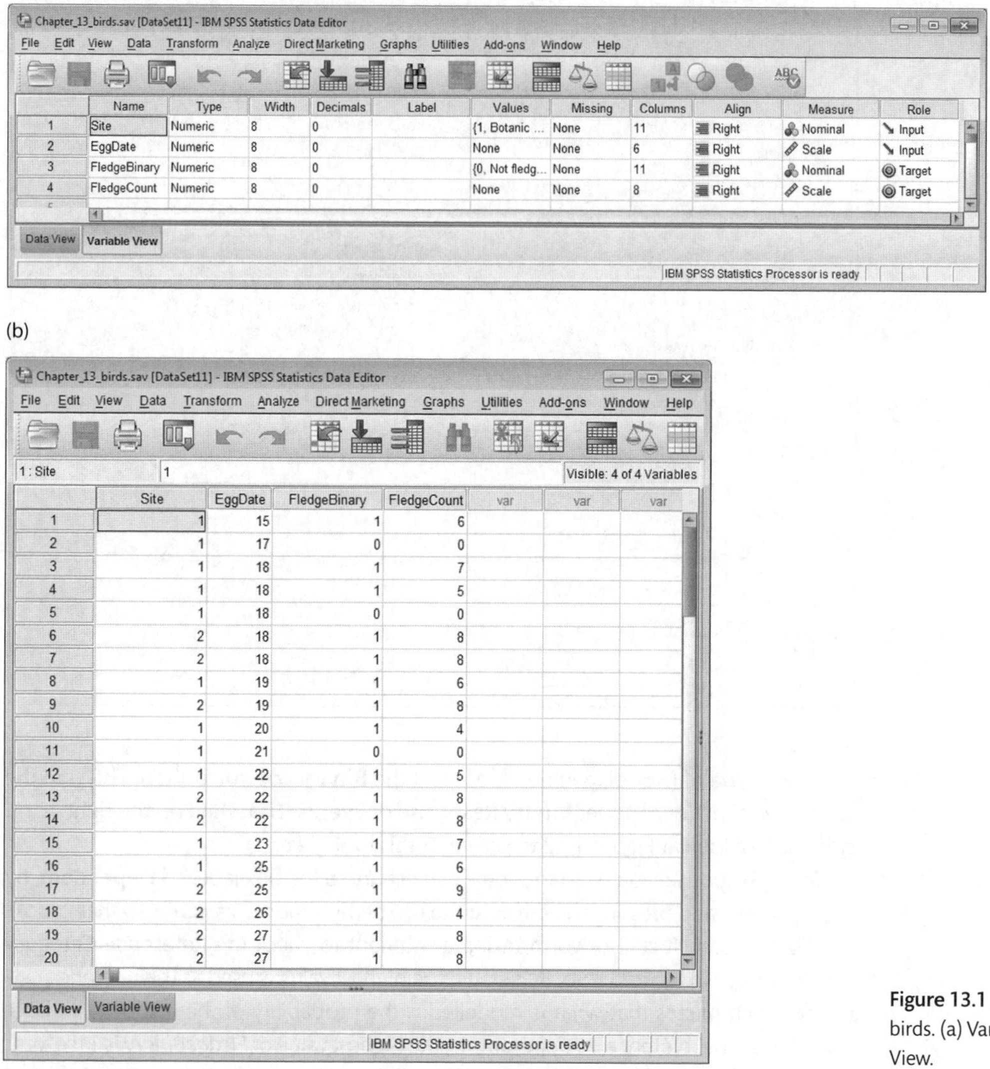

Figure 13.1 Example data in SPSS, birds. (a) Variable View. (b) Data View.

Once you have your data file and your **word equation**, you need to make the following selections:

A **word equation** is a representation of a model using words and simple mathematical symbols. For a GLM, the response variable goes to the left of an equals sign and the explanatory variable(s) to the right. Explanatory variables are separated by plus signs.

Analyze
→Generalized linear models
→Generalized linear models...

A window will appear with nine tabs along the top, we are going to look at just six of them: **Type of Model**, **Response**, **Predictors**, **Model**, **Statistics**, and **Save**. Click on each of these tabs in turn as you go through in the following paragraphs.

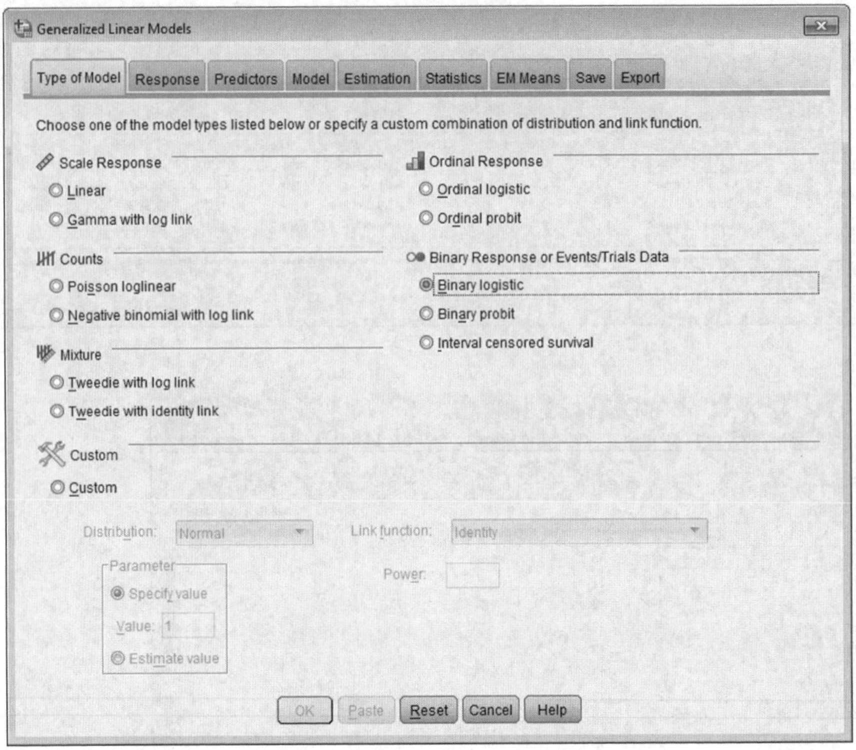

Figure 13.2 Conducting a binary logistic model using SPSS. Dialogue window, Type of Model tab.

Under **Type of Model**, you need to select the **Binary Logistic** option. This is in the second section down under **Binary Response** or **Events/Trial Data** on the right-hand side: Have a look at Figure 13.2 to see what this looks like.

Under **Response,** you need to send *FledgeBinary* to the box labelled **Dependent**, by highlighting it and pressing the arrow at the top of the window **Variable**, so that it looks like Figure 13.3a. It is also worth taking a moment to check the **Reference Category** option, shown in Figure 13.3b. By default, the highest value is used as the reference category, which means that what is modelled is the probability of the response variable using this as the reference. In our example, the highest value of *FledgeBinary* is 1 (i.e. yes), so what will be modelled as default is the probability of it being the lowest value 0 (i.e. no) in comparison with this reference category. In short, if we use the default, we are modelling the probability of *not* fledging. If you would prefer to work with the probability of fledging instead, you just need to change the reference category to the lowest value.

Under **Predictors,** you need to send *Site* to the box labelled **Factors** by highlighting it, then pressing the first arrow down. Then you need to send *EggDate* to the box labelled **Covariates** by highlighting it and pressing the second arrow down. Look at Figure 13.4, this is what you should now see.

Under **Model,** select both variables in the left-hand box, labelled **Factors and Covariates** and click the arrow to send them to the right-hand box, labelled **Model**. In the **Build Term(s)** box, **Type** should be **Main effects**. Once you have done this, it should look like Figure 13.5.

(a) (b)

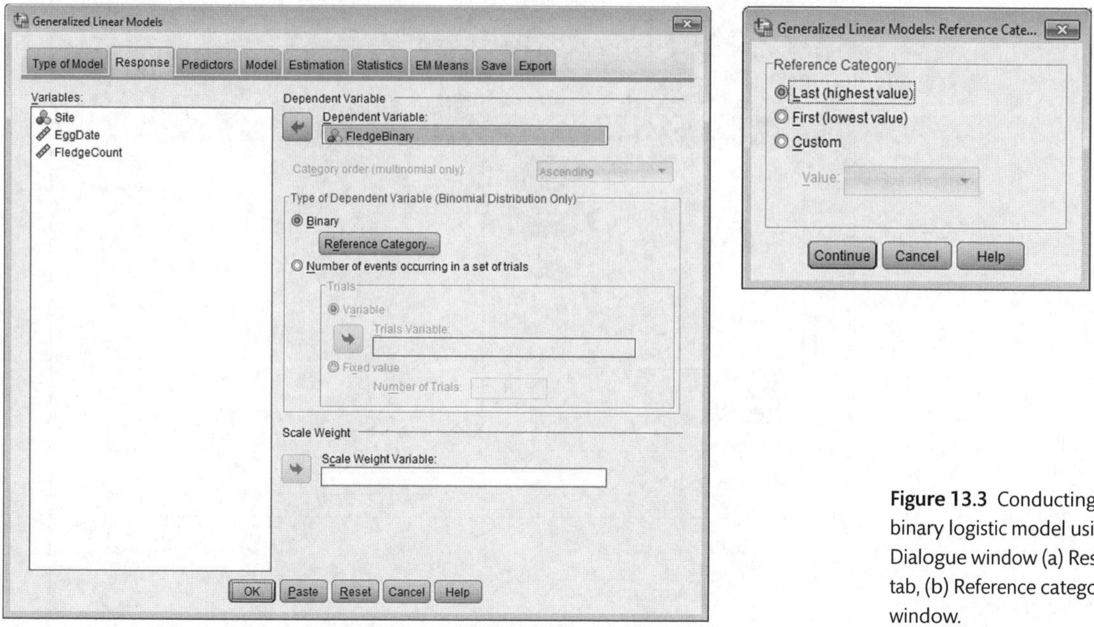

Figure 13.3 Conducting a binary logistic model using SPSS. Dialogue window (a) Response tab, (b) Reference category window.

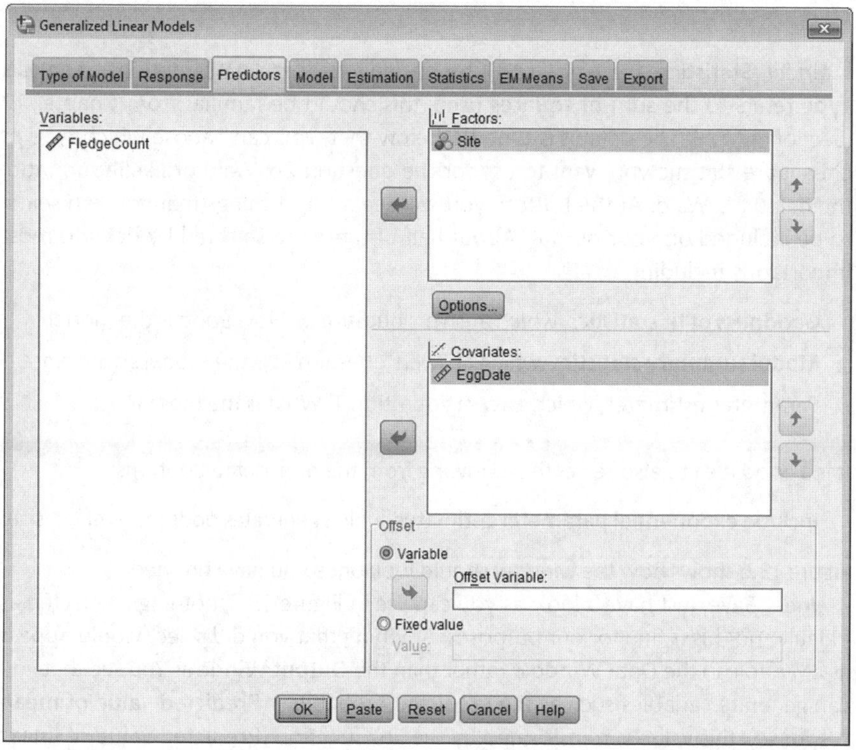

Figure 13.4 Conducting a binary logistic model using SPSS. Dialogue window, Predictors tab.

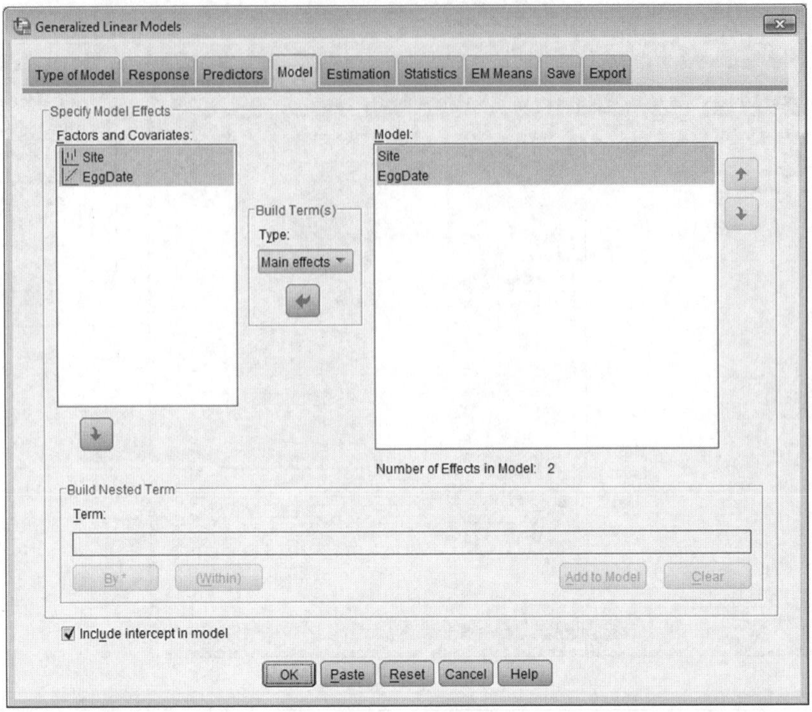

Figure 13.5 Conducting a binary logistic model using SPSS. Dialogue window, Model tab.

Type III (adjusted) sums of squares for an explanatory variable are calculated after all other explanatory variables in the model have been taken into account.

Under **Statistics,** we will largely be sticking to the default options. **Analysis Type** refers to the **sum of squares** used; this should be familiar from Chapter 12 (Section 12.1.5). The default is **type III**. Below that, you can choose which type of chi-square statistic you want to use for the question 2b: Wald or likelihood ratio; the default is Wald. At the bottom you will see a list of things that you can select to be included on your output. About half of these are selected by default, most importantly including:

Goodness of fit statistics, which answers question 3: How good is the model?

Model summary statistics, which answers question 2: Is the model significant?

Parameter estimates, which answers question 1: What is the model?

It is a good idea to also select the following from the non-default options:

Include exponential parameter estimates, which generates odds ratios effect size.

Figure 13.6 shows how the window should look once you have finished.

Under **Save,** just have a look: as you can see in Figure 13.7, nothing is selected as default and this is fine for our purposes. Anything that you did select would appear as a variable in the **Data Window** rather than the **Output Window**. You see that you can generate variables such as the predicted probability (**Predicted value of mean response**), this value is transformed by the link function (**Predicted value of linear**

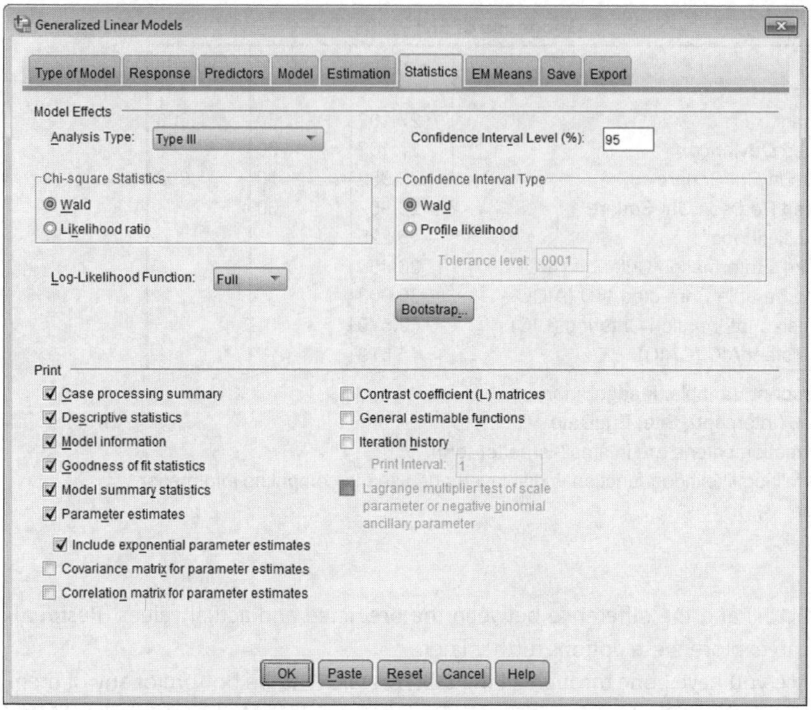

Figure 13.6 Conducting a binary logistic model using SPSS. Dialogue window, Statistics tab.

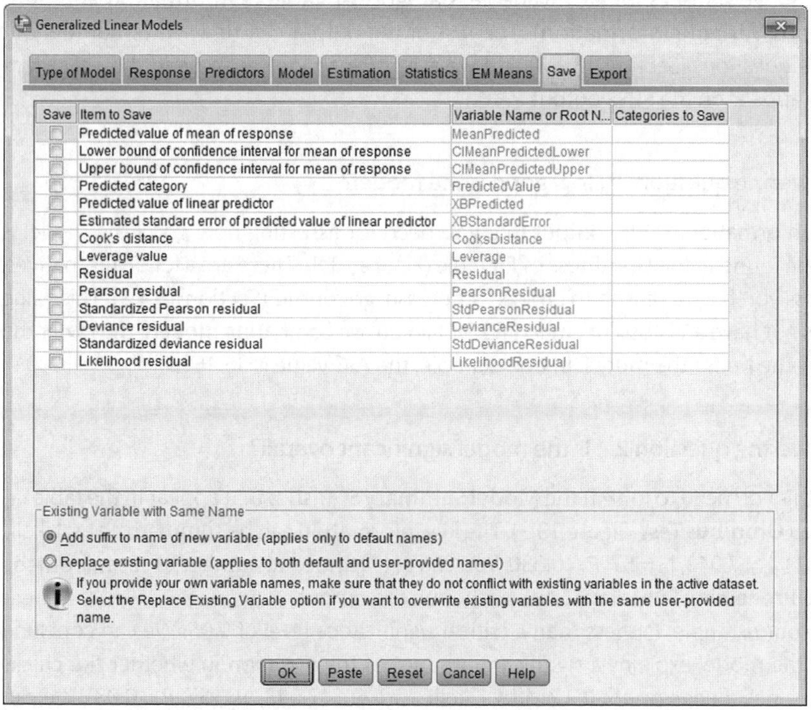

Figure 13.7 Conducting a binary logistic model using SPSS. Dialogue window.

Goodness of Fit[a]

	Value	df	Value/df
Deviance	27.102	30	.903
Scaled Deviance	27.102	30	
Pearson Chi-Square	29.952	30	.998
Scaled Pearson Chi-Square	29.952	30	
Log Likelihood[b]	–15.225		
Akaike's Information Criterion (AIC)	36.450		
Finite Sample Corrected AIC (AICC)	36.903		
Bayesian Information Criterion (BIC)	42.579		
Consistent AIC (CAIC)	45.579		

Dependent Variable: FledgeBinary
Model: (Intercept), Site, EggDate
[a]Information criteria are in small-is-better form.
[b]The full log likelihood function is displayed and used in computing information criteria.

Figure 13.8 SPSS output for a binary logistic model. Goodness of Fit table annotated to show Akaike's Information Criterion (AIC).

predictor) and the difference between the predicted and actual values (**Residual**). You can explore these options further later.

Once you have gone through all the tabs, press **OK** at the bottom of any of them. The output produced starts with some basic information about the model (**Model Information**), a summary of the data used (**Case Processing Summary**), and basic descriptive statistics for each variable (**Categorical Variable Information** and **Continuous Variable Information**). The rest of the output can be used to address our three questions (Section 13.1.3) and we are going to look at these in the order that they appear on the SPSS output.

Answering question 3: How good is the model?

The information on the output that you need for assessing how good the model is can be found in the **Goodness of Fit** table (Figure 13.8). There are a lot of alternatives. Unless you have a reason to choose otherwise, go for AIC (Section 13.1.3). This value does not have any absolute meaning, but if you are comparing models, the lower the value the better the model. In our example, the AIC value is 36.450.

Answering question 2a: Is the model significant overall?

For this you need to look at the **Model Summary Statistics** that appear in the table labelled **Omnibus Test** (Figure 13.9). The key information can be summarized as: Likelihood $X^2_2 = 7.043$, $N = 57$, $P = 0.030$. You can get N, the total number of cases, from the **Case Processing Summary** table further up the output.

In our example, *P* is less than a critical significance level of 0.05. We can conclude that the model explains a significant amount of the variation in whether the chicks from a nest fledge or not (GLZM(b): Likelihood $X^2_2 = 7.043$, $N = 57$, $P = 0.03$). We can

Statistic (likelihood χ^2) **Degrees of freedom** ***P***

Omnibus Test[a]

Likelihood Ratio Chi-Square	df	Sig.
7.043	2	.030

Dependent Variable: FledgeBinary
Model: (Intercept), Site, EggDate
[a]Compares the fitted model against the intercept-only model.

Tests of Model Effects

Source	Type III		
	Wald Chi-Square	df	Sig.
(Intercept)	2.145	1	.143
Site	4.748	1	.029
EggDate	.005	1	.941

Dependent Variable: FledgeBinary
Model: (Intercept), Site, EggDate

Statistic (Wald χ^2) **Degrees of freedom** ***P***

Figure 13.9 SPSS output for a binary logistic model. Model Summary Statistics with Omnibus Test table and Tests of Model Effects table, annotated for key information for the individual explanatory variable *Site*, annotated for key information for overall model.

not tell from this which explanatory variables are associated with the variation in fledging success.

Answering question 2b: Are individual explanatory variables significant?

Having established whether the model overall explains a significant amount of variation in the response variable, we can now look at each explanatory variable to answer whether it explains a significant amount of variation in the response variable. For this we need to look at the **Model Summary Statistics** that appear in the table labelled **Tests of Model Effects** (Figure 13.9). The key information can be summarized as:

Site: Wald $X^2_1 = 4.748$, N = 57, $P = 0.029$.

EggDate: Wald $X^2_1 = 0.005$, N = 57, $P = 0.941$

You can get N, the total number of cases, from the **Case Processing Summary** table further up the output.

In our example, *P* is less than a critical significance level of 0.05 for *Site* but not *EggDate*. We can conclude that the site where the nest is located makes a significant contribution to the amount of the variation in whether or not the chicks from a nest fledge (GLZM(b) for Site: Wald $X^2_1 = 4.748$, N = 57, $P = 0.029$), but that the date that they lay the first egg in their clutch does not (GLZM(b) for Site: Wald $X^2_1 = 0.005$, N = 57, $P = 0.941$).

Intercept (c)

Coefficients for site

Parameter Estimates

Parameter	B	Std. Error	95% Wald Confidence Interval		Hypothesis Test			Exp(B)	95% Wald Confidence Interval for Exp(B)	
			Lower	Upper	Wald Chi-Square	df	Sig.		Lower	Upper
(Intercept)	−3.614	1.8993	−7.336	.109	3.620	1	.057	.027	.001	1.115
[Site=1]	2.466	1.1318	.248	4.685	4.748	1	.029	11.779	1.281	108.275
[Site=2]	0[a]	.	.	.	.	.	.	1	.	.
EggDate	.006	.0793	−.150	.161	.005	1	.941	1.006	.861	1.175
(Scale)	1[b]									

Dependent Variable: FledgeBinary
Model: (Intercept), Site, EggDate
[a]Set to zero because this parameter is redundant.
[b]Fixed at the displayed value.

Coefficient to cut diameter

95% CI for coefficient of EggDate

Wald hypothesis test for EggDate

Figure 13.10 SPSS output for a binary logistic model. Parameter Estimates table with model intercept and key information for model coefficients, annotated for *Site* and *EggDate*.

Answering question 1: What is the model?

For this you need to look at the **Parameter Estimates** table that appears in the table of the same name (Figure 13.10). The second column gives estimates for the coefficients in the model. In our example, if we put these in our equation it would look like this:

$$\textit{Fledging Success} = \begin{array}{l} 2.466 \textit{ for Site 1} \\ 0.000 \textit{ for Site 2} \end{array} + 0.006\ \textit{EggDate} - 3.614$$

The value for the *Fledging Success* produced if you substituted values for the explanatory variables *Site* and *EggDate* into the equation, and calculated it all out, would be in the form of a probability of not fledging, transformed by the **Link Function**. Remember, it's the probability of not fledging rather than fledging, because we used the default **Reference Category** (under the **Response** tab).

As with the general linear model we went through in the previous chapter, the coefficients for the categorical explanatory variables are all relative to the final category, which is set to zero. This is the way SPSS does it by default, but other packages may have different default reference points. **The link function** makes direct interpretation of the coefficients less straightforward, but in our example we can tell that the probability of not fledging is higher at *Site 1* (Botanic Garden) compared with *Site 2* (Brampton Wood), because the coefficient is positive.

Also as for the general linear model, each coefficient is reported along with a confidence interval and a statistic, in this case Wald chi-square. The confidence interval gives you an idea of the precision of a coefficient (Section 4.2.2). For *EggDate*, the coefficient is 0.006 with a 95% confidence interval from –0.150 to 0.161.

For continuous variables, the Wald chi-square can be used as an alternative to the one in the **Tests of Model Effects** table used for answering question 2b for

Parameter Estimates

Parameter	B	Std. Error	95% Wald Confidence Interval		Hypothesis Test			Exp(B)	95% Wald Confidence Interval for Exp(B)	
			Lower	Upper	Wald Chi-Square	df	Sig.		Lower	Upper
(Intercept)	−3.614	1.8993	−7.336	.109	3.620	1	.057	.027	.001	1.115
[Site=1]	2.466	1.1318	.248	4.685	4.748	1	.029	11.779	1.281	108.275
[Site=2]	0[a]	.	.	.	.	.	.	1	.	.
EggDate	.006	.0793	−.150	.161	.005	1	.941	1.006	.861	1.175
(Scale)	1[b]									

Dependent Variable: FledgeBinary
Model: (Intercept), Site, EggDate
[a]Set to zero because this parameter is redundant.
[b]Fixed at the displayed value.

Effect size for EggDate (odds ratio) (annotation pointing to Exp(B) for EggDate)

95% CIs for effect size (annotation pointing to the Lower and Upper values for EggDate)

Figure 13.11 SPSS output for a binary logistic model. Parameter Estimates table with effect size information annotated for *EggDate*.

continuous variables, since they will give the same result. For our example: *EggDate: Wald* $X^2_1 = 0.005$, N = 57, *P* = 0.941.

For categorical variables, the Wald statistic can be used to assess whether each category is significantly different from the reference point which, for SPSS, is the last category. When there are only two categories, again, this can be used as an alternative means of answering question 2b. This applies in our example: *Site 1 compared to Site 2: Wald* $X^2_1 = 4.748$, N = 57, *P* = 0.029. No Wald statistic is reported for the last category, since it's clearly not significantly different from itself.

Effect size

The coefficients in the **Parameter Estimates** table can be used as a measure of effect size, but it's easier to work with odds ratios (Section 5.3), which you can get if you choose the **Include exponential parameter estimates** option on the statistics tab. The **Parameter Estimates** table will then also contain a column displaying the odds ratio (Box 5.3), labelled Exp(B), along with their associated 95% Confidence Intervals. In Figure 13.11, you will see that the odds ratio for *Site 1* (Botanic Garden) is 11.779 (95% CI 1.281 to 108.275), compared with the reference 1 for *Site 2* (Brampton Wood). This means that the odds of not fledging are 11.779 times higher in the Botanic Garden compared with Brampton Wood, controlling for all other variables in the model, in this case just *EggDate*. The odds ratio for EggDate is 1.006 (95% CI 0.861 to 1.175), which is close to 1 (i.e. no effect) and, given the confidence intervals span 1, we can interpret this as no effect.

Checking for overdispersion

If you look at the first and third rows in the **Goodness-of-fit** table, you will see values for **Deviance** and **Pearson Chi-square** with degrees of freedom and, in the final

Goodness of Fit[a]

	Value	df	Value/df
Deviance	27.102	30	.903
Scaled Deviance	27.102	30	
Pearson Chi-Square	29.952	30	.998
Scaled Pearson Chi-Square	29.952	30	
Log Likelihood[b]	−15.225		
Akaike's Information Criterion (AIC)	36.450		
Finite Sample Corrected AIC (AICC)	36.903		
Bayesian Information Criterion (BIC)	42.579		
Consistent AIC (CAIC)	45.579		

Deviance ratio of value to df

Dependent Variable: FledgeBinary
Model: (Intercept), Site, EggDate
[a]Information criteria are in small-is-better form.
[b]The full log likelihood function is displayed and used in computing information criteria.

Figure 13.12 SPSS output for a binary logistic model. Goodness of Fit table annotated to show ratio of deviance to degrees of freedom (used for assessing overdispersion).

column, a ratio of the value to the degrees of freedom (Figure 13.12). Since this ratio is close to 1 in both cases, we don't need to worry about overdispersion in this example.

13.3.4 **Literature link: Death by AMI**

online resource centre

Abildstrom *et al.* (2005) European Heart Journal, 26: 990–995 (OUP).

AMI stands for acute myocardial infarction, or heart attack to the lay person. An increase in troponin, a protein found in muscle, can be used to indicate that an AMI has occurred. Abildstom and colleagues decided to have a look to see if use of this indicator influenced hospitalization rates and mortality in patients in Denmark between 1994 and 2001.

They measured mortality as yes the patient had died within 28 days of being admitted to hospital, or no the patient was still alive 28 days after being admitted to hospital: this is binary. The model they used took into account age, gender, and year, as well as whether troponin levels were used as diagnostic or not.

In word equation form this can be written as:

Death at 28 days (yes/no) = Tropinin + Age + Gender + Year

where Death at 28 days is binary and Troponin, Age, Gender and Year are categorical.

Here is how they phrased it in their methods: 'Trends in case-fatality rates were evaluated using multiple logistic regression analysis with logit as the link function. Age was entered in the model as 10 year age bands (30–39, 40–49,...,90–99 years).'

In their results they conclude: 'The 28 day mortality decreased steadily from 25.9% in 1994 to 17.5% in 2002 and was not affected by the use of troponin measurement. The odds ratio associated with the use of troponin measurement adjusted for age, gender, and calendar year was 1.05 (95% CI 0.98–1.11, P = 0.14)'. This means that the odds of being dead after 28 days were only 0.05 higher if troponin was used as an indicator of AMI compared to if it was not. Since this effect size is small and its confidence intervals are narrow and span 1, this is evidence that the use of troponin as a marker does not influence mortality.

You will notice that these authors are reporting odds ratio effect size rather than the model coefficients or chi-squared statistics associated with the overall model and its explanatory variables. This gives all the information needed in the context and in a form that is useful to decision-making.

13.4 Using the loglinear model

13.4.1 When to use

You can use a loglinear model when you would have a response variable that is measured at scale level in the form of counts which cannot be assumed to be normally distributed, and one or more explanatory variables that are either continuous and/or categorical. Before proceeding you should think whether there is any reason that your data definitely overdispersed (Section 13.1.4). If you do not come up with a reason then you can proceed, but it's good to check for overdispersion as part of the procedure. Issues relating to when and when not to use a GLM were covered in the introduction to this chapter (Section 13.1). They are summarized in Box 13.3.

Box 13.3 When to use a loglinear model

Use this test when:

- You want to assess the influence of *one or more explanatory variables* (independent variables) on a *single response variable* (dependent variable).
- The explanatory variable(s) is(are) measured at *any level*, but the response variable is *scale level in the form of counts*.
- You want to take a *modelling* approach and/or *prediction* would be useful.

Do not use this technique when:

- The response variable is not counts.
- Sample sizes are very small.
- Your data are overdispersed (Section 13.1.4).

13.4.2 Loglinear models and the four steps

As explained in Section 13.3.2, we are not going to explicitly use our four-step approach now that we have graduated to GLZMs. For the equivalent of step 3, the statistics calculated are chi-squares but the likelihood or Wald versions, rather than the Pearson version used in chi-square null hypothesis significance testing (Chapter 6), and we shall still be using 0.05 as our guide when judging statistical significance (Box 13.4).

Box 13.4 Decision using *P* values for the loglinear model

If $P \leq \alpha \rightarrow$ significant contribution to variation in response variable.
If $P > \alpha \rightarrow$ non significant result.

13.4.3 **Worked example: using SPSS**

online resource centre

We are going to conduct a loglinear logistic model analysis on the bird box data (Section 13.2) using SPSS. To do this the data must be entered into SPSS as shown in Figure 13.1. In newer versions of SPSS (version 10 onwards), the **Data Editor** window has a **Variable View** tab and a **Data View** tab. Figure 13.1a shows the former and Figure 13.1b the latter. Appendix I has notes on how to enter data into SPSS.

You should start by writing out the model you are considering. In this example we are going to look at the model:

Fledging Success (number of chicks) = Site + EggDate

where *Fledging Success* is counts and where *Site* is categorical and *EggDate* is continuous.

A **word equation** is a representation of a model using words and simple mathematical symbols. For a GLM, the response variable goes to the left of an equals sign and the explanatory variable(s) to the right. Explanatory variables are separated by plus signs.

Once you have your data file and your **word equation**, you need to make the following selections:

Analyze
→Generalized linear models
→Generalized linear models...

You should see nine tabs along the top; we are going to look at just six of them: **Type of Model**, **Response**, **Predictors**, **Model**, **Statistics**, and **Save**. Click on each of these tabs in turn and do what's described in the following paragraphs.

Under **Type of Model**, you need to select the **Poisson loglinear** option. This is in the second section down under **Counts** on the left-hand side. Have a look at Figure 13.13 to see what this looks like.

Under **Response,** you need to send *FledgeCount* to the box labelled **Dependent**, by highlighting it and pressing the arrow at the top of the window **Variable** so that it looks like Figure 13.14.

Under **Predictors,** you need to send *Site* to the box labelled **Factors**, by highlighting it then pressing the first arrow down. Then you need to send *EggDate* to the box labelled **Covariates**, by highlighting it and pressing the second arrow down. Look at Figure 13.15, this is what you should now see.

Under **Model,** select both variables in the left-hand box labelled **Factors and Covariates** and click the arrow to send them to the right-hand box labelled **Model**.

In the **Build Term(s)** box, **Type** should be **Main effects**. Once you have done this it should look like Figure 13.16.

Under **Statistics, Analysis Type** refers to the **sum of squares** used; this should be familiar from Chapter 12 (Section 12.1.5). The default is **Type III**. Below that you can

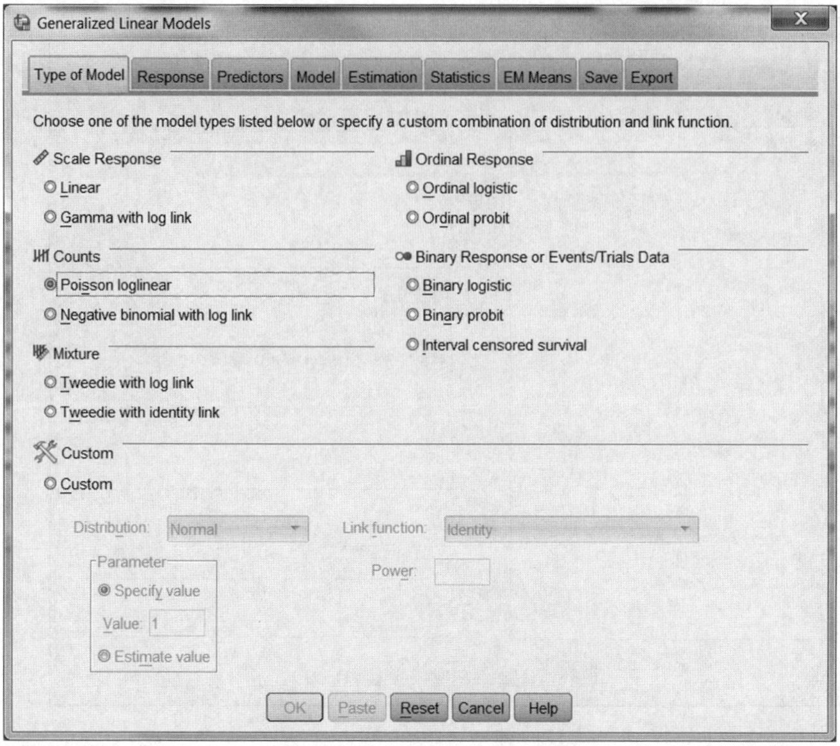

Figure 13.13 Conducting a loglinear model using SPSS. Dialogue window, Type of Model tab.

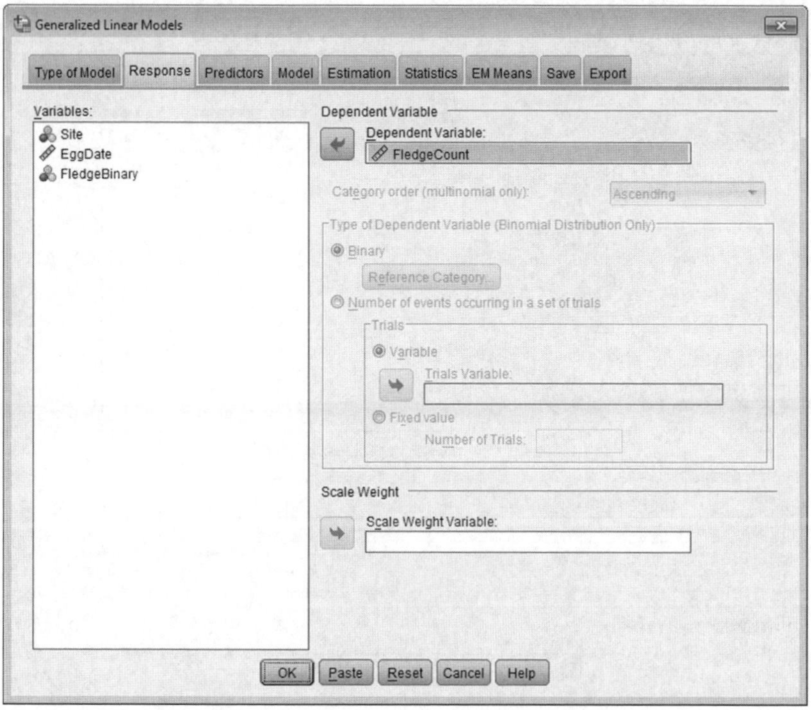

Figure 13.14 Conducting a loglinear model using SPSS. Dialogue window, Response tab.

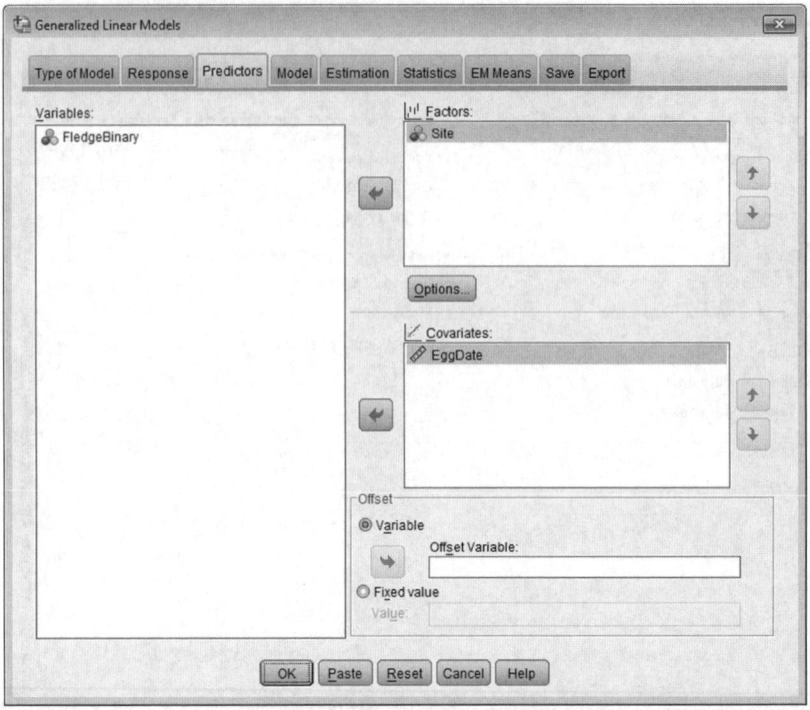

Figure 13.15 Conducting a loglinear model using SPSS. Dialogue window, Predictors tab.

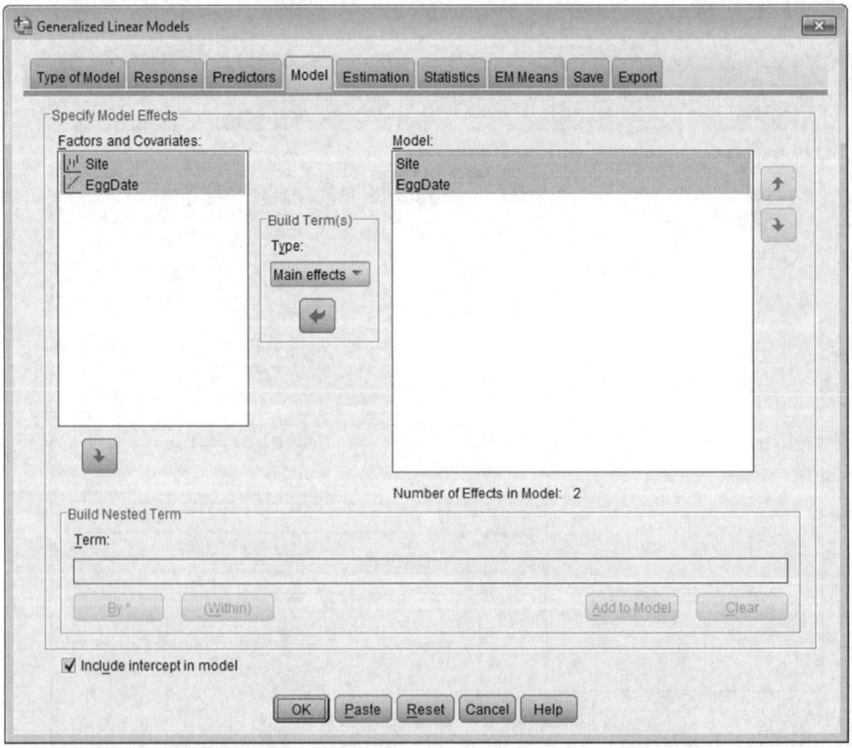

Figure 13.16 Conducting a loglinear model using SPSS. Dialogue window, Model tab.

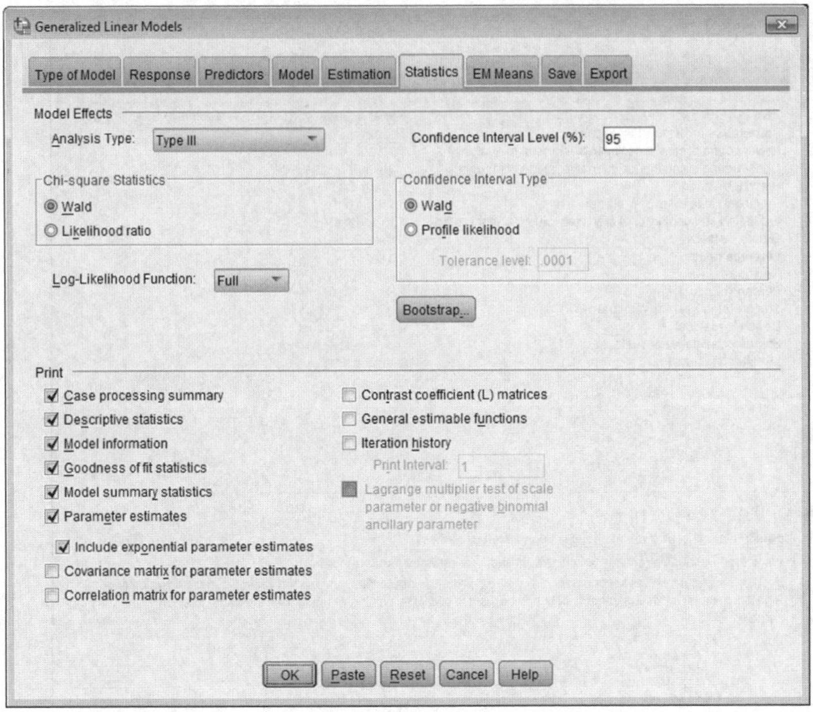

Figure 13.17 Conducting a loglinear model using SPSS. Dialogue window, Statistics tab.

choose which type of chi-square statistic you want to use for question 2b: Wald or likelihood; the default is Wald. At the bottom you will see a list of things that you can select to be included in your output. About half of these are selected by default, most importantly including:

Goodness of fit statistics, which answers question 3: How good is the model?

Model summary statistics, which answers question 2: Is the model significant?

Parameter estimates, which answers question 1: What is the model?

It's a good idea to also select the following from the non-default options:

Include exponential parameter estimates, which generates odds ratios.

Figure 13.17 shows how the window should look once you have finished.

Under **Save**, again just have a look; as you can see in Figure 13.18, nothing is selected as default, and this is fine for our purposes. Anything that you did select would appear as a variable in the **Data Window** rather than the **Output Window**. You see that you can generate variables such as the predicted probability (**Predicted value of mean response**), this value transformed by the link function (**Predicted value of linear predictor**) and the difference between the predicted and actual values (**Residual**). These are all options you can play with later once you have the basic procedure under your belt.

Type I (sequential) sums of squares for an explanatory variable are calculated after any explanatory variables listed before it in the model have been taken into account.

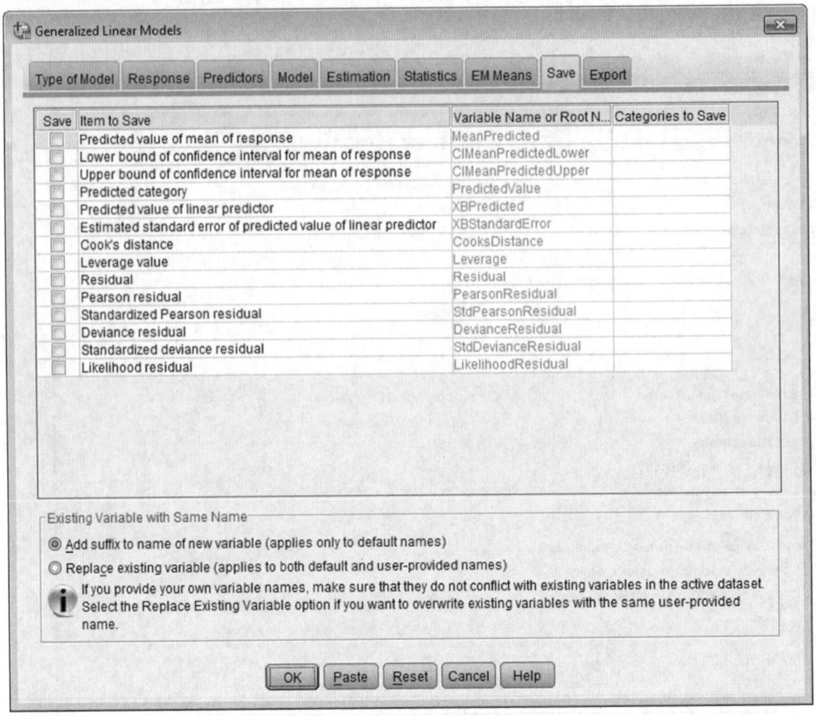

Figure 13.18 Conducting a loglinear model using SPSS. Dialogue window, Save tab.

Once you have gone through all the tabs, press **OK** at the bottom of any of them. The output produced starts with some basic information about the model (**Model Information**), a summary of the data used (**Case Processing Summary**), and basic descriptive statistics for each variable (**Categorical Variable Information** and **Continuous Variable Information**). The rest of the output can be used to address our three questions (Section 13.1.3) and we are going to look at these in the order that they appear on the SPSS output.

Answering question 3: How good is the model?

The information on the output that you need in order to assess how good the model can be, is found is in the **Goodness of Fit** box (Figure 13.19). There are a lot of alternatives. Unless you have a reason to choose otherwise, go for AIC (Section 13.1.3). This value doesn't have any absolute meaning, but if you are comparing models the lower the value the better the model. In our example the AIC value is 283.159.

Answering question 2a: Is the model significant overall?

For this you need to look at the **Model Summary Statistics** that appear in the table labelled **Omnibus Test** (Figure 13.20). The key information can be summarized as:

Goodness of Fit[a]

	Value	df	Value/df
Deviance	86.169	54	1.596
Scaled Deviance	86.169	54	
Pearson Chi-Square	55.582	54	1.029
Scaled Pearson Chi-Square	55.582	54	
Log Likelihood[b]	−138.580		
Akaike's Information Criterion (AIC)	283.159 ← AIC		
Finite Sample Corrected AIC (AICC)	283.612		
Bayesian Information Criterion (BIC)	289.288		
Consistent AIC (CAIC)	292.288		

Dependent Variable: FledgeCount
Model: (Intercept), Site, EggDate
[a]Information criteria are in small-is-better form.
[b]The full log likelihood function is displayed and used in computing information criteria.

Figure 13.19 SPSS output for a loglinear model. Goodness of Fit table annotated to show Akaike's Information Criterion (AIC).

Likelihood $X^2_2 = 39.086$, N = 57, $P = 0.000$. You can get N, the total number of cases, from the **Case Processing Summary** table further up in the output.

In our example, *P* is less than a critical significance level of 0.05. We can conclude that the model explains a significant amount of the variation in the number of chicks that fledged (GLZM(p): Likelihood $X^2_2 = 39.086$, N = 57, $P = 0.000$). We can not tell from this which explanatory variables are associated with the variation in fledging success.

Answering question 2b: Are individual explanatory variables significant?

Having established that the model overall explains a significant amount of variation in the response variable, we can now look at each explanatory variable to answer whether it explains a significant amount of variation in the response variable. For this we need to look at the **Model Summary Statistics** that appear in the table labelled **Tests of Model Effects** (Figure 13.20). The key information can be summarized as:

Site: Wald $X^2_1 = 35.221$, N = 57, $P = 0.000$.

EggDate: Wald $X^2_1 = 0.301$, N = 57, $P = 0.583$.

You can get N, the total number of cases, from the **Case Processing Summary** table further up the output.

In our example, *P* is less than a critical significance level of 0.05 for *Site*, but not *EggDate*. We can conclude that the site where they nest makes a significant contribution to the amount of the variation in the number of chicks fledged (GLZM(p) for Site: Wald $X^2_1 = 35.221$, N = 57, $P = 0.000$), but that when the eggs were laid, in this example at least, does not (GLZM(p) for Site: Wald $X^2_1 = 0.301$, N = 57, $P = 0.583$).

Statistic (likelihood χ^2) | Degrees of freedom | *P*

Omnibus Test[a]

Likelihood Ratio Chi-Square	df	Sig.
39.086	2	.000

Dependent Variable: FledgeCount
Model: (Intercept), Site, EggDate
[a]Compares the fitted model against the intercept-only model.

Tests of Model Effects

Source	Type III		
	Wald Chi-Square	df	Sig.
(Intercept)	111.511	1	.000
Site	35.221	1	.000
EggDate	.301	1	.583

Dependent Variable: FledgeCount
Model: (Intercept), Site, EggDate

Statistic (Wald χ^2) | Degrees of freedom | *P*

Figure 13.20 SPSS output for a loglinear model. Model Summary Statistics with Omnibus Test table and Tests of Model Effects table annotated for key information for the individual explanatory variable *Site*, annotated for key information for the overall model.

Answering question 1: What is the model?

For this you need to look at the **Parameter Estimates** table that appears in the table of the same name (Figure 13.21). The second column gives estimates for the coefficients in the model. In our example, if we put these in our equation it would look like this:

Fledging Success = 0.720 *for Site 1* − 0.005 *EggDate* + 2.192

0.000 *for Site 2*

The value for the *Fledging Success* produced if you substituted values for the explanatory variables *Site* and *EggDate* into the equation, and calculated it all out, would be in the form of a log. To get back to number of eggs you would need to take an anti-log i.e. raise the number known as 'e' (2.718, to three decimal places) to the power of the number calculated.

As for the general linear model we went through in the previous chapter, the coefficients for the categorical explanatory variables are all relative to the final category, which is set to zero. This is the way SPSS does it by default, but other packages may have different default reference points. The involvement of a link function that is not identity complicates the process slightly, but we can tell that the number of chicks fledging is lower at *Site 1* (Botanic Garden) compared with *Site 2* (Brampton Wood), because the coefficient is negative.

Intercept

Coefficients for site

Parameter Estimates

Parameter	B	Std. Error	95% Wald Confidence Interval		Hypothesis Test			Exp(B)	95% Wald Confidence Interval for Exp(B)	
			Lower	Upper	Wald Chi-Square	df	Sig.		Lower	Upper
(Intercept)	2.192	.1807	1.837	2.546	147.151	1	.000	8.949	6.281	12.752
[Site=1]	–.720	.1213	–.958	–.482	35.221	1	.000	.487	.384	.617
[Site=2]	0[a]	.	.	.	.	.	.	1	.	.
EggDate	–.005	.0086	–.022	.012	.301	1	.583	.995	.979	1.012
(Scale)	1[b]									

Dependent Variable: FledgeCount
Model: (Intercept), Site, EggDate
[a]Set to zero because this parameter is redundant.
[b]Fixed at the displayed value.

EggDate coefficient

95% CI of EggDate coefficient

Wald test for EggDate

Figure 13.21 SPSS output for a loglinear model. Parameter Estimates table with model intercept and key information for model coefficients, annotated for *Site* and *EggDate*.

Also, as for GLM, each coefficient is reported along with a confidence interval and a statistic, in this case Wald chi-square. The confidence interval gives you an idea of the precision of a coefficient (Section 4.2.2). For *EggDate*, the coefficient is –0.005 with a 95% confidence interval from –0.022 to 0.012 (Figure 13.21).

For continuous variables, the Wald chi-square can be used as an alternative to the one in the **Tests of Model Effects** table used for answering question 2b for continuous variables, since they will give the same result. For our example: *EggDate: Wald* $X^2_1 = 0.301$, N = 57, $P = 0.583$.

For categorical variables, the Wald statistic can be used to assess whether each category is significantly different from the reference point which, for SPSS is the last category. When there are only two categories, this again can be used as an alternative means of answering question 2b. This applies in our example: *Site 1 compared to Site 2: Site: Wald* $X^2_1 = 35.221$, N = 57, $P = 0.000$. No Wald statistic is reported for the last category, since it's clearly not significantly different from itself.

Effect size

The coefficients in the **Parameter Estimates** table can be used as a measure of effect size, but it's easier to work with odds ratios (Section 5.2), which you can get if you choose the **Include exponential parameter estimates** option on the statistics tab. The **Parameter Estimates** table will then also contain a column displaying the odds ratio (Box 5.3), labelled Exp(B), along with their associated 95% confidence intervals. In Figure 13.22 you will see that the odds ratio for *Site 1* (Botanic Garden) is 0.487 (95% CI –0.958 to –0.482) compared with the reference 1 for *Site 2* (Brampton Wood). This means that for every chick fledged at Brampton Wood, the odds of one fledging at the CUBG were about half (controlling for all other variables in the model, in this case

Parameter Estimates

Parameter	B	Std. Error	95% Wald Confidence Interval		Hypothesis Test			Exp(B)	95% Wald Confidence Interval for Exp(B)	
			Lower	Upper	Wald Chi-Square	df	Sig.		Lower	Upper
(Intercept)	2.192	.1807	1.837	2.546	147.151	1	.000	8.949	6.281	12.752
[Site=1]	–.720	.1213	–.958	–.482	35.221	1	.000	.487	.384	.617
[Site=2]	0[a]	.	.	.	.	.	.	1	.	.
EggDate	–.005	.0086	–.022	.012	.301	1	.583	.995	.979	1.012
(Scale)	1[b]									

Dependent Variable: FledgeCount
Model: (Intercept), Site, EggDate
[a]Set to zero because this parameter is redundant.
[b]Fixed at the displayed value.

Effect size for EggDate (odds ratio)

95% CIs for effect size

Figure 13.22 SPSS output for a loglinear model. Parameter Estimates table with effect size information annotated for *EggDate*.

just *EggDate*). The odds ratio for EggDate is 0.995 (95% CI 0.979 to 1.012), which is close to 1 (i.e. no effect) and, given that the confidence intervals span 1, we can interpret this as no effect.

Checking for overdispersion

If you look at the first and third rows in the **Goodness-of-fit** table (Figure 13.23), you will see values for **Deviance** and **Pearson Chi-square**, with degrees of freedom and in the final column, a ratio of the value to the degrees of freedom. Since this ratio is close to 1 in both cases, we don't need to worry about overdispersion.

Goodness of Fit[a]

	Value	df	Value/df
Deviance	86.169	54	1.596
Scaled Deviance	86.169	54	
Pearson Chi-Square	55.582	54	1.029
Scaled Pearson Chi-Square	55.582	54	
Log Likelihood[b]	–138.580		
Akaike's Information Criterion (AIC)	283.159		
Finite Sample Corrected AIC (AICC)	283.612		
Bayesian Information Criterion (BIC)	289.288		
Consistent AIC (CAIC)	292.288		

Deviance ratio of value to df

Dependent Variable: FledgeCount
Model: (Intercept), Site, EggDate
[a]Information criteria are in small-is-better form.
[b]The full log likelihood function is displayed and used in computing information criteria.

Figure 13.23 SPSS output for a loglinear model. Goodness of Fit table annotated to show ratio of deviance to degrees of freedom (used for assessing overdispersion).

13.4.4 Literature link: sea cows

Burgess *et al.* (2013) Conservation Physiology 1 : 1–14 (OUP).

online resource centre

Sea cows are, as their name suggests, large herbivorous mammals that live in the sea. They are also known as dugongs, are non-distinct grey in colour, tend to live deep in remote turbid waters and are threatened throughout their age range. Adult males and some females have tusks which can cause 'rake' wounds on the other dugongs during copulation and during fights between males. As part of a study to explore the causes of stress in a population off the coast of Australia, Elizabeth Burgess and her colleagues counted the tusk wounds on the middle region of the backs of 158 dugongs using a 'GLM fitted by maximal likelihood', to analyse data on 'tusk rake injury counts (Poisson probability distribution and log-link)'. Their model included reproductive status (pregnant female, non-pregnant female, non-pregnant female with calf, immature female, immature male, and mature male), and season (spring, summer, autumn, winter), as categorical explanatory variables, and they reported Wald (chi-square) statistics and *P* values and interaction is suggested.

13.5 The general, binary logistic, and loglinear models compared

By now you should be happy that when deciding between whether to use a general linear model, a binary logistic model, or loglinear model, it is the characteristics of the response variable that are key. If it is continuous scale data go for the general linear model, if it is binary data go for the binary logistic model, and if it is count data go for the loglinear model.

The way we carried out the general linear model in SPSS in the previous chapter was through the **General linear model** option in the **Analyze** menu, while for the binary logistic and loglinear models in this chapter, we use the **Generalized linear models** choice option. If you selected Linear on the **Model Type** tab in SPSS it would use the identity link and you would get a comparable result to doing a general linear model through the **General linear model** option, except using the chi-squared statistic rather than the F statistic. Since it is more appropriate to use an F statistic with normal distributed data, plus it is easier to include random variables in the **General linear model** option, you should go with that where you can. Table 13.1 compares the three model types we have come across in this and the previous chapter.

13.6 Alternatives and extensions

One of the great things about the GLZM is the range of types of data it can handle. We have covered three of the most common types of biological data, continuous,

Table 13.1 Comparison of binary logistic, loglinear, and general linear models within the generalized linear model, including how to get the answers to our three model questions using SPSS.

GLZM sub-framework	Binary logistic model	Loglinear model	General linear model (see Chapter 12)
Abbreviation	**GLZM(b)**	**GLZM(p)**	**GLZM(n) or GLM**
Response variable characteristics	Nominal level - two categories	Scale level - counts	Scale level - continuous
Response variable assumed distribution	Binary	Poisson	Normal
Key assumptions	Overdispersion and linearity		Normality, homogeneity, and linearity
Explanatory variables	One or more which can be categorical and/or continuous	One or more which can be categorical and/or continuous	One or more which can be categorical and/or continuous
Link function (where x is the value generated by the linear predictor)	Logit Log_e (1-x/x) Where x is probability of getting one category relative to the other (reference) category	Log Log_e (x) Where x is directly comparable to the response variable	Identity x Where x is directly comparable to the response variable
Question 1: What is the model?	Answered from **Parameter Estimates** table		Answered from **Parameter Estimates** table if **Parameter estimates** selected under **Options**
Question 2a: Is the model significant?	Answered using Likelihood X^2 statistic in **Omnibus Test** table		Answered using F statistic for *Corrected Model* in **Tests of Between-Subject Effects** table
Question 2b: Are the individual explanatory variables significant?	Answered using Wald (default) or Likelihood X^2 statistic in **Tests of Model Effects** table		Answered using F statistics for variable in **Tests of Between-Subject Effects** table
Question 3: How good is the model?	Answered by information in **Goodness-of-fit** table e.g. AIC		Answered by R^2 and $R^2_{adjusted}$ values below **Tests of Between-Subject Effects** table
Effect size estimate and 95% confidence intervals	Odds ratios reported in **Parameter Estimates** table if **Include exponential parameter estimates** selected on **Statistics** tab		Answered from **Parameter Estimates** table if **Parameter estimates** selected on **Options** under options

binary, and count, with the general, binary logistic, and loglinear models, but there are lots more options. For example, if you explore the **Type of Model** tab in SPSS, you will see mention of the negative binomial with log-link option, which is good if you have overdispersed count data (Section 13.1.4) or the ordinal logistic model suitable for ordinal data.

One circumstance that we have not covered with GLZM is what to do if your data lack independence, for example, if they are related through being repeated measures, or matched pairs (Section 2.3.1). For this you can turn to the related technique of **general estimating equations (GEE)**. In SPSS you can get to this by choosing the following:

General estimating equations (GEE) are an extension of GLZM appropriate for use where data points are non-independent (e.g. repeated or matched designs).

Analyze
→Generalized linear models
→Generalized Estimating Equations...

You will see that all but one of the tabs are the same as we came across going through the binary logistic and loglinear models in this chapter. Only the first tab, labelled **Repeated**, is new and this is where you specify variables that identify non-independent data.

Checklist of key points

- The binary logistic and loglinear models are sub-frameworks of the generalized linear model (GLZM). Whereas the general linear model (GLM or GLZM(n), Chapter 12) is based on the normal distribution, the binary logistic and loglinear models are based on the binomial and Poisson distributions respectively, and can be abbreviated as GLZM(b) and GLZM(p).
- You should consider using a binary logistic model when you are interested in determining the influence of one or more explanatory (independent) variables on a single response (dependent) variable which is measured at nominal level with only two categories (e.g. alive/dead, healthy/diseased, present/absent, male/female). Explanatory variables can be either continuous or categorical. The link function in a binary logistic model is logit.
- You should consider using a loglinear model when you are interested in determining the influence of one or more explanatory (independent) variables on a single response (dependent) variable, which is measured as counts (e.g. the number of eggs in a nest, the number cells in a petri dish, the number of piglets in a litter). Explanatory variables can be either continuous or categorical. The link function in a loglinear model is log.
- Word equations are a useful way of summarizing your model. They should take the general form:

  Response variable = Explanatory variable 1 + Explanatory variable 2 + ...

  Also state if the response variable is binary or counts and whether each explanatory variable is categorical or continuous.

- The three questions of GLZM apply:
  1. What is the model? To do this you need to find the values of the model coefficients. Coefficients have confidence intervals and chi-square statistics (likelihood or Wald) associated with them. (In SPSS, this information is produced in a table labelled Parameter Estimates.)
  2. Is the model significant? Does the model overall explain a significant amount of variation in the response variable? And, does each explanatory variable make a significant contribution to the model? To do this you need to use the chi-squared statistic (likelihood or Wald). Where *P* values are less than your critical significance level (typically 0.05), this indicates that the model, or one of its component explanatory variables, explains a significant amount of variation in the response variable. If the model is significant, you can use your answer to question 1 for prediction and it is worth asking question 3. (In SPSS this information is produced in the Model Summary Statistics tables, labelled Ominbus Test and Test of Model Effects.)
  3. How good is the model? To do this you calculate the model's goodness-of-fit and look at, for example, the AIC value, which can be used to compare models. Remember it is only worth asking this question if the answer to question 2 is yes.
- Model coefficients can be used to assess effect size, but generating odds ratios, or relative risk, and their associated 95% confidence intervals is a better option.
- In terms of assumptions, the key points to consider are: (1) Are your data independent? If your data are repeated or matched, for example, look at Section 13.6; (2) Is the response variable of the appropriate form (binary for the binary logistic model and counts for the loglinear model)?; (3) Are your data overdispersed (Section 13.1.4)? Having too many zeros is often a cause of this; (4) If you have count data, do you have any zeros?
- You can assess whether your data are overdispersed using the deviance to degrees of freedom ratio: the closer to one the better. You can potentially fix overdispersion by adding variables to your model, or by using the negative binomial distribution.
- If you have non-independent data (e.g. due to a repeated or matched design), consider using general estimating equations (GEE).

Self-help questions

Link back: Section 13.1.1

1. What distributions are the following based on:
   (a) General linear model.
   (b) Binary logistic model.
   (c) Loglinear model.

Link back: Section 13.1.2

2. What are the link functions used in:
   (a) General linear model.
   (b) Binary logistic model.
   (c) Loglinear model.

Link back: Section 13.1.3

3. (a) What statistic is used to assess the statistical significance of the overall model, explanatory variables within the model, and model coeffiecients?

Link back: Section 13.3.3

   (b) What does it indicate if the *P* value associated with these statistics is less than your critical significance level (e.g. 0.05)?

Link back: Section 13.1.3

4. What can you use AIC values to do?

Link back: Sections 5.2, 13.3.3, and 13.4.3

5. What effect size should you report in association with a binary logistic or loglinear model?

Link back: Section 13.1.4

6. What is overdispersion and why do you have to be aware of it?

Link back: Section 13.6

7. When might you find general estimating equations useful?

14 Choosing the right test and graph

Chapter Aims

In Chapters 6 to 13 you have been given a series of recipes to follow for conducting a range of analyses to suit a variety of situations. One aim of this chapter is to show you how to choose between these recipes. A second aim is to give you guidance in which type of graph you might use to help you communicate your findings and to show you how to get SPSS to produce these graphs.

14.1 Introduction to choosing

This chapter is all about choosing the right analyses for your data, and the appropriate charts or tables to help you communicate your findings. In the preceding chapters we have gone through a series of analyses for detecting significant patterns and trends in your data. The procedures have been presented as recipes that are straightforward to follow. The trick to statistical analysis really comes when you have to decide which test to use on a particular set of data. In Section 14.2 I explain how to choose between the various recipes presented in this book, using a flow diagram that I call *The Choosing Chart*. The tests detailed in this book cover the vast majority of scenarios you are likely to meet as an undergraduate. If you cannot find the test you need using this chart you will have to consult one of the books in the Selected further reading section. When you are conducting your own research you should ideally have a particular test in mind before you collect your data as this will help you develop a rigorous design (Chapter 1; Appendix IV). This strategy also gives you the opportunity to keep things simple so that you don't have to look elsewhere.

In Section 14.3 I am going to review some common graphical formats used to support the findings of the procedures covered in Chapters 6 to 13. The presentation of data in a single sample from a single variable was considered in Chapter 3, where histograms, pie charts, and boxplots were introduced as ways of depicting single samples graphically. Chapter 3 also introduced the use of error bars in conjunction with bars or points showing central tendency. The use of points and error bars (errorplots) was further developed in Chapter 4 in relation to estimation and sample comparison involving the standard error and confidence intervals of the mean. In this chapter, we build on this by focusing on the graphical presentation of data from two or more variables. Such techniques are not only to be used to support statistical findings once analyses are complete; they can be used to explore your data visually before employing a statistical test on them. This is a good practice to get into. In fact, in the case of tests of relationship covered in Chapters 10 and 11, you need to do this to make sure that linearity applies.

14.2 Which test?

Before reading this section you should take a good look at Figure 14.1, *The Choosing Chart*. As we proceed, follow the decision-making process on this diagram. Start where it says *The Choosing Chart*. The first thing you need to decide is what sort of test you need.

The first question to ask yourself is:

What type of test do I need?

Start by deciding if you need a test of frequencies. In Chapter 6, I describe how observations are a first stage in data collection but that, from these, frequency distributions can be derived. Tests of frequencies deal with the latter rather than the former.

If you decide that a test of frequencies is for you, you will need to ask:

Do I have one or two sets of categories?

If you have one set, you could use a one-way chi-square test to compare an observed frequency distribution with a theoretical one, based for example, on a Mendelian inheritance ratio. If you have two sets of categories, you could use a two-way chi-square, which looks for an association between the two sets of categories. If you decide you need a test of frequencies, you need to review Chapter 6.

If you eliminate tests of frequencies as a possibility, then ask yourself:

How many independent variables do I have?

If the answer is more than one, you have no further choices to make. The only technique presented in this book that can cope with more than one independent variable is the generalized linear model, covered in Chapters 12 (for dependent variables that are scale level and can be assumed to be normally distributed) and 13 (for dependent variables that are binary (nominal with only two categories) or counts (scale but not normally distrubuted)).

If the answer is that you have only one independent variable, then you can choose between a test of difference and a test of relationship. Ask yourself:

Am I looking for a difference or a relationship?

If you have two or more samples of your dependent variable (measured at the scale or ordinal level), distinguished by different categories of an independent variable, then you should consider a test of difference. These tests are covered in Chapters 7, 8,

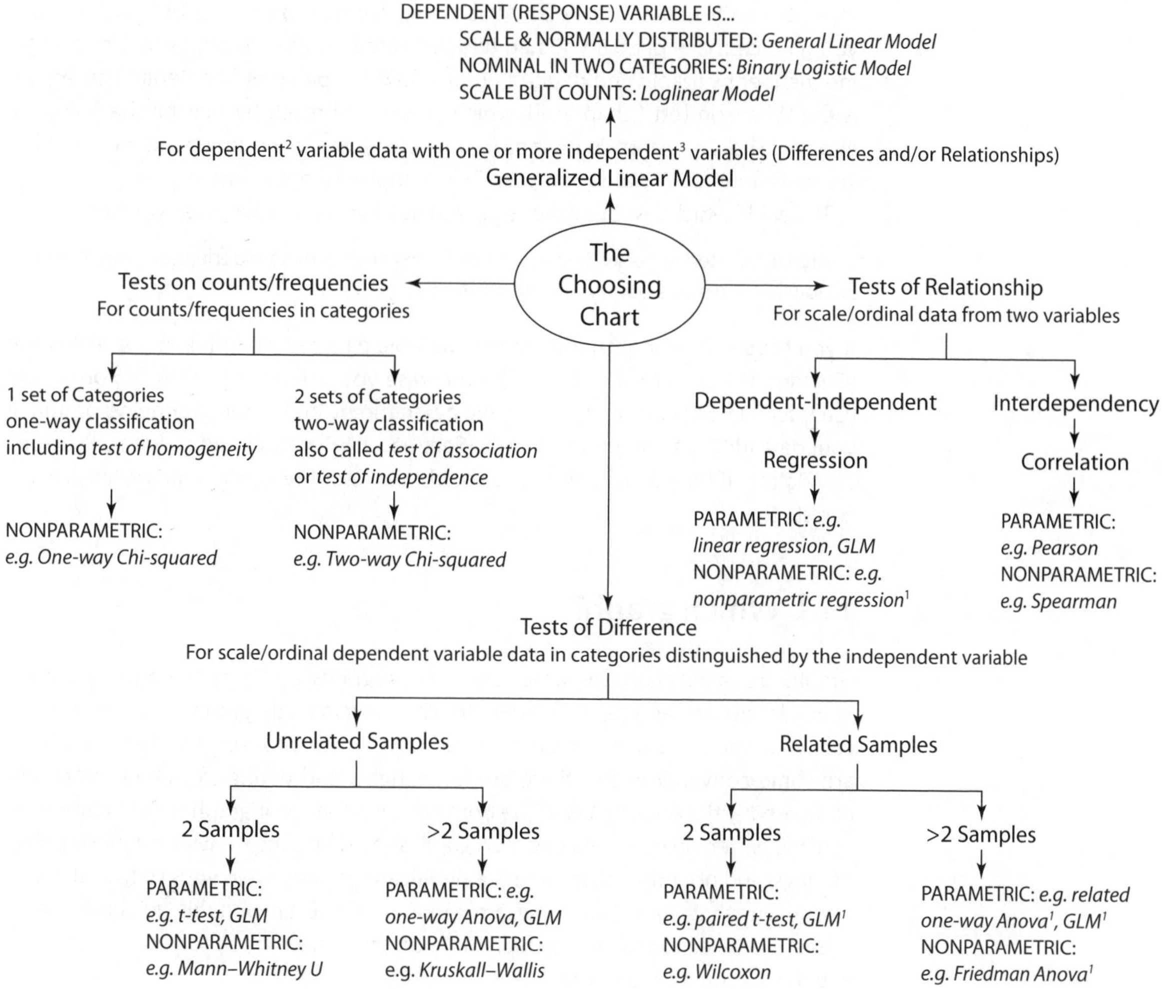

Figure 14.1 The Choosing Chart for selecting the right statistical test. [1]Not covered in detail in this book; [2]response; [3]explanatory.

9, 12, and 13. If you have two samples from two variables then you should consider a test of relationship covered in Chapters 10, 11, 12, or 13.

To decide which test of difference you need, you will have to ask yourself two questions:

How many samples does my independent variable distinguish?

Are the data in my samples related or unrelated?

For example, if you have two unrelated samples distinguished by a single independent variable, then you are looking at a *t*-test or Mann-Whitney U test (Chapter 7). To decide which of these to choose, you need to assess if your data fulfil the parametric criteria for a *t*-test (Section 7.1.3). If they do, then the *t*-test is for you. If not, then it's the Mann-Whitney U test that you should go with. Alternatively, you could look at GLM (Chapter 12). Follow this choice through on Figure 14.1. What would your choices have been if you had two unrelated samples distinguished by a single independent variable and data that didn't fulfil the parametric criteria? The answer is the Wilcoxon test (Chapter 8), which you would reach by making the following choices: The Choosing Chart–not frequencies–One or no independent variable–Test of Difference–Related Samples–Two Samples–Nonparametric.

To decide which test of relationship you need you will have to ask yourself

> Am I interested in how a dependent variable varies with an independent variable or how two variables covary interdependently?

If you have a dependent/independent variable pair you can think about regression (Chapter 10) or GLM (Chapter 12); otherwise you will need to look at correlation (Chapter 11). For correlation you have a parametric and a nonparametric option. If your data fulfil the parametric criteria (Section 11.1.3) you should go with a Pearson's correlation. If they do not, then you should consider the nonparametric Spearman's correlation.

14.3 **Which graph?**

Graphs are called charts by some computer programs, and in papers and reports are generally labelled as figures. The rules for choosing the right graph to help communicate your findings are much less hard and fast than for choosing the right test. There are some conventions but there are fewer rights and wrongs for choosing graphs compared with choosing tests. The bottom line is that your graph should communicate the pattern in your data clearly. A graph should be consistent with your statistical findings, and not imply that there is a significant pattern when your statistical analyses indicate that any apparent pattern is due to chance. Beyond this, you have a lot of scope. However, to help you get a handle on the options open to you I suggest that you think about the following:

- Pie charts or frequency distributions in conjunction with tests of frequencies.
- Boxplots or errorplots (or bar graphs with error bars) to show differences.
- Scatterplots to show relationships.

A good rule to follow when deciding whether to go for a boxplot or graphs with error bars in support of a test of difference is:

- Boxplots to support nonparametric tests.
- Errorplots (or bar graphs with error bars) to support parametric tests.

Lines that best summarize the data can be drawn through the points on a scatterplot. These are known as lines of best fit or fit lines. They are only easy to calculate for regression and it is therefore usually inappropriate to draw fit lines on data you are correlating. Furthermore, I would suggest only using fit lines with regression if the test or model is significant.

One of the most widely adopted graphing conventions assigns dependent and independent variables to whichever axis is appropriate:

- The vertical *y*-axis should be used for dependent variables;
- The horizontal *x*-axis should be used for independent variables.

The only main class of graph we are not going to cover in this book is the line graph. Line graphs look a bit like scatterplots except that the data points are connected by lines. They are typically used to show changes over time. There is a tendency for the inexperienced to use line graphs for everything, but they have quite limited uses.

As an alternative to using a graph of any type, you might consider using a table. Your table might contain raw observations, but more likely frequencies or descriptive statistics. Tables are good if there is a lot of information that you want to summarize. You can also include details from your null hypothesis significance test in a table (the statistic, degrees of freedom and/or sample size(s), and *P* values). You can find examples of the use of tables in the Literature Link sections (e.g. Tables 6.6 and 8.6), and elsewhere in the preceding chapters (e.g. Tables 3.3 and 7.3).

14.4 Worked examples: graphs using SPSS

In this section I am going to use example data introduced in previous chapters to demonstrate how to get SPSS to produce pie charts, boxplots, errorplots, and scatterplots. SPSS works from raw observations rather than frequency distribution or descriptive statistics. If you want to produce graphs directly from frequency tables or descriptive statistics, then you are better off using Microsoft Excel. You should see Excel and SPSS as complimentary in this respect, and note that it is very easy to copy and paste data between these two packages.

The options for SPSS under **Graphs** include **Chart Builder** and **Legacy Dialogs**. You can do most graphs through either route. We met **Chart Builder** in Chapter 3 for producing pie charts and boxplots for single samples, and in Chapter 4 for producing errorplots. We could have done all of these using **Legacy Dialogs** (see Online Resource Centre). In this section, we use **Chart Builder** for pie charts and **Legacy Builder** for boxplots, errorplots, and scatterplots. You could use **Legacy Dialog** for pie charts, but it is slightly more involved to produce two pie charts, as we do here. You could produce exactly the same graphs for boxplot, errorplot, and scatterplot quite easily through **Chart Builder**, except for boxplots for related data.

online resource centre

You can edit the appearance of any graph you produce using SPSS through a **Chart Editor** window, which you call up by double-clicking anywhere on the background of the graph you want to edit in the output window. We touch on this in Section 14.4.4, when adding a fit line to a scatterplot, but otherwise I will leave you to play and learn about this by yourself.

Once you have produced a graph in SPSS, you can incorporate it directly into your report by copying and pasting it into, for example, a Microsoft Word document. To do this you need to select the desired graph in the SPSS output and use the **Edit**, **Copy Object** facility. Once you have done this you just open your Word document and paste the graph in at the appropriate point.

14.4.1 Pie charts

The data I will be using to demonstrate how to produce pie charts are the Mikumi elephant data, described in Section 6.2.2. We are going to do two pie charts that could be used to support the result of the two-way chi-square test:

online resource centre

Graphs
→ Chart Builder...

This will bring up the **Chart Builder** dialogue window (Figure 14.2a). (If a window called **Element Properties** is in the way, move it to one side or close it.) On the **Gallery** tab in the bottom half of the window select **Pie/Polar** from the **Choose from** list. Look at Figure 14.2a, this is what you should see. Now select the simple pie chart (your only option) and drag it to the **Chart preview** area. Next select the **Groups/Points ID** tab (Figure 14.2b). Now select the variable *Group Type* from the **Variables** box on the left and drag it to the area marked **Sliced by?** in the **Chart preview...** area, and the variable *Season* to the area marked **Panel?**. You might have to wiggle the dragged variables around a bit until they 'connect', which is indicated by the blue rectangle marking the area going black. At this point, the display should look like Figure 14.2b. Click **OK**. The output you will produce is shown in Figure 14.3.

In Chapter 6, we concluded from the statistics that *Group Type and Season* were associated (two-way classification chi-square, $X^2_3 = 19.30$, $N = 562$, $P = 0.000$; Figure 14.3). The pie charts show how the proportion of Solitary Bulls and Bull Groups compared with Family Groups (with and without bulls) was higher in the wet season.

14.4.2 Boxplots

Unrelated samples

To demonstrate how to get SPSS to produce boxplots for unrelated samples, I am going to use the bone data described in Section 7.2. After opening the data file you need to choose:

online resource centre

Graphs
→ Legacy Dialogs
→ Boxplot...

This will bring up the window shown in Figure 14.4a. We want a **Simple** boxplot and, since the samples are unrelated, **Summaries for groups of cases** should be selected under **Data in the Chart are**. After checking that these options are selected,

(a)

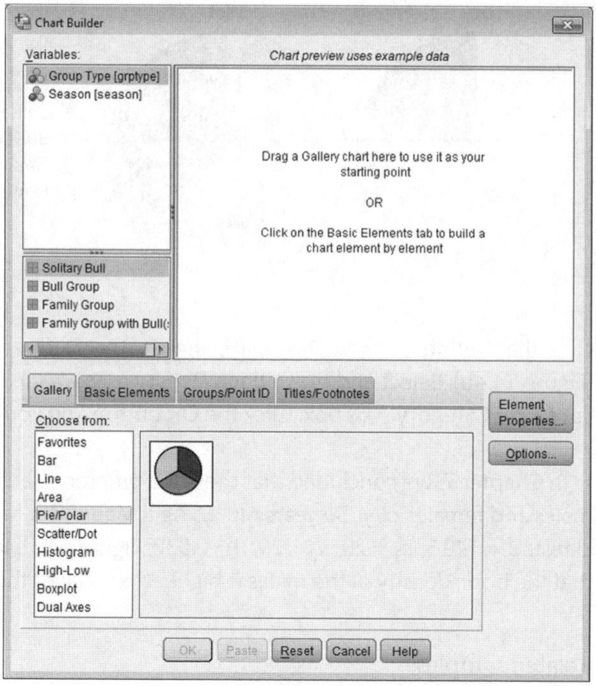

(b)

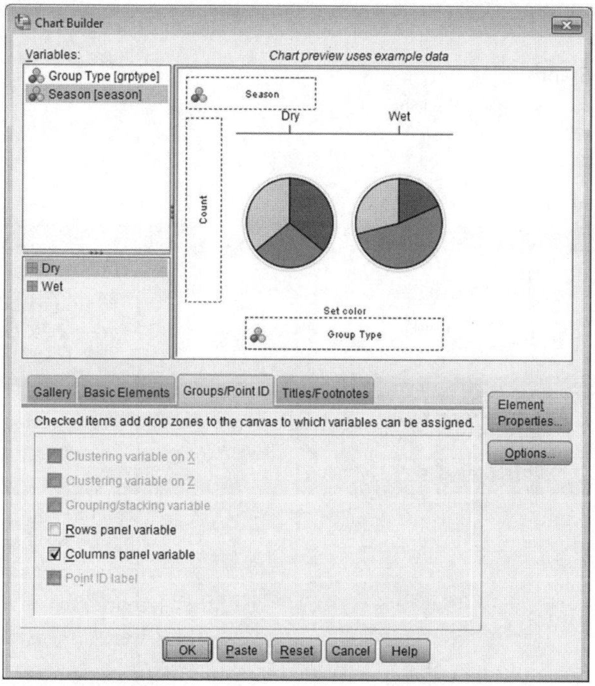

Figure 14.2 Producing pie charts using SPSS. (a) Chart Builder dialogue window. (b) Chart Builder dialogue window with pie chart selections made.

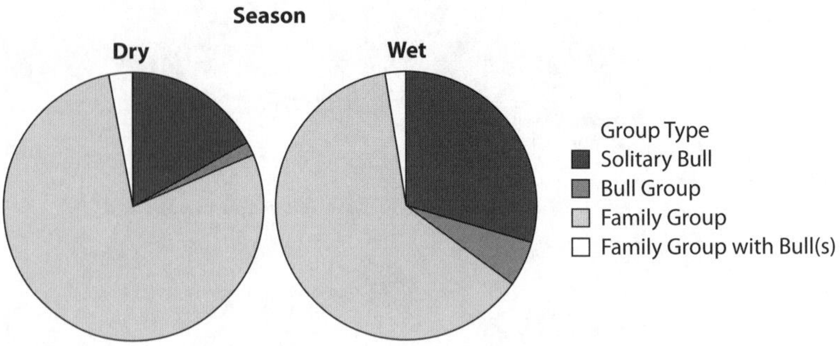

Figure 14.3 Pie charts from SPSS output.

click the **Define** button. This will bring up the **Define Simple Boxplot** window (Figure 14.4b). Select and move *Bone Density Measurement* to the **Variable** box and *Sex* to the **Category Axis** box. Click the **OK** button and you will get the output shown in Figure 14.5.

In Chapter 7, we concluded that there is a difference between the bone density of males and females over 50 years old, using a Mann–Whitney U test (Mann–Whitney U test: $U = 140.5$, $n_1 = 20$, $n_2 = 20$, $P = 0.032$; Figure 14.5). In Figure 14.5, you can see that the bone density of the males is higher than that of the females.

Related samples

To demonstrate how to get SPSS to produce boxplots for related samples, I am going to use the ewe data described in Section 8.2. After opening the data file you need to choose:

Graphs
→ Legacy Dialogs
→ Boxplot

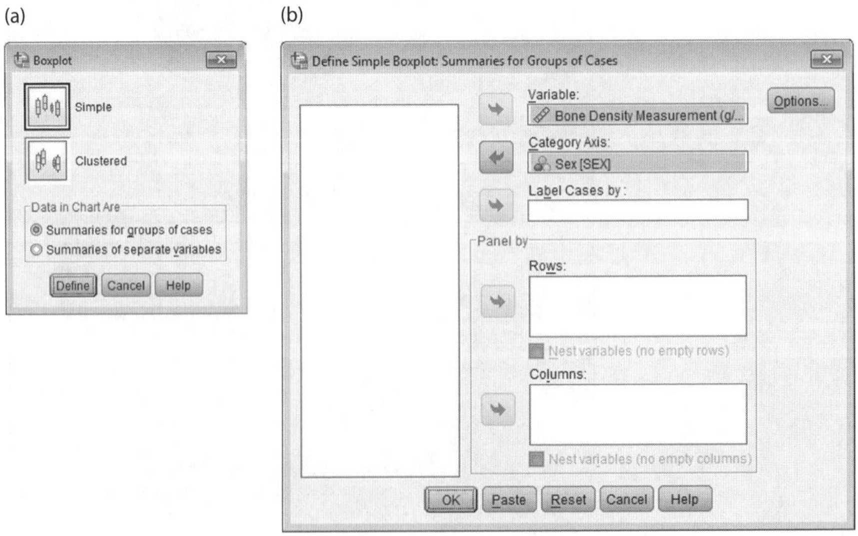

Figure 14.4 Producing boxplots for unrelated data using SPSS. (a) Main dialogue window. (b) Define Simple Boxplot dialogue window.

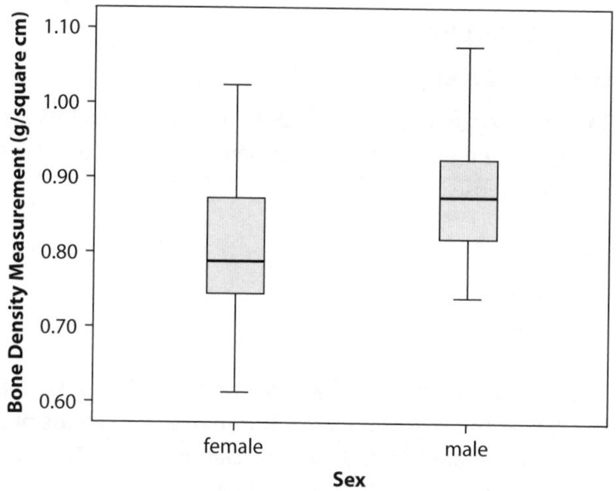

Figure 14.5 Boxplots from SPSS output.

This will bring up the window shown in Figure 14.6a. Again we want the **Simple** boxplot option, and up until this point the procedure is the same as for unrelated samples. However, for related samples you need to select the **Summaries of separate variables** option. Click the **Define** button to bring up the **Define Simple Boxplot** window (Figure 14.6b). Select *Without lamb* and *With lamb* and send them to the **Boxes Represent** box. Clicking the **OK** button will give you output, including boxplots, such as are shown in Figure 8.1 (Chapter 8).

In Chapter 8, we concluded that the time ewes spent grazing was different, depending on whether they did or did not have lambs, by using a Wilcoxon signed-rank test (Wilcoxon signed-rank test: $T=107$, $n=16$, $N=16$, $P=0.044$; Figure 8.1). By referring to Figure 8.1, we can see that the time they spent grazing when they had lambs was higher than when they did not.

(a) (b)

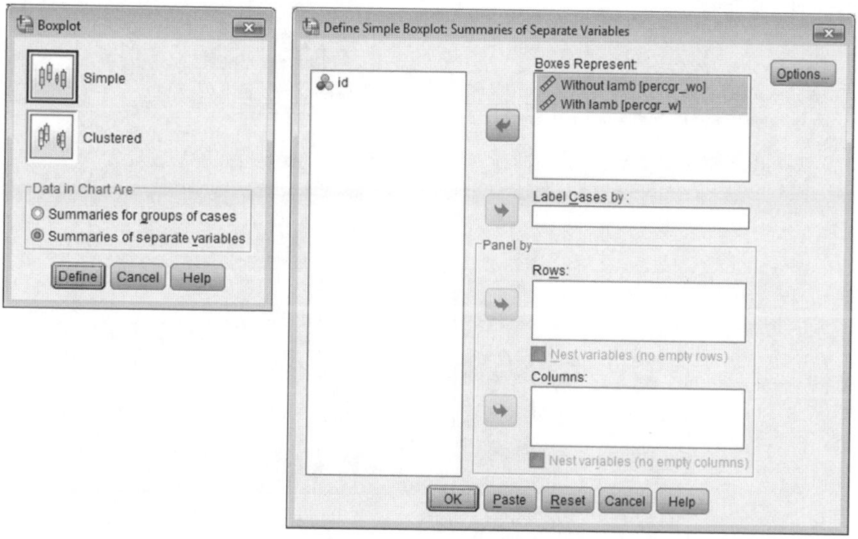

Figure 14.6 Producing boxplots for related data using SPSS. (a) Main dialogue window. (b) Define Simple Boxplot dialogue window.

14.4.3 **Errorplots**

Unrelated samples

To demonstrate how to get SPSS to produce errorplots for unrelated samples, I am going to use the bone data described in Section 7.2. After opening the data file you need to choose:

online resource centre

Graphs
→ Legacy Dialogs
→ Error Bar...

This will bring up the window shown in Figure 14.7a . We want **Simple** errorplots and the data in the chart should be **Summaries for groups of cases**. After checking that these options are selected, click the **Define** button. This will bring up the **Define Simple Error Bar** window (Figure 14.7b). Select and move *Bone Density Measurement* to the **Variable** box and *Sex* to the **Category Axis** box. Check that **Confidence interval for mean** is selected from the drop-down menu under **Bars Represent** with **Level** as **95%**. Click the **OK** button and you will get output, including errorplots, such as that shown in Figure 7.2 (Chapter 7). As an alternative, you could have selected **Standard error of mean** or **Standard deviation** from the **Bars Represent** drop-down menu.

In Chapter 7 we concluded that there is a difference between the bone density of males and females over 50 years old using a *t*-test (*t*-test: $t_{38} = 2.06$, $P = 0.046$; Figure 7.2). Reference to Figure 7.2 indicates that the bone density of the males is higher than that of the females.

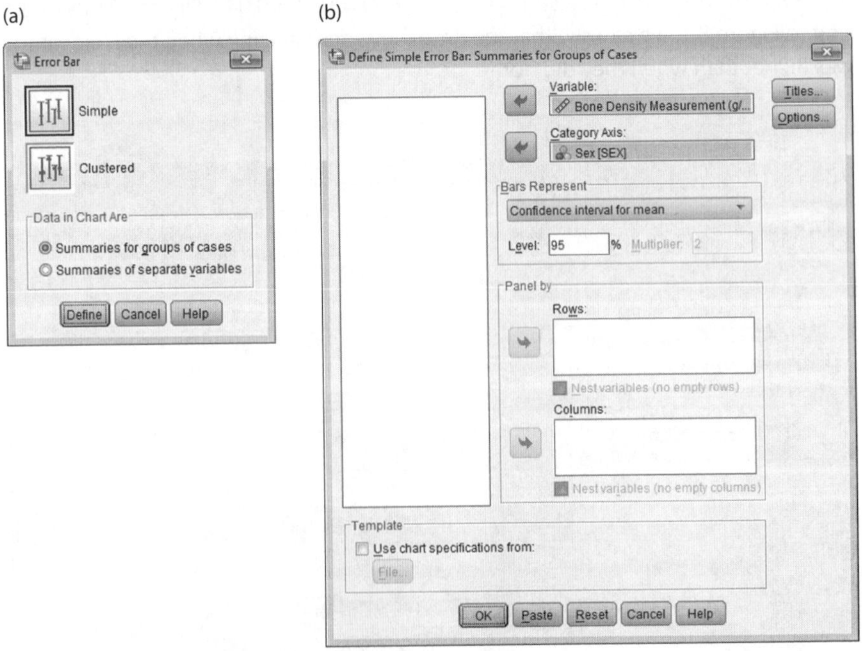

Figure 14.7 Producing errorplots for unrelated data using SPSS.
(a) Main dialogue window.
(b) Define Simple Error Bar dialogue window.

Related samples

To demonstrate how to get SPSS to produce errorplots for related samples, I am going to use the ewe data described in Section 8.2. After opening the data file you need to choose:

Graphs
  → Legacy Dialogs
    →Error Bar...

online resource centre

This will bring up the window shown in Figure 14.8a. We want the **Simple** style and **Summaries of separate variables** for the **Data in Chart Are** options. Click the **Define** button to bring up the **Define Simple Error Bar** window (Figure 14.8b). Select *Without lamb* and *With lamb* and send them to the **Error Bars** box. Check that **Confidence interval for mean** is selected from the drop-down menu under **Bars Represent** with **Level** as **95%**. (The other options open to us in the **Bars Represent** drop-down menu are **Standard error of the mean** or **Standard deviation**.) Look at Figure 14.8b, this is what you should now see. Click the **OK** button and you will get the output shown in Figure 14.9.

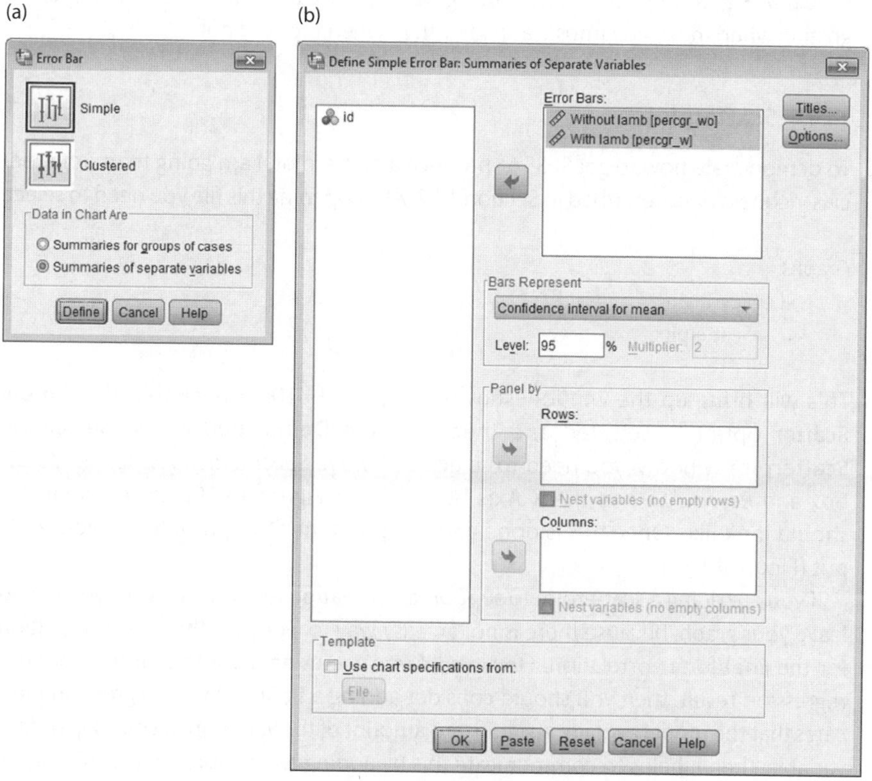

Figure 14.8 Producing errorplots for related data using SPSS. (a) Main dialogue window. (b) Define Simple Error Bar dialogue window.

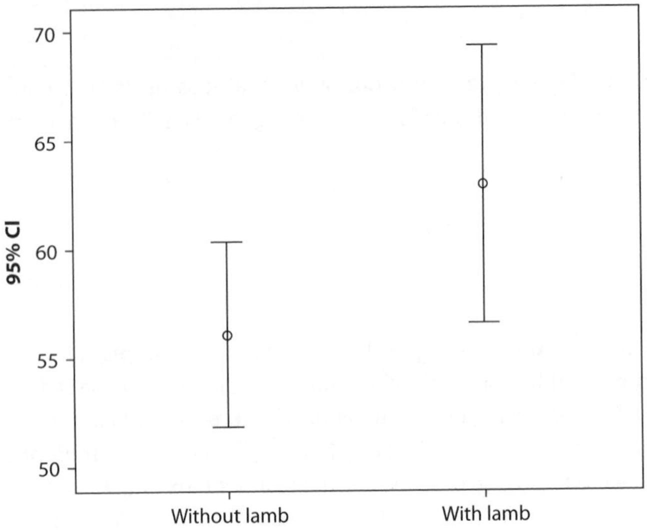

Figure 14.9 Errorplots from SPSS output.

In Chapter 8, we concluded that the time ewes spent grazing was different when they did and did not have lambs using a paired *t*-test (paired *t*-test: $t_{15} = 2.226$, $P = 0.042$; Figure 14.9). By referring to Figure 14.9, we can see that the time they spent grazing when they had lambs was higher than when they did not.

14.4.4 **Scatterplots**

To demonstrate how to get SPSS to produce a scatterplot, I am going to use the species-richness data described in Section 10.2. After opening this file you need to select:

online resource centre

Graphs
→ Legacy Dialogs
→Scatterplot...

This will bring up the window shown in Figure 14.10a. Check that the **Simple Scatter** option is selected and then click the **Define** button. In the **Simple Scatterplot** window, you need to select and move *Species Richness* to the **Y Axis** box and *Reserve Size* to the **X Axis** box. Look at Figure 14.10b to see what this should look like. Once this is done you can press the **OK** button to get your output (Figure 14.11).

If you are doing a scatterplot to support a correlation result, this is how you would leave your graph, because there is not an easy way to work out the model (equation for the line) for a correlation. However, if you are using a scatterplot to support a regression result, then you should consider adding a fit line. If your regression indicates that the model explains a significant amount of the variation in your dependent variable, then it helps to communicate this by adding a line representing this model to your graph.

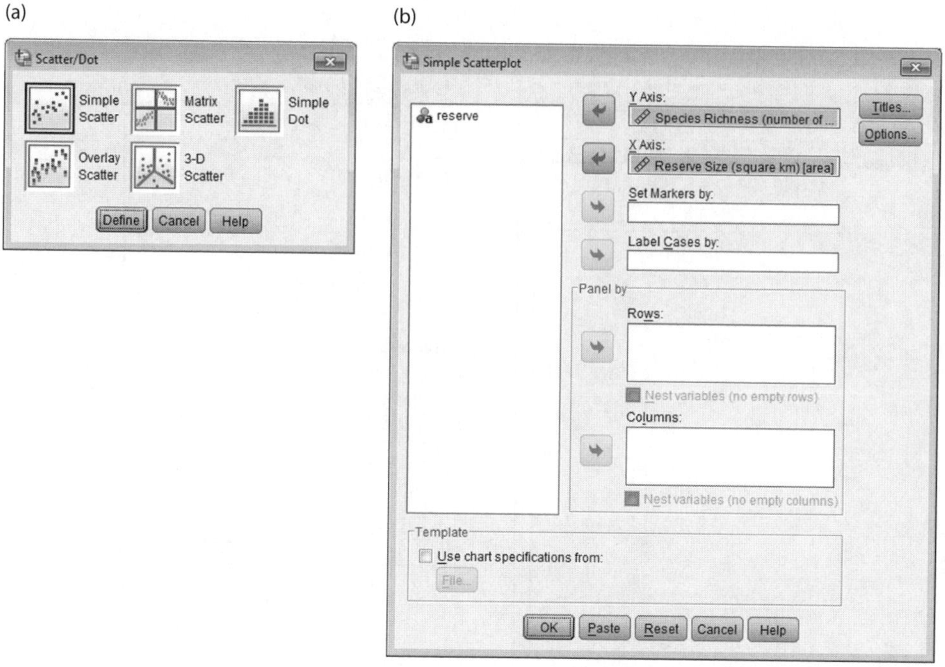

Figure 14.10 Producing a scatterplot using SPSS. (a) Main dialogue window. (b) Simple Scatterplot dialogue window.

To add a fit line to a scatterplot you must first double-click on the chart in your output to launch the **Chart Editor** window (Figure 14.12a). Then, click one of the data points on the graph to select all the data points. Click the **Add Fit Line** icon (Figure 14.12a) to bring up the **Properties** window (Figure 14.12b). Select the **Linear Fit Method**, click the **Apply** and then the **Close** button. This will give you the output shown in Figure 14.13.

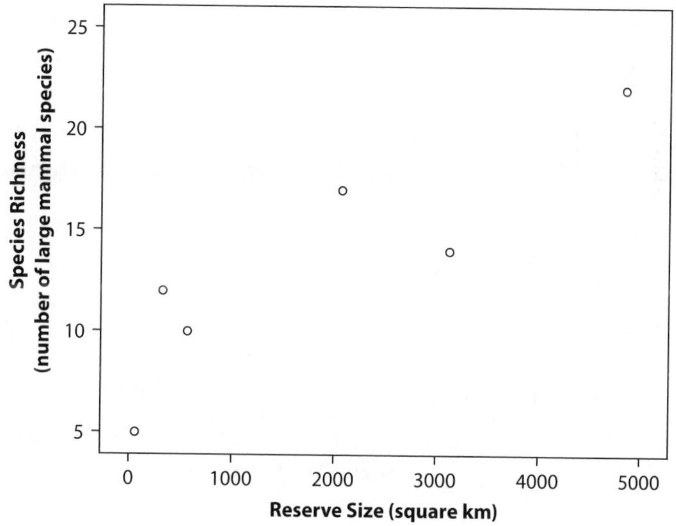

Figure 14.11 Scatterplot from SPSS output.

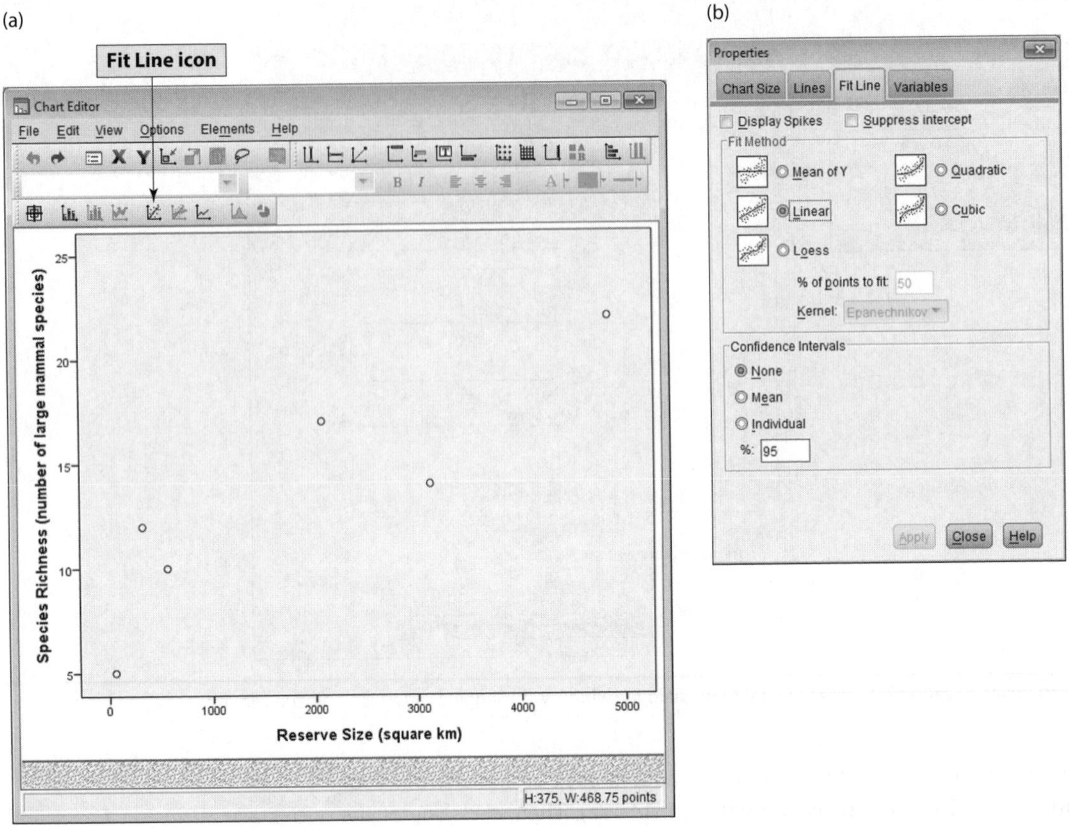

Figure 14.12 SPSS Chart Editor window for a scatterplot. (a) Main window. (b) Properties dialogue window.

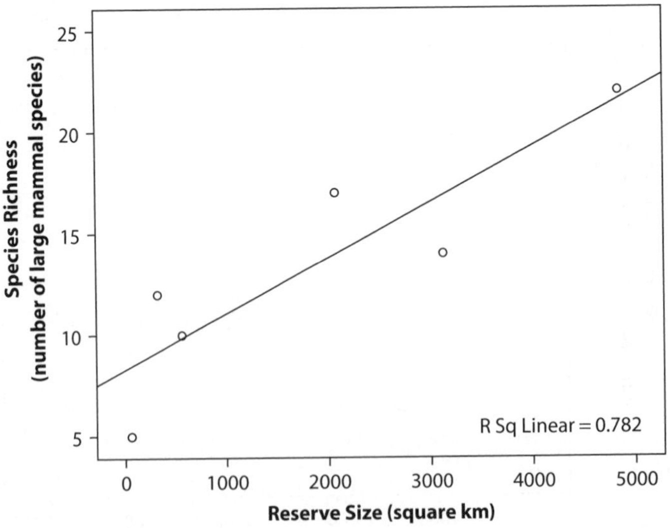

Figure 14.13 Scatterplot with fit line from SPSS output.

The **Linear Fit Method** in SPSS calculates the regression line that best fits the points, using the ordinary least squares method (Section 10.1.2). This line will have the slope (b) and constant (c) presented in the **Coefficients** table on the regression output (Figure 10.6 , Chapter 10).

In Chapter 10, we concluded that species richness is linearly related to reserve size (regression: $y = 0.003x + 8.310$; $F_{1,4} = 14.352$, $P = 0.019$; $R^2 = 0.782$; Figure 14.13). Figure 14.13 is a useful supplement to this result, helping to communicate the nature of the relationship between reserve size and species richness.

Checklist of key points

- Use the choosing chart (Figure 14.1) to help you decide which test you need.
- The first question to ask yourself is what type of test do you need: frequency, difference, or relationship? It is easiest to decide whether you have frequency data first, and if not to then focus on how many independent variables you have, and whether it is a test of difference or a test of relationship.
- For tests on frequencies you have to consider whether you are dealing with one (one-way classification chi-square test) or two (two-way classification chi-square test) sets of categories.
- For tests of difference you have to consider:
  - whether your samples are unrelated (*t*-test, Mann-Whitney U test, one-way Anova, Kruskal-Wallis Anova, GLM) or related (paired *t*-test or Wilcoxon signed-rank test);
  - whether you have two samples (*t*-test, Mann-Whitney U test, paired *t*-test, Wilcoxon signed-rank test, GLM) or more than two samples (one-way Anova, Kruskal-Wallis Anova, GLM).
- For tests of relationship you have to consider whether you are dealing with a dependent/independent variable pair (regression) or interdependency (correlation).
- Your data must fulfil the parametric criteria for a parametric test. If they do not, and you only have one independent variable, you can use the nonparametric alternative.
- The Generalized Linear Model can handle multiple independent (explanatory) variables which are categorical (fixed factors) or continuous (covariates). The appropriate sub-framework to use is determined by the response variable: use the General Linear Model for scale data that can be assumed to be normally distributed; use the binary logistical model for binary data; use the loglinear model for count data.
- As a general guide on which graph type to use:
  - Pie charts are a good option to choose with tests on frequencies;
  - Boxplots, errorplots, or bar graphs with error bars are a good option to choose with tests of difference, or with a categorical independent variable in GLM;
  - Scatterplots are a good option to choose with tests of relationship or with a continuous independent variable in GLM.
- For graph types to go with tests of difference, a general guide is:
  - Boxplots are a good choice to go with nonparametric tests;
  - Errorplots or bar graphs with error bars are a good choice to go with parametric tests.
- Fit lines can be drawn on scatterplots but they are hard to calculate for correlations. In general, you should only consider adding a fit line if you have a significant regression.

- If you have a dependent/independent variable pair then:
  - The dependent variable should go on the *y*-axis;
  - The independent variable should go on the *x*-axis.
- Tables are an alternative way of presenting any sort of data.

Self-help questions

For each of the scenarios described below:

Link back: Section 14.2

(a) State which analysis you think the researchers had in mind for their data when designing their project (include a summary of the path you took through the choosing chart to get to this choice).

Link back: Section 14.3

(b) Suggest which type of graph the researchers could use to help communicate their findings.

1. Many female primates give loud calls during copulation. One function of these calls may be to increase competition between males. If this is the function of the calls, call length would be predicted to increase as ovulation approaches. Biologists studying baboons in Africa collected the data in Table 14.1 to test this prediction.
2. A geneticist had 39 brown, 20 orange, 29 black, and 45 albino corn snakes and wanted to see whether this is significantly different from a 1:1:1:1 ratio of phenotypes.
3. In a study of the effectiveness of a new drug on migraine, sufferers were assigned randomly to taking a course of an old drug, the new drug, or a placebo. After one month, each person was asked to rank how they had felt over the preceding four weeks on a well-being scale from 10 (really great all the time) to 1 (really bad all the time). There were 20 people in the placebo group, 25 people in the old-drug group, and 23 people in the new-drug group.
4. In a study of the effectiveness of a new drug on migraine, sufferers were assigned randomly to taking a course of an old drug or a new drug. After one month, each person was asked to rank how they had felt over the preceding four weeks on a well-being scale from 10 (really great all the time) to 1 (really bad all the time). The treatments were then swapped around so that people previously taking the old drug took the new drug, and vice versa. After a further month they were again asked to score how they had felt over the previous four weeks. There were 100 people involved in the trials; the results from five of them are shown in Table 14.2.
5. The bone density of 100 adults, all over 50 years old, were recorded, along with information on the sex of the person, how much exercise they took (none, low, moderate, or high level), and how tall they were in centimetres.

Table 14.1 Copulation call length and time to ovulation in baboons referred to in self-help question 1.

Call length (seconds)	3.3	2.7	4.2	3.9	5.1	5.9	6.1	5.7
Time to ovulation (days)	7	5	6	2	1	2	1	3

Table 14.2 Selection of records of five people in trials described in self-help question 4.

Person ID code	Well-being score	
	After 1 month on old drug	**After 1 month on new drug**
0345	2	4
0784	3	10
0895	6	9
0661	5	5
0342	6	2

Answers to self-help questions

Chapter 1

1 All.

2 Descriptive questions—answered directly. Questions about cause, mechanism, or function—answered via hypotheses (hypothetico-deductive approach).

3 Generate hypotheses; develop predictions; test predictions.

4 Analysing data to answer descriptive questions directly; testing predictions in an hypothesis-driven study.

5 Developing robust hypotheses and predictions; designing good experiments; taking accurate and reliable measurements; critically evaluating everything; assessing health, safety, and ethical implications.

6 Descriptive statistics for organizing, summarizing, and describing data. Inferential statistics for assessing patterns in data apparent from samples (includes estimation and null hypothesis significance testing procedures).

Chapter 2

1 Quantity of food available, mineral content of food, genetics, physical aspects of habitat, amount of energy diverted to growth, measurement errors, recording errors.

2 (a) Population, sample.

(b) Sample, population, sample error.

(c) Random, bias.

(d) Ordinal.

(e) Scale.

(f) Nominal.

(g) Scale.

(h) Repeated-measures, related.

(i) Looking-up behaviour, room colour.

3 Natural, correlational or observational studies or experiments; alternative is when it's generated artificially (manipulative, planned, designed or experimental studies or experiments).

4 Potential for pseudoreplication / lack of independence of data.

5 Correlated or non-orthogonal.

6 (a) 28 (given by $4 + 0 + 4 + 16 + 4$).

(b) 11 (given by $2 + 0 + 2 + 4 + 2) + 1$.

(c) 15 (given by $3 + 1 + 3 + 5 + 3$).

(d) 53 (given by $9 + 1 + 9 + 25 + 9$).

(e) 100 (given by $2 + 0 + 2 + 4 + 2 = 10$, followed by 10×10).

7 (a) $\Sigma(y - 1)$.

(b) $\Sigma(y + 2)^2$.

Chapter 3

1 (a) 10 (given by $8 + 9 + 11 + 11 + 13 + 12 + 6 + 9 + 11 = 90$ divided by 9).

(b) 11 (since it's the middle value: 6, 8, 9, 9, **11**, 11, 11, 12, 13).

(c) 11 (since it's the most commonly occurring value: 8, 9, **11**, **11**, 13, 12, 6, 9, **11**).

(d) 7 or 6–13 (since the lowest number is 6 and the highest is 13 and the difference between these is 7: **6**, 8, 9, 9, **11**, 11, 11, 12, **13**; $13 - 6 = 7$).

(e) 8.5–11.5 (since it's the middle of the lower quartile, *, to the middle of the upper quartile, **: 6, 8, *, 9, 9, **11**, 11, 11, **, 12, 13).

(f) 4.75 (to two decimal places).

(g) 2.18 (to two decimal places).

(h) 9.

2 (a) Touching.

(b) Yes.

3 (a) Median and interquartile range (range is acceptable) because the data are ordinal.

(b) Median = 3; interquartile range = 2–4; range = 1–5.

4 (a) Gap.

(b) No.

5 Frequency-distribution graph, pie chart, boxplot, bar chart with error bars, errorplot.

Chapter 4

1 Our questions are about populations, but our data are usually only samples from these populations. Apparent patterns in our sample data might be a product of sample error alone.

2 To estimate parameters from statistics (estimation). To make decisions on whether the patterns in data can be accounted for by chance alone (null hypothesis significance testing).

3 0.05.

4 Yes, in fact they can be constructed for any descriptive statistic (although this book only deals with standard errors and confidence intervals of means).

5 That the confidence intervals of 95 out of every 100 samples taken are likely to include the population mean. There is a good chance that the 95% confidence interval of any sample includes the population mean.

6 The sampling distribution of the means has to be normally distributed, which covers almost all circumstances involving scale data, whether they are continuous or discrete, normally distributed or not.

7 The mean for the males is higher than that for the females, but the confidence intervals overlap quite a lot. We cannot tell for definite from this figure whether or not the difference in the means is due to sample error alone.

Chapter 5

1 Null hypothesis significance testing using the Neyman-Pearson (frequentist) approach.

2 Four: construct H_0; choose α; calculate statistic; reject or accept H_0.

3 (d) All of the above.

4 (b) A researcher decides on his/her own significance level.

5 (a) Anytime H_0 is rejected.

6 (c) Decreasing background variation.

7 (a) True.
(b) False.
(c) False (because *P* could be between 0.01 and 0.05).
(d) True.
(e) False.
(f) True.

8 Type I.

9 The level of measurement of the data (has to be scale) and that the parametric criteria of the test are fulfilled (generally to do with data being normally distributed and sample variances being similar).

10 To assess biological significance and non-significance and facilitate meta-analyses.

Chapter 6

1 (a) Yes.
(b) No.
(c) No.
(d) No.
(e) No.
(f) Yes.

2 (a) Neither.
(b) Both.
(c) Both.
(d) Neither.
(e) Neither.
(f) One-way.
(g) Two-way.
(h) Both.
(i) Both.
(j) Both.
(k) Both.
(l) Both.
(m) Neither.
(n) Both.
(o) Neither.
(p) Neither.
(q) Both.

3 Two-way chi-square test: $X^2_3 = 29.9$, $N = 80$, $P = 0.000$.

Chapter 7

1 (a) *t*-test.
(b) Mann-Whitney U test.
(c) Neither.
(d) Mann-Whitney U test.
(e) *t*-test.
(f) Neither.
(g) Neither.
(h) Both.
(i) *t*-test.
(j) Mann-Whitney U test.
(k) *t*-test and Mann-Whitney U test, if higher U used.
(l) Mann-Whitney U test, if lower U used.
(m) Both.

2 (a) *t*-Test: $t_{23} = 3.497$, $P = 0.002$.
(b) For a *t*-test the data must fulfil the parametric criteria (samples normal, variances similar) and she should have checked these, especially as percentages are not usually normally distributed and require transformation before being used in parametric tests.

Chapter 8

1 (a) Paired *t*-test.
(b) Wilcoxon signed-rank test.
(c) Neither.
(d) Wilcoxon.
(e) Neither.
(f) Neither.
(g) Both.
(h) Paired *t*-test.
(i) Wilcoxon.
(j) Paired *t*-test and Wilcoxon when using higher T.
(k) Wilcoxon when using lower T.
(l) Both.

2 Wilcoxon signed-rank test: $T = 2.50$ (or $z = 2.933$), $n = 10$, $N = 9$, $P = 0.017$.

Chapter 9

1 (a) One-way Anova.
(b) Kruskal-Wallis Anova.
(c) Neither.
(d) Kruskal-Wallis Anova.
(e) One-way Anova.
(f) One-way Anova.
(g) Both.
(h) Neither.
(i) Both.
(j) One-way Anova.
(k) Kruskal-Wallis Anova.
(l) Both.
(m) Neither.
(n) Both.

2 (a) One-way Anova: $F_{2,27} = 13.405$, $P = 0.000$.
(b) For a one-way Anova, the data must fulfil the parametric criteria (samples normal, variances similar). She must have been satisfied that her data fulfilled these criteria.

Chapter 10

1 (a) Parametric.
(b) Measurements.
(c) Relationship.
(d) Reject.
(e) Reject.

2 (a) 1. Model; 2. Significance; 3. Amount.
(b) 2. Significance.
(c) 1. Constructing a model that can potentially be used for prediction; 2. Calculating a coefficient of determination to see how much variation is explained by the model.
(d) F.
(e) R^2.

3 (a) Regression: $y = -0.478x + 29.406$, $F_{1,20} = 9.087$, $P = 0.007$, $R^2 = 0.312$.
(b) It is better if the dependent variable is continuous, but this is a grey area and the analysis in general still remains reliable if parametric criteria are assumed.

Chapter 11

1 (a) Pearson.
(b) Spearman.
(c) Neither.
(d) Spearman.
(e) Neither.
(f) Both.
(g) Both.
(h) Pearson.
(i) Spearman.
(j) Both.
(k) Neither.
(l) Both.

2 (a) Spearman correlation: $r_s = -0.295$, $N = 12$, $P = 0.352$.
(b) Call peak frequency and duration are not correlated.
(c) He must not have been convinced that his data fulfilled the parametric criteria. He could have considered transforming his data (Section 5.4.3).

Chapter 12

1 Using single framework covering everything makes things simpler; can be used easily for multiple explanatory variables; can be easily used in multiple model approach; good preparation for learning generalized linear model, which covers an even wider range of techniques.

2 (a) *t*-test or one-way Anova.
(b) Anova.
(c) Multiple regression.

3 (a) What's the model? Is it/each explanatory variable significant? How much is explained by the model?
(b) F statistics for overall model and individual explanatory variables; model coefficients with associated *t* statistics and confidence intervals.
(c) Coefficient of determination indicates how much variation explained; adjusted coefficient of determination takes into account number of coefficients (which in turn is determined by the number of variables and factor levels) in model.
(d) The model accounts for more variation in the response variable than can be accounted for by chance alone.
(e) The explanatory variable accounts for more variation in the response variable (once the other variables have been taken into account if using Type III sums of squares) than can be accounted for by chance alone.

4 For example, Stress hormone = Presence of toy (yes/no) + Breed + Time of day (am/pm) + Body size + Age
where Presence of toy, Breed, and Time of day are categorical and Body size and Age are continuous.

5 Normality, homogeneity, and linearity. Predicted Fitted values and Residuals/Errors.

6 (a) Example from Chapter 9
Anova: $F_{2,12} = 11.950$, $P = 0.001$.
GLM: $F_{2,12} = 11.950$, $P = 0.001$; $R^2 = 0.666$.
Example from Chapter 10
Regression: $y = 0.003x + 8.310$; $F_{1,4} = 14.332$, $P = 0.019$; $R^2 = 0.782$.
GLM: Species richness = 0.003 Reserve area + 8.310; $F_{1,4} = 14.332$, $P = 0.019$; $R^2 = 0.782$.

(b) Nitrogen levels between sites 1 and 3 differed ($t_{12} = -2.725$, $P = 0.018$).

Chapter 13

1 (a) normal.

(b) binomial.

(c) Poisson.

2 (a) identity.

(b) logit.

(c) log.

3 Chi-square–likelihood or Wald.

4 Compare the quality of models based on the same data; the larger value indicates the best model.

5 Odds ratio (with 95% confidence intervals).

6 Overdispersion is when the data are more variable than would be predicted based on the theoretical distribution assumed, and is a common problem when assuming the binomial or Poisson distributions. If your data is overdispersed when using a binary logistic or loglinear model, the results are unreliable.

7 For related data, for example, repeated measures from the same individuals.

Chapter 14

1 (a) Pearson or Spearman correlation: test of relationship (between duration of call and days to ovulation); variables possibly interdependent, not a dependent/independent relationship; if they decided that the data met the parametric criteria for Pearson, then they could do a Pearson test; otherwise the Spearman test. The Choosing Chart–(not frequencies)–One or no independent variable–Test of relationship–Interdependency–Correlation.

(b) Scatterplot (in fact they should draw this before they do the test, just to check that an assumption of linearity is not unreasonable).

2 (a) One-way classification chi-square: the data for the numbers of snakes of different colour are frequencies; there is just one set of categories, the frequencies of which are to be tested against an expected frequency based on the 1:1:1:1 ratio (which makes it a test of homogeneity). The Choosing Chart–Test on counts/frequencies–One set of categories.

(b) Pie chart or frequency distribution chart.

3 (a) Kruskal-Wallis Anova: test of difference between well-being on the different treatments defined by the independent variable; samples are unrelated (different people in each sample); there are more than two samples (placebo, old drug, new drug); the data could not fulfil the parametric criteria since the well-being score is ordinal, so they must have had a Kruskal-Wallis Anova, rather than a one-way Anova, in mind. The Choosing Chart–(not frequencies)–One or no independent variable–Test of difference–Unrelated–Two samples.

(b) Boxplots.

4 (a) Wilcoxon signed-rank test: test of difference between well-being on the different treatments defined by the independent variable; samples are related (repeated-measures design); there are only two samples (old drug, new drug); the data could not fulfil the parametric criteria since the well-being score is ordinal, so they must have had a Wilcoxon signed-rank test, rather than a paired *t*-test, in mind. The Choosing Chart–(not frequencies)–One or no independent variable–Test of difference–Related–Two samples.

(b) Boxplots.

5 (a) General linear model: the dependent variable is bone density and there are three independent variables, sex (categorical), exercise (categorical), and height (continuous). Model assumptions would need to be checked. The Choosing Chart–(not frequencies)–More than two independent variables–GLM.

(b) A number of possibilities, including errorplots to show differences in bone density between males and females, or different exercise levels and scatterplots show the relationship between bone density and height. Other alternatives include using different colours for males and females and different symbols for different exercise levels on a scatterplot of bone density against height.

APPENDIX I

How to enter data into SPSS

When you open SPSS you will see a **Data Editor** window. In newer versions of SPSS (version 10 onwards) this **Data Editor** window has two parts to it, identified by tabs at the bottom of the window: one is labelled **Variable View** and the other **Data View**. You have seen several examples of these in this book (e.g. Figures 11.4a and 11.4b). You can switch between the two views by clicking on the tabs.

Data View is where you enter your data. SPSS uses the terminology variables for columns and cases for rows in **Data View**. However, before entering your data you should define each variable in the **Variable View**, using a separate row for each variable. You need to fill in the information for the eight features listed as column headings. In older versions of SPSS there is only one view in the **Data Editor** window. If you click on 'var' at the top of a column, it will bring up a **Define Variable** dialogue box through which you can define the features for that variable.

The features of a variable that you need to define are listed below, along with notes on how to define them using the **Variable View** table in newer versions of SPSS.

Name

Type in a short name for the variable.

Type, width, and decimals

Click on the three dots at the right side of the **Type** box to bring up the **Variable Type** window. I recommend that you leave type on the default, **Numeric**, but you can use the **Width** and **Decimals** boxes to modify the format of the number if you like.

Label

Write out the full name of your variable, including any units.

Values

It is always a good idea to enter data into SPSS as numbers. If you have recorded observations as text, for example female or male, you should enter these as number codes, for example 1 and 2. You can use the **Values** feature to assign text names to

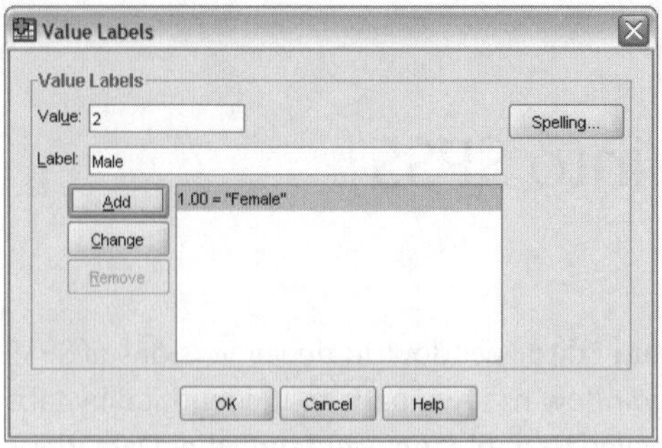

Figure AI.1 SPSS Value Labels dialogue window.

number codes. If you double-click on the three dots to the right side of the box it will bring up the **Value Labels** window (Figure AI.1). Type your first number code in the **Value** box and its word name in the **Value Label** box. Click the **Add** button. Repeat this until all your number codes and their value labels have been entered. Return to the **Variable View** tab by clicking **OK**. In **Data View** you can switch between number codes and value labels through the **Value Labels** option under **View**.

Missing

Click on the three dots at the right side of the box to bring up the **Missing Values** window. You can use this to tell SPSS any numbers you are using to code for missing data. The default is no missing values.

Columns

You can use the up and down arrows to the right of this box to set the width of the column in **Data View**.

Align

This sets the alignment of data in the columns in **Data View**. The default is right. There is a drop-down menu activated by clicking the down arrow to the right of this box, if you want to change it to centre or left.

Measure

You use this to tell SPSS the level of measurement used for the variable. The default is nominal. There is a drop-down menu activated by clicking the down arrow to the right of this box, if you want to change it to ordinal or scale.

Role

You can use this to indicate what role the variable will play in your analyses. The main options are:

- **Input** for an independent variable. This is the default.
- **Target** for a dependent variable.
- **Both** for a variable that can be either independent or dependent.
- **None** if you don't want to assign a role.

APPENDIX II

Statistical tables of critical values

Table AII.1 Critical values of the chi-square (χ^2) distribution. If $X^2 \geq \chi^2$ critical $\rightarrow$ reject $H_0 \rightarrow$ significant result.

χ^2

df	$\alpha = 0.10$	$\alpha = 0.05$	$\alpha = 0.01$
1	2.706	3.841	6.635
2	4.605	5.991	9.210
3	6.251	7.815	11.345
4	7.779	9.488	13.277
5	9.236	11.070	15.086
6	10.645	12.592	16.812
7	12.017	14.067	18.475
8	13.362	15.507	20.090
9	14.684	16.919	21.666
10	15.987	18.307	23.209
11	17.275	19.675	24.725
12	18.549	21.026	26.217
13	19.812	22.362	27.688
14	21.064	23.685	29.141
15	22.307	24.996	30.578
16	23.542	26.296	32.000
17	24.769	27.587	33.409
18	25.989	28.869	34.805
19	27.204	30.144	36.191
20	28.412	31.410	37.566

df	$\alpha = 0.10$	$\alpha = 0.05$	$\alpha = 0.01$
21	29.615	32.671	38.932
22	30.813	33.924	40.289
23	32.007	35.172	41.638
24	33.196	36.415	42.980
25	34.382	37.652	44.314
26	35.563	38.885	45.642
27	36.741	40.113	46.963
28	37.916	41.337	48.278
29	39.087	42.557	49.588
30	40.256	43.773	50.892
31	41.422	44.985	52.191
32	42.585	46.194	53.486
33	43.745	47.400	54.775
34	44.903	48.602	56.061
35	46.059	49.802	57.342
36	47.212	50.998	58.619
37	48.363	52.192	59.893
38	49.513	53.384	61.162
39	50.660	54.572	62.428
40	51.805	55.758	63.691

Source: values generated by Toby Carter using the CHIINV() function in Microsoft Excel.

Table AII.2 Critical values of the *t* (two-tailed) distribution. If $t \geq t_{critical} \rightarrow$ reject $H_0 \rightarrow$ significant result.

t

df	$\alpha = 0.10$	$\alpha = 0.05$	$\alpha = 0.01$
1	6.314	12.706	63.656
2	2.920	4.303	9.925
3	2.353	3.182	5.841
4	2.132	2.776	4.604
5	2.015	2.571	4.032
6	1.943	2.447	3.707
7	1.895	2.365	3.499
8	1.860	2.306	3.355
9	1.833	2.262	3.250
10	1.812	2.228	3.169
11	1.796	2.201	3.106
12	1.782	2.179	3.055
13	1.771	2.160	3.012
14	1.761	2.145	2.977
15	1.753	2.131	2.947
16	1.746	2.120	2.921
17	1.740	2.110	2.898
18	1.734	2.101	2.878
19	1.729	2.093	2.861
20	1.725	2.086	2.845

df	$\alpha = 0.10$	$\alpha = 0.05$	$\alpha = 0.01$
21	1.721	2.080	2.831
22	1.717	2.074	2.819
23	1.714	2.069	2.807
24	1.711	2.064	2.797
25	1.708	2.060	2.787
26	1.706	2.056	2.779
27	1.703	2.052	2.771
28	1.701	2.048	2.763
29	1.699	2.045	2.756
30	1.697	2.042	2.750
31	1.696	2.040	2.744
32	1.694	2.037	2.738
33	1.692	2.035	2.733
34	1.691	2.032	2.728
35	1.690	2.030	2.724
36	1.688	2.028	2.719
37	1.687	2.026	2.715
38	1.686	2.024	2.712
39	1.685	2.023	2.708
40	1.684	2.021	2.704

Source: values generated by Toby Carter using the TINV() function in Microsoft Excel.

Table AII.3a Critical values of the (two-tailed) Mann–Whitney U distribution (with $\alpha = 0.05$), where U is the lower value of U_1 or U_2. If $U \leq U_{critical} \rightarrow$ reject $H_0 \rightarrow$ significant result.

U

n_2 \ n_1	1	2	3	4	5	6	7	8	9	10	11	12	13	14	15	16	17	18	19	20
1																				
2								0	0	0	0	1	1	1	1	1	2	2	2	2
3					0	1	1	2	2	3	3	4	4	5	5	6	6	7	7	8
4				0	1	2	3	4	4	5	6	7	8	9	10	11	11	12	13	13
5			0	1	2	3	5	6	7	8	9	11	12	13	14	15	17	18	19	20
6			1	2	3	5	6	8	10	11	13	14	16	17	19	21	22	24	25	27
7			1	3	5	6	8	10	12	14	16	18	20	22	24	26	28	30	32	34
8		0	2	4	6	8	10	13	15	17	19	22	24	26	29	31	34	36	38	41
9		0	2	4	7	10	12	15	17	20	23	26	28	31	34	37	39	42	45	48
10		0	3	5	8	11	14	17	20	23	26	29	33	36	39	42	45	48	52	55
11		0	3	6	9	13	16	19	23	26	30	33	37	40	44	47	51	55	58	62
12		1	4	7	11	14	18	22	26	29	33	37	41	45	49	53	57	61	65	69
13		1	4	8	12	16	20	24	28	33	37	41	45	50	54	59	63	67	72	76
14		1	5	9	13	17	22	26	31	36	40	45	50	55	59	64	67	74	78	83
15		1	5	10	14	19	24	29	34	39	44	49	54	59	64	70	75	80	85	90
16		1	6	11	15	21	26	31	37	42	47	53	59	64	70	75	81	86	92	98
17		2	6	11	17	22	28	34	39	45	51	57	63	67	75	81	87	93	99	105
18		2	7	12	18	24	30	36	42	48	55	61	67	74	80	86	93	99	106	112
19		2	7	13	19	25	32	38	45	52	58	65	72	78	85	92	99	106	113	119
20		2	8	13	20	27	34	41	48	55	62	69	76	83	90	98	105	112	119	127

Table AII.3b Critical values of the (two-tailed) Mann-Whitney U distribution (with $\alpha = 0.05$), where U is the higher value of U_1 or U_2. If $U \geq U_{critical} \rightarrow$ reject $H_0 \rightarrow$ significant result.

U

n_2 \ n_1	1	2	3	4	5	6	7	8	9	10	11	12	13	14	15	16	17	18	19	20
1	–	–	–	–	–	–	–	–	–	–	–	–	–	–	–	–	–	–	–	–
2	–	–	–	–	–	–	–	16	18	20	22	23	25	27	29	31	32	34	36	38
3	–	–	–	–	15	17	20	22	25	27	30	32	35	37	40	42	45	47	50	52
4	–	–	–	16	19	22	25	28	32	35	38	41	44	47	50	53	57	60	63	66
5	–	–	15	19	23	27	30	34	38	42	46	49	53	57	61	65	68	72	76	80
6	–	–	17	22	27	31	36	40	44	49	53	58	62	67	71	75	80	84	89	93
7	–	–	20	25	30	36	41	46	51	56	61	66	71	76	81	86	91	96	101	106
8	–	16	22	28	34	40	46	51	57	63	69	74	80	86	91	97	102	108	114	119
9	–	18	25	32	38	44	51	57	64	70	76	82	89	95	101	107	114	120	126	132
10	–	20	27	35	42	49	56	63	70	77	84	91	97	104	111	118	125	132	138	145
11	–	22	30	38	46	53	61	69	76	84	91	99	106	114	121	129	136	143	151	158
12	–	23	32	41	49	58	66	74	82	91	99	107	115	123	131	139	147	155	163	171
13	–	25	35	44	53	62	71	80	89	97	106	115	124	132	141	149	158	167	175	184
14	–	27	37	47	57	67	76	86	95	104	114	123	132	141	151	160	169	178	188	197
15	–	29	40	50	61	71	81	91	101	111	121	131	141	151	161	170	180	190	200	210
16	–	31	42	53	65	75	86	97	107	118	129	139	149	160	170	181	191	202	212	222
17	–	32	45	57	68	80	91	102	114	125	136	147	158	169	180	191	202	213	224	235
18	–	34	47	60	72	84	96	108	120	132	143	155	167	178	190	202	213	225	236	248
19	–	36	50	63	76	89	101	114	126	138	151	163	175	188	200	212	224	236	248	261
20	–	38	52	66	80	93	106	119	132	145	158	171	184	197	210	222	235	248	261	273

Source: values derived from Milton (1964), following Zar (1999).

Table AII.4a Critical values of the (two-tailed) Wilcoxon T distribution, where T is the lower value of T^+ or T^- and only non-zero differences are ranked. N is the number of non-zero differences. If $T \leq T_{critical} \rightarrow$ reject $H_0 \rightarrow$ significant result.

T

N	α = 0.10	α = 0.05	α = 0.01	N	α = 0.10	α = 0.05	α = 0.01
1				**21**	67	58	42
2				**22**	75	65	48
3				**23**	83	73	54
4				**24**	91	81	61
5	0			**25**	100	89	68
6	2	0		**26**	110	98	75
7	3	2		**27**	119	107	83
8	5	3	0	**28**	130	116	91
9	8	5	1	**29**	140	126	100
10	10	8	3	**30**	151	137	109
11	13	10	5	**31**	163	147	118
12	17	13	7	**32**	175	159	128
13	21	17	9	**33**	187	170	138
14	25	21	12	**34**	200	182	148
15	30	25	15	**35**	213	195	159
16	35	29	19	**36**	227	208	171
17	41	34	23	**37**	241	221	182
18	47	40	27	**38**	256	235	194
19	53	46	32	**39**	271	249	207
20	60	52	37	**40**	286	264	220

Source: values derived from McCornack (1965), following Zar (1999).

Table AII.4b Critical values of the (two-tailed) Wilcoxon T distribution, where T is the higher value of T^+ or T^- and only non-zero differences are ranked. N is the number of non-zero differences. If $T \geq T_{critical} \rightarrow$ reject $H_0 \rightarrow$ significant result.

T

N	α = 0.10	α = 0.05	α = 0.01
1			
2			
3			
4			
5	15		
6	19	21	
7	25	26	
8	31	33	10
9	37	40	17
10	45	47	25
11	53	56	36
12	61	65	51
13	70	74	70
14	80	84	92
15	90	95	119
16	101	107	150
17	112	119	187
18	124	131	230
19	137	144	278
20	150	158	333

N	α = 0.10	α = 0.05	α = 0.01
21	164	173	395
22	178	188	464
23	193	203	541
24	209	219	625
25	225	236	718
26	241	253	821
27	259	271	932
28	276	290	1054
29	295	309	1185
30	314	328	1327
31	333	349	1481
32	353	369	1646
33	374	391	1823
34	395	413	2013
35	417	435	2215
36	439	458	2430
37	462	482	2660
38	485	506	2904
39	509	531	3162
40	534	556	3435

Source: values derived from Table II.4a using sum of all rank minus crtitical value, when T is the lower value of T^+ or T^-

Table AII.5 Critical values of the (two-tailed) F distribution (with $\alpha = 0.05$). Note: df1 applies to the numerator, or between-treatment, variance; df2 applies to the denominator, or within-treatment, variance. If $F \geq F_{critical} \rightarrow$ reject $H_0 \rightarrow$ significant result. (A numerator is the top bit of a fraction and the denominator is the bottom bit i.e. $\frac{\text{numerator}}{\text{denominator}}$.)

F

df 2 \ df 1	1	2	3	4	5	6	7	8	9	10
1	647.793	799.482	864.151	899.599	921.835	937.114	948.203	956.643	963.279	968.634
2	38.506	39.000	39.166	39.248	39.298	39.331	39.356	39.373	39.387	39.398
3	17.443	16.044	15.439	15.101	14.885	14.735	14.624	14.540	14.473	14.419
4	12.218	10.649	9.979	9.604	9.364	9.197	9.074	8.980	8.905	8.844
5	10.007	8.434	7.764	7.388	7.146	6.978	6.853	6.757	6.681	6.619
6	8.813	7.260	6.599	6.227	5.988	5.820	5.695	5.600	5.523	5.461
7	8.073	6.542	5.890	5.523	5.285	5.119	4.995	4.899	4.823	4.761
8	7.571	6.059	5.416	5.053	4.817	4.652	4.529	4.433	4.357	4.295
9	7.209	5.715	5.078	4.718	4.484	4.320	4.197	4.102	4.026	3.964
10	6.937	5.456	4.826	4.468	4.236	4.072	3.950	3.855	3.779	3.717
11	6.724	5.256	4.630	4.275	4.044	3.881	3.759	3.664	3.588	3.526
12	6.554	5.096	4.474	4.121	3.891	3.728	3.607	3.512	3.436	3.374
13	6.414	4.965	4.347	3.996	3.767	3.604	3.483	3.388	3.312	3.250
14	6.298	4.857	4.242	3.892	3.663	3.501	3.380	3.285	3.209	3.147
15	6.200	4.765	4.153	3.804	3.576	3.415	3.293	3.199	3.123	3.060
16	6.115	4.687	4.077	3.729	3.502	3.341	3.219	3.125	3.049	2.986
17	6.042	4.619	4.011	3.665	3.438	3.277	3.156	3.061	2.985	2.922
18	5.978	4.560	3.954	3.608	3.382	3.221	3.100	3.005	2.929	2.866
19	5.922	4.508	3.903	3.559	3.333	3.172	3.051	2.956	2.880	2.817
20	5.871	4.461	3.859	3.515	3.289	3.128	3.007	2.913	2.837	2.774
21	5.827	4.420	3.819	3.475	3.250	3.090	2.969	2.874	2.798	2.735
22	5.786	4.383	3.783	3.440	3.215	3.055	2.934	2.839	2.763	2.700
23	5.750	4.349	3.750	3.408	3.183	3.023	2.902	2.808	2.731	2.668
24	5.717	4.319	3.721	3.379	3.155	2.995	2.874	2.779	2.703	2.640
25	5.686	4.291	3.694	3.353	3.129	2.969	2.848	2.753	2.677	2.613
26	5.659	4.265	3.670	3.329	3.105	2.945	2.824	2.729	2.653	2.590
27	5.633	4.242	3.647	3.307	3.083	2.923	2.802	2.707	2.631	2.568
28	5.610	4.221	3.626	3.286	3.063	2.903	2.782	2.687	2.611	2.547
29	5.588	4.201	3.607	3.267	3.044	2.884	2.763	2.669	2.592	2.529
30	5.568	4.182	3.589	3.250	3.026	2.867	2.746	2.651	2.575	2.511
31	5.549	4.165	3.573	3.234	3.010	2.851	2.730	2.635	2.558	2.495
32	5.531	4.149	3.557	3.218	2.995	2.836	2.715	2.620	2.543	2.480
33	5.515	4.134	3.543	3.204	2.981	2.822	2.701	2.606	2.529	2.466
34	5.499	4.120	3.529	3.191	2.968	2.808	2.688	2.593	2.516	2.453
35	5.485	4.106	3.517	3.179	2.956	2.796	2.676	2.581	2.504	2.440
36	5.471	4.094	3.505	3.167	2.944	2.785	2.664	2.569	2.492	2.429
37	5.458	4.082	3.493	3.156	2.933	2.774	2.653	2.558	2.481	2.418
38	5.446	4.071	3.483	3.145	2.923	2.763	2.643	2.548	2.471	2.407
39	5.435	4.061	3.473	3.135	2.913	2.754	2.633	2.538	2.461	2.397
40	5.424	4.051	3.463	3.126	2.904	2.744	2.624	2.529	2.452	2.388

Source: values generated by Toby Carter using the FINV() function in Microsoft Excel.

Table AII.6 Critical values of the Kruskal–Wallis *H* distribution (with $\alpha = 0.05$). If $H \geq H_{critical} \rightarrow$ reject $H_0 \rightarrow$ significant result.

H

n_1	n_2	n_3	*H*
2	2	2	–
3	2	1	–
3	2	2	4.714
3	3	1	5.143
3	3	2	5.361
3	3	3	5.600
4	2	1	–
4	2	2	5.333
4	3	1	5.208
4	3	2	5.444
4	3	3	5.791
4	4	1	4.967
4	4	2	5.455
4	4	3	5.598
4	4	4	5.692

n_1	n_2	n_3	*H*
5	2	1	5.000
5	2	2	5.160
5	3	1	4.960
5	3	2	5.251
5	3	3	5.648
5	4	1	4.985
5	4	2	5.273
5	4	3	5.656
5	4	4	5.657
5	5	1	5.127
5	5	2	5.338
5	5	3	5.705
5	5	4	5.666
5	5	5	5.780

Source: values derived from Iman *et al.* (1975), following Zar (1999).

Table AII.7 Critical values of the correlation coefficient, *r*. If $r \geq$ or $= r_{critical} \rightarrow$ reject $H_0 \rightarrow$ significant result.

r

df	α = 0.10	α = 0.05	α = 0.01
1	0.988	0.997	1.000
2	0.900	0.950	0.990
3	0.805	0.878	0.959
4	0.729	0.811	0.917
5	0.669	0.754	0.875
6	0.621	0.707	0.834
7	0.582	0.666	0.798
8	0.549	0.632	0.765
9	0.521	0.602	0.735
10	0.497	0.576	0.708
11	0.476	0.553	0.684
12	0.458	0.532	0.661
13	0.441	0.514	0.641
14	0.426	0.497	0.623
15	0.412	0.482	0.606
16	0.400	0.468	0.590
17	0.389	0.456	0.575
18	0.378	0.444	0.561
19	0.369	0.433	0.549
20	0.360	0.423	0.537
21	0.352	0.413	0.526
22	0.344	0.404	0.515
23	0.337	0.396	0.505
24	0.330	0.388	0.496
25	0.323	0.381	0.487
26	0.317	0.374	0.479
27	0.311	0.367	0.471
28	0.306	0.361	0.463
29	0.301	0.355	0.456
30	0.296	0.349	0.449
31	0.291	0.344	0.442
32	0.287	0.339	0.436
33	0.283	0.334	0.430
34	0.279	0.329	0.424
35	0.275	0.325	0.418
36	0.271	0.320	0.413
37	0.267	0.316	0.408
38	0.264	0.312	0.403
39	0.260	0.308	0.398
40	0.257	0.304	0.393

Source: values derived by Toby Carter from the critical values of the *t* distribution using the equation:

$$r_{\alpha,\mathrm{df}} = \sqrt{\frac{t^2_{\alpha,\mathrm{df}}}{t^2_{\alpha,\mathrm{df}} + v}}$$

where $v = n - 2$.

Table AII.8 Critical values of the Spearman rank-order correlation coefficient, r_s. If $r_s \geq r_{s\,critical} \rightarrow$ reject $H_0 \rightarrow$ significant result.

r_s

N	α = 0.10	α = 0.05	α = 0.01
1			
2			
3			
4	1.000		
5	0.900	1.000	
6	0.829	0.886	1.000
7	0.714	0.786	0.929
8	0.643	0.738	0.881
9	0.600	0.700	0.833
10	0.564	0.648	0.794
11	0.536	0.618	0.755
12	0.503	0.587	0.727
13	0.484	0.560	0.703
14	0.464	0.538	0.679
15	0.446	0.521	0.654
16	0.429	0.503	0.635
17	0.414	0.485	0.615
18	0.401	0.472	0.600
19	0.391	0.460	0.584
20	0.380	0.447	0.570

N	α = 0.10	α = 0.05	α = 0.01
21	0.370	0.435	0.556
22	0.361	0.425	0.544
23	0.353	0.415	0.532
24	0.344	0.406	0.521
25	0.337	0.398	0.511
26	0.331	0.390	0.501
27	0.324	0.382	0.491
28	0.317	0.375	0.483
29	0.312	0.368	0.475
30	0.306	0.362	0.467
31	0.301	0.356	0.459
32	0.296	0.350	0.452
33	0.291	0.345	0.446
34	0.287	0.340	0.439
35	0.283	0.335	0.433
36	0.279	0.330	0.427
37	0.275	0.325	0.421
38	0.271	0.321	0.415
39	0.267	0.317	0.410
40	0.264	0.313	0.405

Source: values from Owen (1962), de Jonge and Van Montford (1972), Franklin (1988), and Olds (1938), following Zar (1999).

APPENDIX III

Summary guidance on reporting statistical results

Descriptive statistics

Whenever you report a measure of central tendency you should report the relevant sample size and ordinal and scale data, you should also report a measure of variability. Remember:

1. Central tendency.
2. Variability.
3. Sample size.

For example, Hatchwell *et al.* (2004) report the number of helper long-tailed tits per nest as means with standard deviation: 'Rivelin Valley, 1.66 ± 0.78 helpers, $n = 44$; Melton Wood, 3.90 ± 1.66, $n = 10$; Ecclesall Wood, 1.54 ± 0.78, $n = 13$'. We know that the measure of variability reported is standard deviation because they tell us in the methods section of their paper that 'Means are reported $\pm$ 1 SD, unless otherwise stated.'

Look back to Section 3.2 for information on descriptive statistics and Section 9.4.4 for information on the example.

Statistical tests

For each statistical test you should give the following information:

1. The name of the test.
2. The value of the test statistic.
3. The degrees of freedom and/or sample size.

Here are some further tips:

- Degrees of freedom are best presented as a subscript to the test statistic, and remember that F values have two degrees of freedom.
- Wherever possible you should give exact *P* values, otherwise quote them as below your critical significance value (e.g. $P < 0.05$). Typically, you can do the former when using statistical software, but will have to do the latter when calculating by hand.

- When using regression or GLM it's a good idea to report the R^2 value for the overall model, as well as the F or t statistic or when using GLZM iAIC value for the overall model as well as the likelihood X^2. You need to decide whether to report information on the model as well, depending on the context. Look back to Sections 10.3.5, 12.3.4, and 12.6.2 for more guidance on reporting the results of regressions and GLZMs.
- When you get a P value of 0.000 on a computer printout, this means the value is less than 0.0005 and therefore rounds up, to three decimal places, to 0.000. People may prefer you to write $P < 0.0005$ or $P < 0.001$, rather than $P = 0.000$, when reporting such a result.

Below are some examples taken from the chapters of this book. The information below should appear in brackets to support written statements in the results section of your write-up. You should refer back to the relevant chapters to see examples of this (look out for the reporting tip icon) and for more context for each test.

one-way classification chi-square: $X^2_3 = 0.618$, $N = 110$, $P = 0.892$
two-way classification chi-square: $X^2_3 = 19.30$, $N = 562$, $P < 0.001$
t-test: $t_{38} = 2.06$, $P = 0.046$
Mann-Whitney U test: $U = 120.5$, $n_1 = 20$, $n_2 = 20$, $P = 0.032$
paired t-test: $t_{15} = 2.226$, $P = 0.042$
Wilcoxon signed-rank test: $T = 29$, $n = 16$, $N = 16$, $P = 0.044$
one-way Anova: $F_{2,12} = 11.791$, $P = 0.001$
Kruskal-Wallis: $X^2_2 = 9.260$, $n_1 = 5$, $n_2 = 5$, $n_3 = 5$, $P = 0.010$
regression: $F_{1,4} = 14.332$, $P = 0.019$, $R^2 = 0.782$
regression: $y = 0.003x + 8.310$; $F_{1,4} = 14.332$, $P = 0.019$, $R^2 = 0.782$
Pearson correlation: $r_{18} = -0.642$, $P = 0.002$
GLM: $F_{5,57} = 10.016$, $P = 0.000$; $R^2 = 0.468$
GLM: $F_{3,57} = 5.193$, $P = 0.000$
GLM: $F_{1,57} = 15.498$, $P = 0.000$ or GLM: $t_{57} = 3.937$, $P = 0.000$
GLZM(b): Likelihood $X^21 = 7.043$, $N = 57$, $P = 0.03$
GLZM(p): Likelihood $X^22 = 39.086$, $N = 57$, $P = 0.000$
Spearman correlation: $r_s = -0.576$, $N = 20$, $P = 0.002$

Effect size

You should report effect size with its associated 95% CIs, for example, with the t-test in Chapter 7, the effect size is:

difference (female-male) = −0.065 (95% CI −0.129 to −0.001).

APPENDIX IV

Statistics and experimental design

The following is a list of experimental designs described in Chapter 4 of Ruxton and Colegrave (2010), with their implications for statistical analysis. You should consult their books for examples.

Designs with control groups

This is where one of the categories of a categorical independent variable is a reference group. You would use a test of difference to investigate. If you plan to use different subjects in one control plus one treatment group look at Chapter 7. If you are thinking of using one control plus one or more treatment groups, then look at Chapter 9, Chapter 12, or Chapter 13. If you plan to use the same subject in your control and treatment groups, check out Chapter 8.

Balanced and unbalanced design

An unbalanced design means you have samples of different sizes. It means that your data are almost certainly unrelated and so you will be looking at techniques in Chapters 7, 9, 12, or 13. If your samples are balanced then this is not the case. Your statistical test will work better if your design is balanced and your results will be easier to interpret.

Completely randomized designs

This is the random allocation of sampling units (subjects) to categories of the independent variable. Since a categorical independent variable is involved, this leads to tests of differences being useful. Randomization is required for the techniques in Chapters 7, 9, 12, and 13 to work reliably.

One-way or one factor designs

This is a design with just one categorical independent variable. The test this will lead you to will depend on the number of levels of the independent variable, for two see Chapter 7 and for two or more see Chapter 7, Chapter 12 or Chapter 13. If it has two levels and the data are related, consult Chapter 8.

Multi-way or multi-factor designs

This is a design with more than one categorical independent variable. For example a two-way design involves two categorical independent variables. The only technique explained in any detail in this book that can deal with such situations is a GLZM, covered in Chapters 12 and 13.

Fully crossed designs

A design must be at least two-way for it to have the possibility of being fully crossed. Fully crossed designs means that all possible combinations of categories between different independent variables are sampled. The GLZM technique covered in Chapters 12 and 13 doesn't work well if the design is not fully crossed.

Incomplete designs

The opposite of a fully crossed design is an incomplete design. The statistics for these is complicated and they are best avoided.

Blocking

Designs involving blocking include a categorical independent variable which is not central to the question being addressed but which is included to reduce background variation. Since such situations involve at least two independent variables, the only technique explained in any detail in this book that can be used is GLZM, covered in Chapters 12 and 13.

Paired design

This is where data in two categories of a categorical independent variable are related through matching data pairs (Section 2.3.1). The techniques in Chapter 8 can be used to analyse data of this kind.

Covariate

A covariate can be thought of as a blocking variable that is continuous rather than categorical. Including a covariate in your design will lead you to GLZMs in Chapters 12 and 13.

Within-subject designs

This is when the data in different categories of a categorical independent variable are related through repeated measures (Section 2.3.1). If your independent variable only has two categories, the tests in Chapter 8 can be used to look for differences.

Split-plot designs

This is where sampling of one categorical independent variable is done within the categories of a second categorical independent variable. Split-plot designs are generally only used because of logistical constraints. They are complicated to analyse statistically and are not covered by any of the techniques in this book, although there are types of GLZM analyses that can cope.

Selected further reading

After studying this book, one of my greatest hopes is that you will want to read others and learn more. I have provided a brief description of a few texts that you might check out. I've divided them into introductory and intermediate level (under Next steps...) and more advanced texts (under For when you are feeling stronger...), plus resources on the internet (under Online) outside the online resource centre for this book <**http://global.oup.com/uk/orc/biosciences/maths/hawkins3e/**>. You should also not forget to use the help that comes with statistical software. SPSS and R, for example, have built-in help and tutorials as well as online support through the websites <**www.spss.com**> and <**http://www.r-project.org**>, respectively. I find the printed SPSS manuals helpful too.

Next steps...

Experimental Design for the Life Sciences
Ruxton, G.D. and Colegrave, N. (2010), 3rd Edition, Oxford University Press, Oxford.

If you are planning your own project, you really should read Ruxton and Colegrave's book. It is not on statistics *per se*, but on how to design research so that it is efficient and effective. This will be informed by your understanding of statistics and will have implications on your choice of data analysis technique.

Choosing and Using Statistics: a Biologist's Guide
Dytham, C. (2010), 3rd Edition, Blackwell Science, Oxford.

Dytham covers more tests in less detail than this book. It has a chapter which introduces techniques such as principal components analysis, factor analysis, and cluster analysis, which are useful when generating rather than testing research hypotheses. Some guidance on the use of computer packages (SPSS, Minitab, and Excel) is given and it has a useful glossary.

Discovering Statistics Using IBM SPSS Statistics
Field, A. (2013) 4th Edition, SAGE, London.

This book covers more material in more detail than any of the others in this section, but its light-hearted style makes it a potential next step. It's written by a psychologist for psychologists, so will probably appeal more to whole behavioural biologists than biomedics. There is a sister book based on R.

Generalized Linear Models and Generalized Estimating Equations
Garson, D. (2013) Statistical Associates Publishing, Asheboro, NC.

Written by a statistician but accessible to biologists, this book will take you further with GLZM and introduce you to generalized estimating equations.

Medical Statistics from Scratch
Bowers, D. (2008), 2nd Edition, John Wiley & Sons, Chichester.

Bowers is a good book for health-care professionals, biomedics, and students of human biology. It covers the same techniques as this book, but in less detail, but its examples are exclusively medically related.

Modern Statistics for the Life Sciences
Grafen, A. and Hails, R. (2002), Oxford University Press, Oxford.

Grafen and Hails is the book to read if you want to learn more about the general linear model. It emphasizes the overall conceptual framework that connects all the parametric tests.

Nonparametric Statistics for the Behavioural Sciences
Siegel, S. and Castellan, Jr, N.J. (1988), 2nd Edition , McGraw-Hill International Editions, New York.

Siegel and Castellan is a marvellous book. It only covers nonparametric techniques in detail, but its introductory chapters are relevant to parametric procedures as well. It is designed specifically for the behavioural sciences.

Practical Statistics for Field Biology
Fowler, J., Cohen, L., and Jarvis, P. (1998), 2nd Edition, John Wiley & Sons, Chichester.

Fowler, Cohen, and Jarvis is a good book for ecologists. As for Dytham, it introduces principal component analysis, factor analysis, and cluster analysis, which are useful in helping to generate research hypotheses rather than testing them. It has a useful appendix listing common symbols and a good section on transformation.

Statistics: An Introduction using R.
Crawley, M.J. (2005) John Wiley & Sons, Chicester.

Crawley's clear insightful text is worth reading, whether or not you plan to learn R.

The Essential Guide to Effect Sizes
Ellis, P.D. (2010) Cambridge University Press, Cambridge.

This book is written by a social scientist, but will appeal to a wide readership and provides a great introduction to statistical power, meta-analysis, and effect size and the link between these three.

For when you are feeling stronger...

Experimental Design and Analysis for Biologists
Quinn, G.P. and Keough, M.J. (2002) Cambridge University Press, Cambridge.

Quinn and Keough is an extensive text covering a lot of ground. In addition to detailing statistical procedures, it has good sections on experimental design, power analysis, the scientific method, and alternatives to null hypothesis significance testing.

Biometry
Sokal, R.R. and Rohlf, F.J. (2012) 4th Edition, W. H. Freemen & Company, New York.

Sokal and Rohlf has traditionally been the statistical bible of professional biologists. It is not for the faint-hearted but if you start to get into statistics at all you will need a copy on your shelves. I use it as a reference source, reading specific bits I've looked up in the index or contents list when I have a particular issue that I am trying to resolve. New material in the 4th Edition includes greater coverage of model selection, effect size, power analyses, and meta-analyses.

Biostatistical Analysis
Zar, J.H. (1999) 4th Edition, Prentice-Hall, New Jersey.

This overlaps with Sokal and Rohlf, but I find that some topics are explained more clearly. It has a particularly good chapter on transformation, for example.

Model Selection and Multimodel Inference: A practical information-theoretic approach
Burnham, K.P. and Anderson, D.R. (2002), 2nd Edition, Springer Science

If you want to go the model selection and comparison route this book is a must read. It talks a lot about AIC values amid much much more.

Online...

Handbook of Biological Statistics
<http://udel.edu/~mcdonald/statintro.html>

I recently came across this site, by John MacDonald of the University of Delaware. It covers a lot of useful topics and is designed specifically for Bioscientists.

Khan Academy
<http://www.khanacademy.org/>

The Khan Academy evolved from when Shia Khan started helping his cousin with maths over the internet. Now this amazing global video-based resource covers a range of disciplines, but it's still a great place to go for help and inspiration with maths and stats support.

NUMBAT
<http://numeracy-bank.net>

NUMBAT as a resource is designed to support numeracy and statistics for students in further- and higher-education, especially in Biosciences. It has been developed by myself and colleagues from Anglia Ruskin University.

References

Here is a full listing of all the references cited in the text in alphabetical order by author. Those asterisked (*) are available as PDFs on the companion website.

*Abildstrom, S.Z., Rasmussen, S. and, Madsen, M. (2005) Changes in hospitalization rate and mortality after acute myocardial infarction in Denmark after diagnostic criteria and methods changed. *European Heart Journal*, **26**: 990–995.

Barnard, C.J., Gilbert, F.S., and MacGregor, P.K. (2007) *Asking Questions in Biology: a Guide to Hypothesis Testing, Experimental Design and Presentation in Practical Work and Research Projects*, 3rd Edition. Pearson Education, Harlow.

*Beasley, C.R., Túry, E., Vale, W.G., and Tagliaro, C.H. (2000) Reproductive cycle, management and conservation of *Paxyodon syrmatophorus* (Bivalvia: Hyriidae) from the Tocantins River, Brazil. *Journal of Molluscan Studies*, **66**: 393–402.

Bowers, D. (2008) *Medical Statistics from Scratch*, 2nd Edition. John Wiley & Sons, Chichester.

Brashares, J.S., Arcese, P., and Sam, M.K. (2001) Human geography and reserve size predict wildlife extinction in West Africa. *Proceedings of the Royal Society of London, Series B*, **268**: 2473–2478.

*Burgess, E.A., Brown, J.L., and Lanyon, J.M. (2013) Sex, scarring, and stress: understanding seasonal costs in a cryptic marine mammal. *Conservation Physiology*, **1**: 1–14.

Burnham, K.P. and Anderson, D.R. (2002) *Model Selection and Multimodel Inference: A Practical Information-Theoretic Approach*, 2nd Edition. Springer Science.

Clements, D.A.V. (2000) The ecology of stem boring mainscot moths in *Phragmites australis* reedbeds. PhD Thesis, Anglia Polytechnic University.

Crawley, M.J. (2005) *Statistics: An Introduction using R.* John Wiley & Sons, Chichester.

de Jonge, C. and Van Montford, M.A.J. (1972) The null distribution of Spearman's S when n = 12. *Statistica Neerlandica*, **26**: 15–17.

Dobson, A.J. (2001) *An introduction to Generalized Linear Models*, 2nd Edition. Chapman and Hall, London.

Doll, R. and Hill, A.H. Smoking and carcinoma of the lung. *British Medical Journal*, 30th September 1950, 739–748.

Dytham, C. (2010) *Choosing and Using Statistics: a Biologist's Guide*, 3rd Edition. Blackwell Science, Oxford.

Ellis, P.D. (2010) *The Essential Guide to Effect Sizes.* Cambridge University Press, Cambridge.

Ereckson, A. (2001) Demographic and behavioural effects of elephant poaching in Mikumi National Park. MPhil Thesis, Anglia Ruskin University.

Field, A. (2013) *Discovering Statistics Using IBM SPSS Statistics*, 4th Edition. SAGE, London.

Fowler, J., Cohen, L., and Jarvis, P. (1998) *Practical Statistics for Field Biology*, 2nd Edition. John Wiley & Sons, Chichester.

Franklin, L.A. (1988) The complete exact null distribution of Spearman's rho for n = 12 (1)18. *Journal of Statistical Computation and Simulation*, **29**: 255–269.

Garson, D. (2013) *Generalized Linear Models and Generalized Estimating Equations.* Statistical Associates Publishing, Asheboro, NC.

*Gavin, L.A, Wamboldt, M.Z., Sorokin, N., Levy, S., and Wamboldt, F.S (1999) Treatment alliance and its association with family functioning. *Journal of Pediatric Psychology*, **24**: 355–365.

Grafen, A. and Hails, R. (2002) *Modern Statistics for the Life Sciences.* Oxford University Press, Oxford.

*Greseth, S.L., Cope, W.G., Rada, R.G., Waller, D.L., and Bartsch, M.R. (2003) Biochemical composition of three species of unionid mussels after emersion. *Journal of Molluscan Studies*, **69**: 101–106.

*Grube, R.C., Brennan, E.B., and Ryder, E.J. (2003) Characterization and genetic analysis of a lettuce (*Lactuca sativa* L.) mutant, *weary*, that exhibits reduced gravotropic response in hypocotyls and inflorescence stems. *Journal of Experimental Botany*, **54**: 1259–1268.

*Hatchwell, B.J., Russell, A.F., MacColl, A.D.C., Ross, D.J., Fowlie, M.K., and McGowan, A. (2004) Helpers increase long-term but not short-term productivity in cooperatively breeding long-tailed tits. *Behavioural Ecology*, **15**: 1–10.

*Hattori, T., Inanaga, S., Tanimoto, E., Lux, A., Luxová, M., and Sugimoto, Y. (2003) Silicon-induced changes in viscoelastic properties of sorghum root cell walls. *Plant and Cell Physiology*, **44**: 743–749.

Hawkins, D., Gallacher, E., and M. Gammell (2013) Statistical power, effect size and animal welfare:

recommendations for good practice. *Animal Welfare*, **22**: 339–344.

*Heitger, M.H., Anderson, T.J., Jones, R.D., Dalrymple-Alford, J.C., Frampton, C.M., and Ardagh, M.W. (2004) Eye movement and visuomotor deficits following mild closed head injury. *Brain*, **127**: 575–590.

Hughes, F.M.R., Johansson, M.E., Xiong, S., Carlborg, E., Svedmark, M., Hawkins, D., Hayes, A., Goodall, A., Richards, K.S.R., and Nilsson, C. (2009) The influence of hydrological regimes on sex ratios in dioecious riparian shrub species in northern Sweden. *Plant Ecology*, **208**: 77–92.

*Hutchinson, K.J., Gómez-Pinilla, F., Crowe, M.J., Ying, Z., and Basso, M.D. (2004) Three exercise paradigms differentially improve sensory recovery after spinal cord contusion in rats. *Brain*, **127**: 1403–1414.

Iman, R.L., Quade, D., and Alexander, D.A. (1975) Exact probability levels for the Kruskal-Wallis test. In *Selected Tables in Mathematical Statistics*, vol. III (Harter, H.L. and Owen, D.B., eds), pp. 329–384, American Mathematical Society, Providence, RI.

Jones, A., Reed, R., and Weyers, J. (2012) *Practical Skills in Biology*, 5th Edition. Pearson, Harlow.

*Jordaens, K., Gielen, H., Van Houtte, N., Bernon, G., and Backeljau, T. (2003) The response of the terrestrial slug. *Journal of Molluscan Studies*, **69**: 285–288.

*Lohar, D.P. and Bird, D.M. (2003) *Lotus japonicus*: A model to study root-parasitic nematodes. *Plant and Cell Physiology*, **44**: 1176–1184.

Mackenzie, J. (2010) Impact of floral origin, floral composition and structural fragmentation on breeding success in blue tits (*Cyanistes caeruleus*) and great tits (*Parus major*). PhD thesis, Anglia Ruskin University.

Martin, P. and Bateson, P. (2007) *Measuring Behaviour: an Introductory Guide*, 3rd Edition. Cambridge University Press, Cambridge.

McCornack R.L. (1965) Extended tables of the Wilcoxon matched pairs signed rank statistic. *Journal of the American Statistical Association*, **60**: 864–871.

Milton, R.C. (1964) An extended table of critical values for the Mann-Whitney (Wilcoxon) two-sample statistic. *Journal of the American Statistical Association*, **59**: 925–934.

*Morton, B. and Britton, J.C. (2002) Holothurian feeding by *Nassarius doratus* on a beach in western Australia. *Journal of Molluscan Studies*, **68**: 187–189.

*Morvan-Bertrand, A., Boucaud, J., and Prud'homme, M. (1999) Influence of initial levels of carbohydrates, fructans, nitrogen, and soluble proteins on the regrowth of *Lolium perenne* L. cv. Bravo following defoliation. *Journal of Experimental Botany*, **50**: 1817–1826.

*Mosimann, U.P., Felblinger, J., Ballinari, P., Hess, C.W., and Müri, R.M. (2004) Visual exploration behaviour during clock reading in Alzheimer's disease. *Brain*, **127**: 431–438.

Neuhaus, P., Broussand, D.R., Murie, J.O., and Dobson, F.S. (2004) Age of primiparity and implications of early reproduction of life history in female Columbian ground squirrels. *Journal of Animal Ecology*, **73**: 36–43.

Olds E.G. (1938) Distributions of the sums of squares of rank differences for small numbers of individuals. *Annals of Mathematical Statistics*, **9**: 133–148.

Owen, D.B. (1962) *Handbook of Statistical Tables*. Addison-Wesley, Reading, MA.

*Pfennig, D.W., Collins, J.P., and Ziemba, R.E. (1999) A test of alternative hypotheses for kin recognition in cannibalistic tiger salamanders. *Behavioral Ecology*, **10**: 436–443.

Pinakin, G. (2002) The influence of corneal dimensions on measurements related to glaucoma and ocular hypertension. PhD Thesis, Anglia Polytechnic University.

Quinn, G.P. and Keough, M.J. (2002) *Experimental Design and Analysis for Biologists*. Cambridge University Press, Cambridge.

Ruckstuhl, K.E. and Festa-Bianchet, M. (1998) Do reproductive status and lamb gender affect the foraging behavior of bighorn ewes? *Ethology*, **104**: 941–954.

Ruckstuhl, K.E. and Neuhaus, P. (2002) Sexual segregation in ungulates: a comparative test of three hypotheses. *Biological Review*, **77**: 77–96.

Ruxton G.D. and Colegrave, N. (2010) *Experimental Design for the Life Sciences*, 3rd Edition. Oxford University Press, Oxford.

*Sackey, M., Weigel, M., and Armijos, R.X. (2003) Predictors and nutritional consequences of intestinal parasitic infections in rural Ecuadorian children. *Journal of Tropical Pediatrics*, **49**: 17–23.

Siegel, S. (1956) *Nonparametric statistics for the behavioral sciences*. McGraw-Hill, New York.

Siegel, S. and Castellan, Jr, N.J. (1988) *Nonparametric Statistics for the Behavioural Sciences*, 2nd Edition. McGraw-Hill International Editions, New York.

Smith, M.D. and Conway, C.J. (2007) Use of mammal manure by nesting burrowing owls: a test of four functional hypotheses. *Animal Behaviour*, **73**: 65–73.

*Snow, L.S.E. and Andrade, M.C.B. (2004) Pattern of sperm transfer in redback spiders: implications for sperm competition and male sacrifice. *Behavioural Ecology*, **15**: 785–792.

Sokal, R.R. and Rohlf, F.J. (2012) *Biometry*, 4th Edition. W.H. Freeman & Company, New York.

Thomas, A. (2003) *Introducing Genetics: from Mendel to Molecule*. Nelson Thornes, Cheltenham.

*Van Osselaer, C. and Tursch, B. (2000) Variability of the genital system *of Helix pomatia* L., 1758 and *H. lucorum* L., 1758 (Gastropoda: Stylommatophora). *Journal of Molluscan Studies*, **66**: 499–515.

*Voisin, A.S., Salon, C., Jeudy, C., and Warembourg, F.R. (2003) Symbiotic N_2 fixation activity in relation to C economy of *Pisum sativum* L. as a function of plant phenology. *Journal of Experimental Botany*, **54**: 2733–2744.

*Wilhelm, J., Frieling, H., Hillemacher, T., , Degner, D., Kornhuber, J., and Bleich, S. (2008) Hippocampal volume loss in patients with alcoholism is influenced by the consumed type of alcoholic beverage. *Alcohol and Alcoholism*, **43**: 296–299.

Zar, J.H. (1999) *Biostatistical Analysis*, 4th Edition. Prentice-Hall, New Jersey.

Index

Page numbers in **bold** indicate that the word or phrase appears also in a marginal note.
Page numbers in *italics* indicate that a box, figure or table is particularly relevant.

V

W

X

Y